ENCYCLOPAEDIA OF
PLANT DISEASES

ENCYCLOPAEDIA OF PLANT DISEASES

Satyavir Singh

ANMOL PUBLICATIONS PVT. LTD.
NEW DELHI - 110 002 (INDIA)

ANMOL PUBLICATIONS PVT. LTD.
H.O.: 4374/4B, Ansari Road, Daryaganj
New Delhi - 110 002
Ph.: 23278000, 23261597
B.O.: No. 1015, Ist Main Road, BSK III Stage
III Phase, III Block
Bangalore - 560 085 (India)
Visit us at: www.anmolpublications.com

Encyclopaedia of Plant Diseases

First Published, 2006

ISBN 81-261-2923-9

PRINTED IN INDIA

Printed at Mehra Offset Press, Delhi.

Preface

This Encyclopaedia is the perfect guide for those studying Plant diseases, either at university or anyone interested in Plant diseases, such as plant pathologists. It provides a comprehensive coverage of Plant diseases. The aim of this encyclopaedia is to provide teachers, students and laymen interested in Plant diseases with clear, concise, and correct definitions and descriptions of the Plant diseases terms, throughout.

In this volume all the branches of Plant diseases are represented. This volume covers whole realm of Plant diseases and includes various topics such as main plant pathogens, fungi, viruses, bacteria, nematodes and viriods. It contains descriptions and thorough supporting references. There are also entries for the names of diseases and disorders, crops and their pathology, fungicides, toxins and vectors. The explanations of entries range from brief definitions of a few lines to expository discussions of various lengths. The Encyclopaedia is to satisfy the immediate technical needs of the user. Concepts are presented in a clear and lucid manner. In terms of its broader objectives, the present book may be viewed as an outline or compendium of Plant diseases ideas correlated with a large collection of specialized items.

The aim of this volume is to provide a concise repository of number of major concepts in this field. In preparation of this book, the author has freely consulted large number of books and journals so no authenticity is claimed. Author is especially thankful to M/S Anmol Publications for shaping this book in its final form. Suggestions for further improvement of this book are not only welcome but also greatly appreciated.

Author

A

A horizon Topsoil.

Aabaxial *Of the side or surface of an organ,* facing away from the axis.

Aabstriction Separation along a septum, often involving constriction.

Aadanal (nematology) Pertaining to a bursa that does not envelop the entire tail.

Aaecidium An aecium with a cup-like outer wall.

Aalternate host One of two kinds of plants on which a parasitic fungus (e.g., a rust) must develop to complete its life cycle.

Abaxial Directed away from the stem of a plant; pertaining to the lower surface of a leaf.

Abdomen The posterior section of the three sections (head-thorax-abdomen) of an insect's body. The abdomen contains the reproductive and excretory organs, and the bulk of the digestive system. The abdomen bears no functional legs in the adult stage.

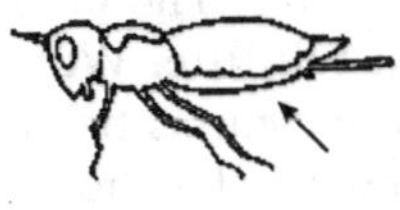

Fig. Abdomen

Abiotic Nonliving factors, such as wind, water, temperature, or soil type or texture.

Abiotic disease A disease caused by a nonliving factor (i.e., hail, cold, nutrient, etc.)

Abiotic disorder A disease caused by factors other than a pathogen; physiological disorder.

Abiotic pathogen A nonliving, disease-, causing entity, e.g., drought, salt, air pollutants.

Abrasive Fine particles of a material such as charcoal, carborundum or diatomaceous earth ('celite') incorporated in the inoculum or dusted on to leaves before inoculation to facilitate the mechanical transmission of a plant virus.

Abrupt Appearing as if cut off transversely; truncate.

Abscise To separate from a plant, as leaves, flowers, and fruits do when they fall.

Abscission Of plants: The shedding of leaves or other parts as the result of physical weakness in a specialized layer of cells (abscission layer) that develops at the base.

Absidia sp An allergen, *Absidia* is commonly found in decaying plant debris, soils, foods, and can often cause food spoilage. The genus contains approximately 21 species, a common isolate being *A. corymbifera* (the only recognized pathogen of *Absidia* species). *A. corymbifera* is a principal cause of zygomycosis or mucormycosis. Zygomycosis can involve the rhino-facial-cranial area, lungs, gastrointestinal tract, and skin, especially in the immunocompromised. The disease is associated with acidotic diabetes, malnourished children, severely burned patients, immunosuppressive therapy, use of cytotoxins and corticosteroids, and diseases such as leukemia and lymphoma. These fungi show an inclination for vessel invasion resulting in embolism and subsequent necrosis. This fungus is a zygomycete and its identification is based on its asexual reproduction, which is similar to *Rhizopus* and *Mucor*. Morphological characteristics are the production of sporangiospores, which are diminutive, round, thick walled bodies resistant to heat and drought, causing the fungi to be thermophilic. They are produced in large numbers in globular envelopes (sporangia) at the tip of special hyphae (sporangiophores). Identification is based on the way the sporangia are formed. Culture - Potato dextrose agar or Malt extract agar, 37°C, 7 - 10 days for *A. corymbifera*; most other *Absibia* sp. are unable to grow at 37°C, but do grow at 20 - 25°C.

Absorb To obtain food by taking up water and dissolved substances across a membrane; this is how fungi operate. This is also called osmotrophic nutrition .

Absorption The uptake of liquids by solids or the passage of a substance into the tissues of an organism; e.g., the movement of a chemical into the bloodstream following exposure.

Acaricide A pesticide used specifically to manage or prevent damage caused by mites. Sometimes generalized to be synonymous with pesticides.

Acceptable daily intake (ADI) A measure of the level of a chemical, i.e. a pesticide residue, that is believed to be able to be consumed on a daily basis over a lifetime without harm.

Acerose Needle-shaped, like pine needles.

Acervulus A more or less saucer-shaped conidioma in which a hymenium of conidiogenous cells develops on the floor of the cavity beneath a covering of host tissue which ruptures at maturity.

Acetylcholinesterase An enzyme present in nerve tissue, muscles and red blood cells that catalyzes the hydrolysis of acetylcholine to choline and acetic acid, allowing neural transmission across synapses to occur. This mechanism is used for neural transmission in insects, humans and other mammals, and other living creatures.

Acetylcholinesterase inhibitor A substance that interferes with the normal transmission of nerve impulses by inhibiting the acetylcholinesterase enzyme. This is the primary mode of action of carbamate and organophosphate insecticides.

Achene A simple, one-seeded fruit in which the seed is attached to the ovary wall at only one point, such as the "seed" on the surface of a strawberry.

Acicular Needle-shaped and stiff.

Acid Having a pH of less than 7.

Acid precipitation Precipitation (fog, rain, snow) with a low pH, due to the presence of nitric and sulfuric acid formed by the reaction of air pollutants ($N0_x$ and $S0_2$) with water add image from acid rain definition.

Acid rain Acid rain is mainly caused by sulpher dioxide, nitrogen oxide and other pollutants when fossil fuels are burnt. Acid rain rarely falls near the source of the pollutant because smoke, fumes etc can be carried many miles by air currents. So the Scandinavia, for example, receives acid rain from British and German industry. The main effects of acid rain are on wildlife. Water in lakes also becomes very clear as fish and microscopic life are killed off. It is thought that acid rain is responsible for the death of some trees, specifically conifers, which gradually lose all their leaves and die. Acid rain is normally considered to be a by-product of modern atmospheric pollution. Even in a pure, uncontaminated world, however, it is likely that the rainfall would be acidic. Carbonic acid (H_2CO_3) is formed when carbon dioxide (CO_2) is absorbed by atmospheric water. Nitric acid (HNO_3) is created during thunderstorms and sulphuric acid (H_2SO_4) is formed from the sulphur dioxide (SO_2) released during volcanic eruptions and forest fires or from the sulphur (S) emitted by phytoplankton during their seasonal bloom period. All of these are natural processes which contribute to make normal rain acid. Current concern over acid rain is not with the naturally produced variety, but rather with that which results from modern industrial activity and the acid gases it produces. Considerable amounts of sulphur dioxide are released into the atmosphere as a by-product of metal smelting, particularly when non-ferrous ores are involved. The burning of coal and oil to provide energy for space heating or to fuel thermal electric power stations also produces sulphur dioxide. The continuing growth of transportation systems using the internal combustion engine contributes to acid rain through the release of oxides of nitrogen (NO_x) into the atmosphere.. Acid rain is normally considered to be a by-product of modern atmospheric pollution. Even in a pure, uncontaminated world, however, it is likely that the rainfall would be acidic. Carbonic acid (H_2CO_3) is formed when carbon dioxide (CO_2) is absorbed by atmospheric water. Nitric acid (HNO_3) is created during thunderstorms and sulphuric acid (H_2SO_4) is formed from the sulphur dioxide (SO_2) released during volcanic eruptions and forest fires or from the sulphur (S) emitted by phytoplankton during their seasonal bloom period. All of these are natural processes which contribute to make normal rain acid. Current concern over acid rain is not with the naturally produced variety, but rather with that which results from modern industrial activity and the acid gases it produces. Considerable amounts of sulphur dioxide are released into the atmosphere as a by-product of metal smelting, particularly when non-ferrous ores are involved. The burning of coal and oil to provide energy for space heating or to fuel thermal electric power stations also produces sulphur dioxide. The continuing growth of transportation systems using the internal combustion engine contributes to acid rain through the release of oxides of nitrogen (NO_x) into the atmosphere.

Acid precipitation produced by human activities differs from natural acid precipitation not only in its origins, but also in its quality. Anthropogenically produced acid rain tends to be many times more acidic than the natural variety. The difference is commonly of the order of 1.0 to 1.5 points on the pH scale used to measure acidity. In North America, for example, naturally acid rain has a pH of about 5.6, whereas measurements of rain falling in southern Ontario, Canada frequently provide values in the range of 4.5 to 4.0. To put these values in perspective, vinegar has a pH of 2.7 and milk a pH of 6.6. Similar values for background levels of acidic rain are indicated by studies in Europe with the average annual pH of rain over Britain between 1978 and 1980 being within the range of 4.5 to 4.2.

Acid precipitation produced by human activities differs from natural acid precipitation not only in its origins, but also in its quality. Anthropogenically produced acid rain tends to be many times more acidic than the natural variety. The difference is commonly of the order of 1.0 to 1.5 points on the pH scale used to measure acidity. In North America, for example, naturally acid rain has a pH of about 5.6, whereas measurements of rain falling in southern Ontario, Canada frequently provide values in the range of 4.5 to 4.0. To put these values in perspective, vinegar has a pH of 2.7 and milk a pH of 6.6. Similar values for background levels of acidic rain are indicated by studies in Europe with the average annual pH of rain over Britain between 1978 and 1980 being within the range of 4.5 to 4.2.

The quality of the rain is determined by a series of chemical processes set in motion when acidic materials are released into the atmosphere. Some of the sulphur dioxide and oxides of nitrogen emitted will return to the surface quite quickly, and close to their source, as dry deposition. The remainder will be carried up into the atmosphere, to be converted into sulphuric and nitric acid, which will eventually return to earth as acid rain. The processes involved are fundamentally simple. Oxidation converts the gases into acids, in either a gas or liquid phase reaction, at a rate which depends upon such variables as the concentration of heavy metals in airborne particulate matter, the presence of ammonia and the intensity of sunlight. The metals and ammonia appear to act as catalysts, while the sunlight provides the energy required by the chemical reactions.

The quality of the rain is determined by a series of chemical processes set in motion when acidic materials are released into the atmosphere. Some of the sulphur dioxide and oxides of nitrogen emitted will return to the surface quite quickly, and close to their source, as dry deposition. The remainder will be carried up into the atmosphere, to be converted into sulphuric and nitric acid, which will eventually return to earth as acid rain. The processes involved are fundamentally simple. Oxidation converts the gases into acids, in either a gas or liquid phase reaction, at a rate which depends upon such variables as the concentration of heavy metals in airborne particulate matter, the presence of ammonia and the intensity of sunlight. The metals and ammonia appear to act as catalysts, while the sunlight provides the energy required by the chemical reactions.

Acid rain remained mainly a local problem in the past. The introduction of the tall stacks policy in an attempt to reduce local pollution, however, increased its geographical extent. The release of pollutants at the greater altitudes possible with the taller smokestacks placed them outside the boundary layer circulation and into the larger scale atmospheric circulation system with its potential for much greater dispersal through the mechanisms involved in the Long Range Transportation of Air Pollution (LRTAP). Local problems were mitigated at the expense of the larger environment.

Acid rain remained mainly a local problem in the past. The introduction of the tall stacks policy in an attempt to reduce local pollution, however, increased its geographical extent. The release of pollutants at the greater altitudes possible with the taller smokestacks placed them outside the boundary layer circulation and into the larger scale atmospheric circulation system with its potential for much greater dispersal through the mechanisms involved in the Long Range Transportation of Air Pollution (LRTAP). Local problems were mitigated at the expense of the larger environment.

The main sources of acid rain are to be found in the industrialized areas of the northern hemisphere. North-eastern North America, Britain and Western Europe have received most attention, although their output of sulphur dioxide has been declining. Levels of oxides of nitrogen have not yet experienced significant decline and continue to rise in some areas. The emission of acid gases from Eastern Europe and the republics of the former USSR – Russia, Ukraine and Kazakhstan – remains high. In Asia, Japanese industries emit large quantities of sulphur dioxide, while the industrial areas of China are also major contributors. Acid emissions remain limited outside of the major industrial nations, but concern has been expressed over growing levels of air pollution, often associated with urban automobile exhaust emissions, which may already have provided a base for acid rain in some Third World countries.

The main sources of acid rain are to be found in the industrialized areas of the northern hemisphere. North-eastern North America, Britain and Western Europe have received most attention, although their output of sulphur dioxide has been declining. Levels of oxides of nitrogen have not yet experienced significant decline and continue to rise in some areas. The emission of acid gases from Eastern Europe and the republics of the former USSR - Russia, Ukraine and Kazakhstan - remains high. In Asia, Japanese industries emit large quantities of sulphur dioxide, while the industrial areas of China are also major contributors. Acid emissions remain limited outside of the major industrial nations, but concern has been expressed over growing levels of air pollution, often associated with urban automobile exhaust emissions, which may already have provided a base for acid rain in some Third World countries.

Once the acid gases have been released into the atmosphere, they are at the mercy of the prevailing circulation patterns. With almost all of the areas currently producing large amounts of acidic pollution located within the mid-latitude westerly wind belt, emissions are normally carried eastwards, or perhaps north-eastwards, often for several hundred kilometres before being redeposited. In this way, pollutants originating in the US Midwest cause acid rain in Ontario, Quebec and the New England states. Emissions from the smelters and power stations of the English Midlands and the Ruhr contribute to the acidity of precipitation in Scandinavia, and acidity in the Arctic originates as far as 8000 km away to the south in North America and Eurasia. Thus the problem of acid rain transcends national boundaries, introducing political overtones to the problem and creating the need for international cooperation if a solution is to be found.

Once the acid gases have been released into the atmosphere, they are at the mercy of the prevailing circulation patterns. With almost all of the areas currently producing large amounts of acidic pollution located within the mid-latitude westerly wind belt, emissions are normally carried eastwards, or perhaps north-eastwards, often for several hundred kilometres before being redeposited. In this way, pollutants originating in the US Midwest cause acid rain in Ontario, Quebec and the New England states. Emissions from the smelters and power stations of the English Midlands and the Ruhr contribute to the acidity of precipitation in Scandinavia, and acidity in the Arctic originates as far as 8000 km away to the south in North America and Eurasia. Thus the problem of acid rain transcends national boundaries, introducing political overtones to the problem and creating the need for international cooperation if a solution is to be found.

The impact of acid rain on the environment depends not only on the level of acidity in the rain, but also on the nature of the environment itself. Areas underlain by acidic rocks such as granitic or quartzitic bedrock, for example, are particularly susceptible to damage, lacking as they do the ability to 'buffer' or neutralize additional acidity from the precipitation. Acid levels therefore rise, the environmental balance is disturbed and serious ecological damage is the inevitable result. In contrast, areas which are geologically basic - underlain by limestone or chalk, for example - are much less sensitive, and may even benefit from the additional acidity. The highly alkaline soils and water of areas underlain by chalk or limestone ensure that the acid added to the environment by the rain is very effectively neutralized.

The impact of acid rain on the environment depends not only on the level of acidity in the rain, but also on the nature of the environment itself. Areas underlain by acidic rocks such as granitic or quartzitic bedrock, for example, are particularly susceptible to damage, lacking as they do the ability to 'buffer' or neutralize additional acidity from the precipitation. Acid levels therefore rise, the environmental balance is disturbed and serious ecological damage is the inevitable result. In contrast, areas which are geologically basic - underlain by limestone or chalk, for example - are much less sensitive, and may even benefit from the additional acidity. The highly alkaline soils and water of areas underlain by chalk or limestone ensure that the acid added to the environment by the rain is very effectively neutralized.

The impact of acid rain on the environment was first recognized in the lakes and rivers of these areas. Reduced pH values which indicated the rising acidity were accompanied by low levels of calcium (Ca) and magnesium (Mg), elevated sulphate concentrations and an increase in the amounts of potentially toxic metals such as aluminum (Al). When aquatic communities were examined in areas as far apart as New York State, Nova Scotia, Norway and Sweden there was clear evidence that increased surface water acidity had adverse effects on fish. In some cases the acidity was sufficiently high that mature fish died, but, more commonly, fish populations began to decline because of the effects of the increasing acidity on reproduction. Damage to the eggs during spawning and the inability of the young fry to survive the higher acidity, particularly during the spring flush, ensured that the older fish were not replaced as they died. As a result, fish populations in many rivers and lakes in eastern North America, Britain and Scandinavia have declined noticeably in the last two to three decades and hundreds of lakes are now completely devoid of fish. Although the impact on the fish is most obvious, all aquatic organisms will decline in number and variety during progressive acidification. Waterbodies that have lost or are in the process of losing their flora and fauna are often described as 'dead' or 'dying'. This is not strictly correct, however, for some organisms are remarkably acid-tolerant, and even the most acid lakes have some life in them. Species of protozoans are found at pH levels as low as 2.0, for example.

The impact of acid rain on the environment was first recognized in the lakes and rivers of these areas. Reduced pH values which indicated the rising acidity were accompanied by low levels of calcium (Ca) and magnesium (Mg), elevated sulphate concentrations and an increase in the amounts of potentially toxic metals such as aluminum (Al). When aquatic communities were examined in areas as far apart as New York State, Nova Scotia, Norway and Sweden there was clear evidence that increased surface water acidity had adverse effects on fish. In some cases the acidity was sufficiently high that mature fish died, but, more commonly, fish populations began to decline because of the effects of the increasing acidity on reproduction. Damage to the eggs during spawning and the inability of the young fry to survive the higher acidity, particularly during the spring flush, ensured that the older fish were not replaced as they died. As a result, fish populations in many rivers and lakes in eastern North America, Britain and Scandinavia have declined noticeably in the last two to three decades and hundreds of lakes are now completely devoid of fish. Although the impact on the fish is most obvious, all aquatic organisms will decline in number and variety during progressive acidification. Waterbodies that have lost or are in the process of losing their flora and fauna are often described as 'dead' or 'dying'. This is not strictly correct, however, for some organisms are remarkably acid-tolerant, and even the most acid lakes have some life in them. Species of protozoans are found at pH levels as low as 2.0, for example.

There is growing evidence that those areas in which the waterbodies have already succumbed to acidification must also face the effects of increasing acid stress on their forests and soils. The threat is not universally recognized, however, and there remains a great deal of controversy over the amount of damage directly attributable to acid rain. Reduction in forest growth in Sweden, physical damage to trees in West Germany, and the death of sugar maples in Quebec and Vermont have all been blamed on the increased acidity of the precipitation in these areas. Many of the impacts, such as the thinning of annual growth rings, reduction in biomass and damage to fine root systems, are only apparent after detailed examination, but others are more directly obvious and have been described as dieback. This involves the gradual wasting of the tree inwards from the outermost tips of its branches. The process is cumulative over several years until the tree dies. The symptoms of dieback have been recognized in the maple groves and red spruce forests of north-eastern North America, and across fifteen countries and some 70,000 km2 of forest in Europe. In Germany, where it was first linked with acid rain, dieback, or Waldsterben, was particularly extensive and in the 1980s was seen as a major threat to the survival of the German forests. More recent research has indicated that forest decline in Europe is a multifaceted process in which acid precipitation may be only one of a series of contributors along with tree harvesting practices, drought and fungal attacks, although its exact role remains controversial.

There is growing evidence that those areas in which the waterbodies have already succumbed to acidification must also face the effects of increasing acid stress on their forests and soils. The threat is not universally recognized, however, and there remains a great deal of controversy over the amount of damage directly attributable to acid rain. Reduction in forest growth in Sweden, physical damage to trees in West Germany, and the death of sugar maples in Quebec and Vermont have all been blamed on the increased acidity of the precipitation in these areas. Many of the impacts, such as the thinning of annual growth rings, reduction in biomass and damage to fine root systems, are only apparent after detailed examination, but others are more directly obvious and have been described as dieback. This involves the gradual wasting of the tree inwards from the outermost tips of its branches. The process is cumulative over several years until the tree dies. The symptoms of dieback have been recognized in the maple groves and red spruce forests of north-eastern North America, and across fifteen countries and some 70,000 km2 of forest in Europe. In Germany, where it was first linked with acid rain, dieback, or Waldsterben, was particularly extensive and in the 1980s was seen as a major threat to the survival of the German forests. More recent research has indicated that forest decline in Europe is a multifaceted process in which acid precipitation may be only one of a series of contributors along with tree harvesting practices, drought and fungal attacks, although its exact role remains controversial.

Present concern over acid rain is concentrated mainly on its effect on the natural environment, but acid rain also contributes to deterioration in the built environment: it attacks limestone and marble used as building stone, for example. The faceless statues and crumbling cornices of the world's famous palaces, castles, abbeys and cathedrals, from the Parthenon in Greece to the Taj Mahal in India, attest to its power. By attacking the fabric of these buildings it not only causes physical and economic damage, but threatens the world's cultural heritage.

Present concern over acid rain is concentrated mainly on its effect on the natural environment, but acid rain also contributes to deterioration in the built environment: it attacks limestone and marble used as building stone, for example. The faceless statues and crumbling cornices of the world's famous palaces, castles, abbeys and cathedrals, from the Parthenon in Greece to the Taj Mahal in India, attest to its power. By attacking the fabric of these buildings it not only causes physical and economic damage, but threatens the world's cultural heritage.

Acid rain in the form of aerosols or attached to smoke particles can cause respiratory problems in humans, as was the case with the infamous London Smog of 1952. There is no evidence that wet deposition is directly damaging to human health, although in its ability to mobilize metals, from lead (Pb) or copper (Cu) pipes, for example, it may have important indirect effects.

Acid rain in the form of aerosols or attached to smoke particles can cause respiratory problems in humans, as was the case with the infamous London Smog of 1952. There is no evidence that wet deposition is directly damaging to human health, although in its ability to mobilize metals, from lead (Pb) or copper (Cu) pipes, for example, it may have important indirect effects.

Solutions to the problem of acid rain are deceptively simple. In theory, a reduction in the emission rate of acid forming gases is all that is required to slow down and eventually stop the damage being caused by the acidification of the environment. Translating that concept into reality has proved difficult, however. One of the first approaches involved adaptation, such as the addition of lime to land or lakes to counteract the increased acidity, and to allow the recovery mechanisms to work more effectively. Experiments involving the liming of lakes in Canada and Sweden, for example, have provided encouraging results. Artificial buffering of lakes in this way is only a temporary measure which may be likened to the use of antacid to reduce acid indigestion. The neutralizing effects of the lime may last longer than those of the antacid, but they do wear off in three to five years and re-liming is necessary as long as acid loading continues. Most of the current proposals for dealing with acid rain tackle the problem at its source. They attempt to prevent, or at least reduce, emissions of acid gases into the atmosphere, and since sulphur dioxide makes the greatest contribution to acid rain in North America and Europe, it has received most attention in the development of abatement procedures. Sulphur dioxide is formed when coal and oil are burned to release energy, and the technology to control it may be applied before, during or after combustion. The exact timing will depend upon such factors as the amount of acid reduction required, the type and age of the system and the cost-effectiveness of the particular process.

Solutions to the problem of acid rain are deceptively simple. In theory, a reduction in the emission rate of acid forming gases is all that is required to slow down and eventually stop the damage being caused by the acidification of the environment. Translating that concept into reality has proved difficult, however. One of the first approaches involved adaptation, such as the addition of lime to land or lakes to counteract the increased acidity, and to allow the recovery mechanisms to work more effectively. Experiments involving the liming of lakes in Canada and Sweden, for example, have provided encouraging results. Artificial buffering of lakes in this way is only a temporary measure which may be likened to the use of antacid to reduce acid indigestion. The neutralizing effects of the lime may last longer than those of the antacid, but they do wear off in three to five years and re-liming is necessary as long as acid loading continues. Most of the current proposals for dealing with acid rain tackle the problem at its source. They attempt to prevent, or at least reduce, emissions of acid gases into the atmosphere, and since sulphur dioxide makes the greatest contribution to acid rain in North America and Europe, it has received most attention in the development of abatement procedures. Sulphur dioxide is formed when coal and oil are burned to release energy, and the technology to control it may be applied before, during or after combustion. The exact timing will depend upon such factors as the amount of acid reduction required, the type and age of the system and the cost-effectiveness of the particular process.

Flue gas desulphurization (FGD) is the most common approach to sulphur dioxide reduction. It uses scrubbers to neutralize the acidity of the flue gases before they are released into the atmosphere. With sulphur dioxide removal rates of between 80 and 95 per cent, FGD is more effective than methods such as fuel switching or fuel desulphurization applied prior to combustion and it is technically simpler than fluidized bed combustion (FBC) and lime injection multi-stage burning (LIMB) incorporated in the combustion process itself. Although FGD systems do little to reduce emissions of oxides of nitrogen they are the preferred method of dealing with acid rain, and future requirements to reduce sulphur dioxide levels will be met in large part by thc installation of FGD equipment. Flue gas desulphurization (FGD) is the most common approach to sulphur dioxide reduction. It uses scrubbers to neutralize the acidity of the flue gases before they are released into the atmosphere. With sulphur dioxide removal rates of between 80 and 95 per cent, FGD is more effective than methods such as fuel switching or fuel desulphurization applied prior to combustion and it is technically simpler than fluidized bed combustion (FBC) and lime injection multi-stage burning (LIMB) incorporated in the combustion process itself. Although FGD systems do little to reduce emissions of oxides of nitrogen they are the preferred method of dealing with acid rain, and future requirements to reduce sulphur dioxide levels will be met in large part by the installation of FGD equipment.

The study of acid rain has declined remarkably since the 1980s, when it was viewed by many as the major environmental problem facing the northern hemisphere. It is one environmental issue in which abatement programmes have met with some success. Sulphur dioxide levels continue to decline, the rain in many areas is considerably less acid than it was a decade ago, and the transboundary disputes which absorbed large amounts of time, energy and money in the 1980s have been resolved. There are indications that lakes and forests are showing some signs of recovery in the most vulnerable areas of North America and Europe, although this is still a matter of dispute. The study of acid rain has declined remarkably since the 1980s, when it was viewed by many as the major environmental problem facing the northern hemisphere. It is one environmental issue in which abatement programmes have met with some success. Sulphur dioxide levels continue to decline, the rain in many areas is considerably less acid than it was a decade ago, and the transboundary disputes which absorbed large amounts of time, energy and money in the 1980s have been resolved. There are indications that lakes and forests are showing some signs of recovery in the most vulnerable areas of North America and Europe, although this is still a matter of dispute.

Developments such as these have created the perception that the acid rain problem is being solved. It would be more accurate, however, to say that it has changed in nature and in geographical extent. In the developed nations, oxides of nitrogen are gradually making a larger contribution to acidity as levels of sulphur dioxide decline. In less developed areas, from Eastern Europe to China, the full impact of acid rain may still be in the future, and the research activities currently winding down in Western Europe and North America may have to be revived to deal with it. Developments such as these have created the perception that the acid rain problem is being solved. It would be more accurate, however, to say that it has changed in nature and in geographical extent. In the developed nations, oxides of nitrogen are gradually making a larger contribution to acidity as levels of sulphur dioxide decline. In less developed areas, from Eastern Europe to China, the full impact of acid rain may still be in the future, and the research activities currently winding down in Western Europe and North America may have to be revived to deal with it.

Acid soil A soil with a pH value of less than 7.0.

Acid tolerance The weathering of soils can result in acidification, limiting the types of plant or microbes that will grow. Repeated fertilization with ammonia can also lead to acidification. To overcome such problems the soil must be limed, or plant species and microbes used that are tolerant of acidity.

Acidification Process of becoming more acid. pH 7 is neutral, above this is alkaline below pH 6 is acidic. i.e Acidification of the soil.

Acidifier An additive used to lower the pH of the water being mixed with a pesticide; pesticides often break down more slowly if the spray water is slightly acid.

Acorn Fruit from an oak tree (nut).

Acquisition access time The length of time that a test vector is given access to a virus source in transmission tests. It is not implied that the vector feeds during all or any of this time. Hence the term is not synonymous with acquisition feeding time.

Acquisition feeding The feeding of a vector on a virus source in transmission tests.

Acquisition feeding time The time that a vector feeds on a virus source intransmission tests. Not synonymous with acquisition access time.

Acquisition The entry of virus into, or attachment to, a vector.

Acquisition threshold period The minimum time necessary for a vector to spend on a virus source in order to obtain an infective charge of virus.

Acre A section of land containing 4,840 square yards or 43,560 square feet.

Acre foot The volume of water equal to the volume of a prism one foot high with a base one acre in area; 43,560 cubic feet.

Acremonium sp *Acremonium* is a common type I & III allergen. Naturally found in soils, decaying organic matter, and plant debris, it is also an agricultural contaminant. This genus can be parasitic or saprophytic to plants and other living fungi, and some species cause vascular wilts in trees. *Acremonium* is the asexual state of *Emericellopsis*, *Chaetomium*, and *Nectripsis*. It has been known to produce a toxin from the trichothecene group and may also give off an unpleasant odor due to the production of volatile organic compounds (VOCs). This fungus is associated with occupant complaints such as nausea, vomiting and diarrhea. *Acremonium* has been isolated from cases of mycetoma, onychomycosis, mycotic keratitis, infection of the cornea, and infections of artificial implants. Morphological characteristics are the production of conidiophores and long, slender phialides; conidia are hyaline, 1-celled, and are collected in a slime drop. Colonies grow fast and are compact and moist, becoming overgrown with loose, cottony hyphae which are white, gray or rose in color. Indoor growth requires extremely wet conditions, and does not grow very well at 37°C. Culture - Potato dextrose agar or Malt extract agar, 20° - 25°C, 7 - 10 days.

Acrogenous *Of conidia*, developing at the apex.

Acropetal Upward from the base to the apex of a shoot of a plant; in fungi, the production of spores in succession in the direction of the apex so that the apical spore is the youngest.

Actinomycetes Filamentous bacteria that have sometimes been classified as Fungi Imperfecti. Actinomycetes typically are saprobes (especially in soil) but a few are pathogenic to man, animals, and plants.

Actinostele A type of protostele that looks like a star in cross section. The walls of the stele are heavily reinforced with lignin and are relatively thick. Xylem occupies the center but the phloem occupies the indentations in the xylem. This generates the star-like pattern.

Action threshold Pest population level at which action needs to be taken to prevent the population from causing economic damage.

Active ingredient The chemical in a pesticide formulation that kills or otherwise controls a pest or a weed. The remainder of a formulated pesticide is one or more inert ingredients.

Active transport Transport across a cell membrane which requires energy generated by the hydrolysis of some energy carrier (usually ATP). The selective active transport of ions (usually sodium) out of the cell is often used as a secondary energy storage mechanism.

Aculeate Prickly; with narrow spines.

Acuminate Tapering gradually to a protracted point.

Acute 1. Developing suddenly, severe (with reference to disease symptoms).

Acute health effects Having rapid onset, severe symptoms and a short course; not chronic.

Acute injury Injury, usually involving necrosis, which develops within several hours to a few days after a high dose exposure to a pollutant; expressed as fleck, scorch, bifacial necrosis, etc.

Acute phase A short, sharp, and relatively severe course of a disease; not chronic.

Acute toxicity The potential of a substance to cause injury or illness after a single exposure or shortterm exposure. One measure of acute toxicity is the lethal dose, or LD50.

Adaptation Non-inherited change in an organism towards increased fitness. Also commonly, but incorrectly, used to indicate an inherited change towards greater fitness.

Adaxial *Of the side or surface of an organ*, facing towards the axis.

Addison's disease A disease marked by the atrophy or destruction of the adrenal cortex. Most cases of Addison's disease appear to involve an autoimmune process.

Additive effects Combined effect of two or more chemicals equal to the sum of their individual effects.

Additives Substances added to feedstuffs for preservation or flavour.

Adherence The ability of a crop protection material to stick to a given surface.

Adjuvant Material added to improve some chemical or physical property (e.g., of a plant protectant) or a biological property (e.g., to improve antibody response to an antigen).

Adnate (mycology) Of gills or tubes broadly attached to the stipe; attached by nearly the entire width of the gills or layer of tubes.

Adpressed Closely flattened down or pressed against a surface.

Adsorption Adhesion of substances to the surface of solids or liquids.

Adsorption coefficient (Koc) A (unitless) measure of a material's tendency to adsorb to soil particles. High Koc values indicate a tendency for the material to be adsorbed by soil particles rather than remain dissolved in the soil solution. Strongly adsorbed molecules will not leach or move unless the soil particle to which they are adsorbed moves (as in erosion). Koc values of less than 500 indicate little or no adsorption and a potential for leaching. Concentration adsorbed/concentration dissolved Koc = % organic carbon in the soil

Adult The final stage of an animal that goes through metamorphosis (cycle of change). With most insects, it is the fourth stage, after it comes out of a pupa. For example, a butterfly is an adult after it comes out of a chrysalis. With frogs, it is an adult after it has changed from a tadpole.

Adventitious Arising from other than the usual place, as roots from a stem rather than as branches of a root.

Adventitious septum A septum formed in the absence of, or independently of, nuclear division, especially in association with the movement of cytoplasm from one part of the fungus to another; characteristic of lower fungi but found in all groups; retraction septum; secondary septum.

Aecial stage (aecium) A spore stage of the rust fungi; a cuplike structure bearing aeciospores.

Aeciospore In Uredinales, a unicellular, non-repeating vegetative spore, produced in an aecium, usually resulting from dikaryotisation, which germinates to give a dikaryotic mycelium.

Aecium In Uredinales, a spore-producing structure of variable morphology, often cup-shaped and with chains of aeciospores; usually found associated with pycnia.

Aerial Occurring in the air.

Aerial photography Photos taken from an airplane, used to help identify variations with a field or other management unit.

Aerification The act of infusing or forcing air into, for example, soil.

Aero aquatic fungi Pond-inhabiting fungi producing elaborate floating propagules which are on hand to colonize autumn-shed leaves as they fall into the water, then 'condition' them in near-anaerobic conditions at the bottom of the pond.

Aerobe An organism that requires free oxygen for respiration.

Aerobic Living or active only in the presence of oxygen.

Aerobic digestion Decomposition of organic matter by micro organisms in the presence of oxygen

Aerobiology The study of fungal (and other) propagules in the atmosphere. This has particular relevance for plant pathologists and for people with respiratory allergies.

Aerosol A material stored in a container under pressure. Fine droplets are produced when the material dissolved in a liquid carrier is released into the air from the pressurized container. Particulates, alternately referred to as particulate matter (PM), aerosols or fine particles are tiny particles of solid or liquid suspended in a gas. They range in size from less than 10 nanometres to more than 100 micrometres in diameter. This range of sizes represents scales from a gathering of a few molecules to the size where the particles no longer can be carried by the gas. Some aerosols occur naturally, originating from volcanoes, dust storms, forest and grassland fires, living vegetation, and sea spray. Human activities, such as the burning of fossil fuels also generate aerosols. Averaged over the globe, anthropogenic aerosols—those made by human activities—currently account for about 10 percent of the total amount of aerosols in our atmosphere.

There are both natural and human sources of atmospheric particulates. The biggest natural sources are wind-blown dust, volcanoes, and forest fires. Sea spray is also a large source of particles though most of these fall back to the ocean close to where they were emitted. The biggest human sources of particles are combustion sources, mainly the burning of fossil fuel in internal combustion engines in automobiles and power plants, and wind blown dust from construction sites and other land areas where the water or vegetation has been removed. Some of these particles are emitted directly to the atmosphere (primary emissions) and some are emitted as gases and form particles in the atmosphere (secondary emissions). The composition of fine particles depends on the source. Wind-blown dust tends to be made of mineral salts and other crustal earth material. Primary emissions from combustion sources are made primarily of unburned fuel (hydrocarbons), elemental carbon (soot), elemental sulfur, mineral salts, and often contain traces of toxic metals. Secondary emissions are a combination of ammonia with either sulfuric acid or nitric acid and water, together with a complex mixture of the organic oxidation products of VOCs. Several classes of chemical compounds are known or believe to contribute to radiative forcing. These are sulfate, sea-salt, carbonaceous, and mineral dust aerosols.

Sulfate aerosols consists of the sulfate anion existing in various chemical states: sulfuric acid, ammonium bisulfate, ammonium sulfate, or as a dissociated anion in aqueous solution. Sea-salt aerosol, the second largest contributor the global aerosol budget, consists mainly of sodium chloride salt from seawater. Other components of seawater include magnesium chloride and organic compounds, which may influence its chemistry. The range of non-mineral, carbon-containing compounds that constitute the carbonaceous aerosol fraction are usually divided into two subclasses: organic carbon and elemental carbon. Elemental carbon is often referred to as black carbon, which is a major consituent of soot. The elemental portion of carbon aerosols are strongly light-absorbing while organic carbon consists of both light-scattering and light-absorbing compounds. Mineral dust refers to soil, dust and other windblown material from the earth's surface. This makes up the majority of particulate matter less than 10 micrometers in size across the globe. In general, the smaller and lighter a particle is, the longer it will stay in the air. Larger particles (greater than 10 micrometres in diameter) tend to settle to the ground by gravity in a matter of hours whereas the smallest particles (less than 1 micrometre) can stay in the atmosphere for weeks and are mostly removed by precipitation.

Aerosols, natural and anthropogenic, can affect the climate by changing the way radiation is transmitted through the atmosphere. Direct observations of the effects of aerosols are quite limited so any attempt to estimate their global effect necessarily involves the use of computer models. The Intergovernmental Panel on Climate Change, IPCC, says: While the radiative forcing due to greenhouse gases may be determined to a reasonably high degree of accuracy... the uncertainties relating to aerosol radiative forcings remain large, and rely to a large extent on the estimates from global modelling studies that are difficult to verify at the present time.

Sulfate aerosol has two main effects, direct and indirect. The direct effect, via albedo, is to cool the planet: the IPCC's best estimate of the radiative forcing is -0.4 watts per square meter with a range of -0.2 to -0.8 W/m^2 but there are substantial uncertainties. The effect varies strongly geographically, with most cooling believed to be at and downwind of major industrial centres. Modern climate models attempting to deal with the attribution of recent climate change need to include sulfate forcing, which appears to account (at least partly) for the slight drop in global temperature in the middle of the 20th century. The indirect effect (via the aerosol acting as cloud condensation nuclei, CCN, and thereby modifying the cloud properties) is more uncertain but is believed to be a cooling.

Black carbon (BC), also called soot rather loosely is one of the most important absorbing aerosol species in the Atmosphere. BC from fossil fuels is estimated by the IPCC in the Third Assessment Report of the IPCC, TAR, to contribute a global mean radiative forcing of +0.2 W/m^2 (was +0.1 W/m^2 in the Second

Assessment Report of the IPCC, SAR), with a range +0.1 to +0.4 W/m². All aerosols both absorb and scatter solar (Sun's) and terrestrial (Earth) radiation. When we say, a species is absorbing, it only means that it dominantly absorbs than scatters radiation. A term called Single Scattering Albedo (SSA) is rather used to explain this. SSA is the ratio of scattering to extinction (Extinction includes both scattering and absorption) of radiation by a particle. High SSA implies the aerosol species of interest mainly scatters radiation. Lower SSA implied absorbing aerosols. The effects of inhaling particulate matter has been widely studied in humans and animals and include asthma, lung cancercardiovascular issues, and premature death. The size of the particle determines where in the body the particle will come to rest if inhaled. Larger particles are generally filtered by small hairs in the nose and throat and do not cause problems, but particulate matter smaller than about 10 micrometres, referred to as PM_{10}, can settle in the bronchies and lungs and cause health problems. Particles smaller than 2.5 micrometres, $PM_{2.5}$, can penetrate directly into the lung, whereas particles smaller than 1 micrometer PM_1 can penetrate into the alveolar region of the lung and tend to be the most hazardous when inhaled. In particular, a study published in the Journal of the American Medical Association, indicates that $PM_{2.5}$ leads to high plaque deposits in arteries, causing vascular inflammation and atherosclerosis -- a hardening of the arteries that reduces elasticity, which can lead to heart attacks and other cardiovascular problems. Researchers suggest that even short-term exposure at elevated concentrations could significantly contribute to heart disease.

There is evidence that particles smaller than 100 nanometres can pass through cell membranes. For example, particles can migrate into the brain. It has been suggested that particulate matter can cause similar brain damage as that found in Alzheimer patients. Particles emitted from modern diesel engines (commonly referred to as Diesel Particulate Matter, or DPM) are typically in the size range of 100 nanometres (0.1 micrometres). In addition, these soot particles also carry carcinogenic components like benzopyrenes adsorbed on their surface. In view of these deposition mechanisms, it is becoming increasingly clear that the legislative limits for engines, which are in terms of emitted mass, are not a proper measure of the health hazard. One particle of 10 µm diameter has approximately the same mass as 1 million particles of 100 nm diameter, but it is clearly much less hazardous, as it probably never enters the human body - and if it does, it is quickly removed. Proposals for new regulations exist in some countries, with suggestions to limit the particle surface area or the particle number.

Aestivation Insect dormancy during summer or periods of high temperature or a dry season.

Aethalium In myxomycetes, a sessile, pulvinate fruit-body, often large (to several centimetres) and sometimes with an outer crust.

Aetiological fraction The reduction in disease when a risk factor is removed. If I* is the number of people that a risk factor is responsible for and I is the total number of cases over the same period, then the aetiological fraction is I*/I. Equivalently, the aetiological fraction is (I-Io)/I. where Io is the number of cases in the absence of the risk factor. Also known as the attributable fraction.

Aetiology The causes of disease.

In medicine and in plant pathology, the science of the causes of disease. Interpretations of the meaning of this term vary widely: some consider that study of the aetiology of a disease consists merely in identifying the causal agent; others take the view that it concerns the whole sequence of events that leads up to the expression of symptoms, and the study of all those factors which may modify the sequence. Its use in the broad sense is recommended, hence: the science of the causes of disease; the study of the causal factor, its nature and its relations with the host.

By extension, the term is applicable to the body of knowledge derived from such studies.

Aflatoxin Virulent toxin produced by the Hyphomycetes, Aspergillus flavus and A. parasiticus, growing on foodstuffs, esp. nuts: highly carcinogenic.

Ag consultant A person trained in agriculture and management science to provide information to land owners/managers for a fee related to the farming operation.

Agar Mixture of polysaccharides derived from red algae that forms a gel at temperatures below about 40C. Used as a support medium, when supplemented by appropriate buffers and/or nutrients and other ingredients, for the production of microbial cultures, overlaying tissue culture cells, electrophoresis, etc.

Agaric Common name for a fungus which produces gilled basidiomata, usually a member of the Agaricales.

Agarose One of the constituents of agar. Often used in preference to agar because it gels at a lower temperature and does not contain the inhibitors of virus growth frequently present in agar. It is also used widely in gel electrophoresis because it has a more uniform pore size than that of agar.

Age structured model A mathematical model which take into consideration the division of the host population into different age classes. Such models can used to consider the consequences of such factors as age-dependent infection, morbidity or mortality rates or of age-specific vaccination schedules.

Agent of disease An organism or abiotic factor that causes disease; a pathogen.

Agent of inoculation That which transports inoculum from its source to or into the infection court (e.g., wind, splashing rain, insects, humans).

Agglutinate Fixed together as if with glue.

Agglutination The formation of insoluble aggregates following the combination of antibodies with cells or other particulate antigens or with soluble antigens bound to cells or other particles or following the combination of soluble (or particulate) antigens with cell-bound or particle-bound antibodies.

Aggregation A distribution of parasites amongst hosts is said to be aggregated, or overdispersed, if parasites are found to co-occur in particular hosts more often than if the parasites were distributed at random amongst all hosts. A more precise test would be if the distribution differed significantly from a Poisson distribution; a good heuristic is whether the variance to mean ratio is bigger than unity (in appropriate units). Macroparasites are typically aggregated in their host populations, so that the majority of hosts harbour few or no parasites while a few hosts harbour large parasite burdens. Aggregated distributions are often well described empirically by the negative binomial distribution. structure.

Aggressiveness J. E. van der Plank recommended that races of a pathogen that differ in the severity of their pathological effects but do not interact differentially with host varieties should be said to differ in aggressive- ness; when there is a differential interaction with varieties the races should be said to differ in virulence,

Robinson accepted this distinction, but recommended using the terms horizontal pathogenicity and vertical pathogenicity respectively, on the grounds that Pathogenicity can be described in terms of the kind of resistance it is able to overcome'. We doubt whether it is necessary or desirable in the present state of knowledge to devise separate terms to describe different kinds of pathogenicity corresponding to various kinds of resistance, but if so it would perhaps be preferable to use precisely defined but 'neutral' terms, such as pathogenicity, rather than applying words which may have connotations that are immediately inappropriate,or become inappropriate in the light of increasing knowledge of host-parasiterelations.

We have not accepted van der Plank's limited definition of virulence, nor can we recommend the use of aggressiveness or aggressivity, either in the restricted sense of van der Plank, or in the less precisely defined usages of other authors, for whom the terms appear to be synonymous with pathogenicity or virulence as defined here.

Agro forestry Cultivated mixtures of trees, crops and/or livestock.

Agronomy The study of crop production and soil management.

AIDS (acquired immune deficiency syndrome) An ultimately fatal condition produced when a virus destroys the T cells of the body's immune system; AIDS patients are attacked and sometimes killed by opportunistic fungi.

Air pollution Any air contaminant that causes undesirable effects on living organisms or materials.

Alae (nematology) Expansions or projections formed by a longitudinal thickening of the cuticle of a nematode. Cervical alae are confined to the anterior region of nematodes parasitic in animals. Caudal alae occur in the posterior region of males in a number of genera. Longitudinal alae, usually four, extend the length of the body sublaterally.

Alanine aminotransferase Aminotransferases are enzymes that facilitate the conversion of one amino acid into another, thus helping to maintain a balanced supply of amino acid building blocks for protein synthesis. Elevated alanine amino transferase activity provides a useful indicator for liver disorders.

Albedo White, spongy inner part of citrus fruit rind.

Albinism Complete absence of green colour, resulting from genetic factors or chimaeral growth.

Albino White or light-colored; having a marked deficiency in pigmentation.

Albumin A type of protein widely distributed in the tissues and fluids of plants and animals. It is the single most abundant protein in blood. Albumin acts as a carrier for numerous substances in the blood.

Alelopathy When one plant releases chemicals which cause certain other plants around it to grow slowly.

Aleurioconidium A terminal conidium, often thick-walled and pigmented, but sometimes thin-walled and hyaline, developed at the end of a conidiogenous cell or hypha, and from which it is not shed or shed only with difficulty; aleuriospore.

Algae Small organisms, mostly living in water, some of which are considered plants and some of which are considered protists. Many of them only have one cell. Algae is eaten by many aquatic animals. Terrestrial (land) algae is shelter and food for small animals like springtails and mites.

Algae (singular alga) encompass several different groups of living organisms that capture light energy through photosynthesis, converting inorganic substances into simple sugars using the captured energy. Algae have been traditionally regarded as simple plants, and some are closely related to the higher plants. Others appear to represent different protist groups, alongside other organisms that are traditionally considered more animal-like (that is, protozoa). Thus algae do not represent a single evolutionary direction or line, but a level of organization that may have developed several times in the early history of life on earth. Algae range from single-celled organisms to multi-cellular organisms, some with fairly complex differentiated form and (if marine) called seaweeds. All lack leaves, roots, flowers, and other organ structures that characterize higher plants. They are distinguished from other protozoa in that they are photoautotrophic, although this is not a hard and fast distinction as some groups contain members that are mixotrophic, deriving energy both from photosynthesis and uptake of organic carbon either by osmotrophy, myzotrophy, or phagotrophy. Some unicellular species rely entirely on external energy sources and have reduced or lost their photosynthetic apparatus.

All algae have photosynthetic machinery ultimately derived from the cyanobacteria, and so produce oxygen as a by-product of photosynthesis, unlike non-cyanobacterial photosynthetic bacteria. Algae are usually found in damp places or bodies of water and thus are common in terrestrial as well as aquatic environments. However, terrestrial algae are usually rather inconspicuous and far more common in moist, tropical regions than dry ones, because algae lack vascular tissues and other adaptations to live on land. Algae can endure dryness and other conditions in symbiosis with a fungus as lichen.

The various sorts of algae play significant roles in aquatic ecology. Microscopic forms that live suspended in the water column—called phytoplankton—provide the food base for most marine food chains. In very high densities (so-called algal blooms) these algae may discolor the water and outcompete or poison other life forms. Seaweeds grow mostly in shallow marine waters. Some are used as human food or harvested for useful substances such as agar or fertilizer. The study of algae is called phycology or algology.

Traditionally the cyanobacteria have been included among the algae, referred to as the cyanophytes or Blue-green Algae, (the term "algae" refers to any aquatic organisms capable of photosynthesis), though some recent treatises on algae specifically exclude them. Cyanobacteria are some of the oldest organisms to appear in the fossil record, dating back about 3.8 billion years (Precambrian). Ancient cyanobacteria likely produced much of the oxygen in the Earth's atmosphere.

Cyanobacteria can be unicellular, colonial, or filamentous. They have a prokaryotic cell structure typical of bacteria and conduct photosynthesis directly within the cytoplasm, rather than in specialized organelles. Some filamentous blue-green algae have specialized cells, termed heterocysts, in which nitrogen fixation occurs. All other algae are eukaryotes and conduct photosynthesis within membrane-bound structures (organelles) called chloroplasts. Chloroplasts contain DNA and are similar in structure to cyanobacteria, presumably representing reduced cyanobacterial endosymbionts. The exact nature of the chloroplasts is different among the different lines of algae, reflecting different endosymbiotic events. There are three groups (Archaeplastida) that have primary chloroplasts: 1. Green algae, together with higher plants

2. Red algae

3. Glaucophytes

In these groups, the chloroplast is surrounded by two membranes and probably developed through a single endosymbiosis. The chloroplasts of red algae have a more or less typical cyanobacterial pigmentation, while those of the green alga have chloroplasts with chlorophyll a and b, the latter found in some cyanobacteria and not most. Higher plants are pigmented similarly to green algae and probably developed from them. Two other groups of algae have green chloroplasts containing chlorophyll b: 1. Euglenids and

2. Chlorarachniophytes.

3. These are surrounded by three and four membranes, respectively, and were probably retained from an ingested green alga. Those of the chlorarchniophytes contain a small nucleomorph, which is the remnant of the alga's nucleus. It has been suggested that the euglenid chloroplasts only have three membranes because they were acquired through myzocytosis rather than phagocytosis. The remaining algae all have chloroplasts containing chlorophylls a and c. The latter chlorophyll type is not known from any prokaryotes or primary chloroplasts, but genetic similarities with the red algae suggest a relationship there. These groups include:

4. Heterokonts
5. Haptophytes
6. Cryptomonads
7. Dinoflagellates

In the first three of these groups (Chromista), the chloroplast has four membranes, retaining a nucleomorph in cryptomonads, and they likely share a common pigmented ancestor. The typical dinoflagellate chloroplast has three membranes, but there is considerable diversity in chloroplasts among the group, as some members have acquired theirs from different sources. The Apicomplexa, a group of closely related parasites, also have plastids though not actual chloroplasts, which appear to have a common origin with those of the dinoflagellates. Note many of these groups contain some members that are no longer photosynthetic. Some retain plastids, but not chloroplasts, while others have lost them entirely.

Most of the simpler algae are unicellular flagellates or amoeboids, but colonial and non-motile forms have developed independently among several of the groups. Some of the more common organizational levels, more than one of which may occur in the life cycle of a species, are: 1. Colonial - small, regular groups of motile cells

2. Capsoid - individual non-motile cells embedded in mucilage

3. Coccoid - individual non-motile cells with cell walls

4. Palmelloid - non-motile cells embedded in mucilage

5. Filamentous - a string of non-motile cells connected together, sometimes branching

6. Parenchymatous - cells forming a thallus with partial differentiation of tissues

There are the brown algae—some of which may reach 70 m in length (kelps)—the red algae, and the green algae. The most complex forms are found among the green algae, in a lineage that eventually led to the higher land plants. The point where these non-algal plants begin and algae stop is usually taken to be the presence of reproductive organs with protective cell layers, a characteristic not found in the other alga groups.

Some species of algae form symbiotic relationships with other organisms. In these symbioses, the algae supply photosynthates (organic substances) to the host organism providing protection to the algal cells. The host organism derives some or all of its energy requirements from the alga. Examples include: 1. lichens - a fungus is the host, usually with a green alga or a cyanobacterium as its symbiont. Both fungal and algal species found in lichens are capable of living independently, although habitat requirements may be greatly different from those of the lichen pair.

2. corals - algae known as zooxanthellae are symbionts with corals. Notable amongst these is the dinoflagellate Symbiodinium, found in many hard corals. The loss of Symbiodinium, or other zooxanthellae, from the host is known as coral bleaching.

3. sponges

Algae are used by man in a great many ways. Because many species are aquatic and microscopic, they are cultured in clear tanks or ponds and either harvested or used to treat effluents pumped through the ponds. Algaculture on a large scale is an important type of aquaculture in some places.

Algicide A pesticide that is used to kill or inhibit algae.

Algicolous Living on algae.

Alimentary toxic aleukia (ATA) A mycotoxicosis caused by T-2 toxin which killed many people in Russia during and after World War II.

Alkaline Having basic (nonacidic) properties; pH greater than 7.

Alkaline phosphatase A group of enzymes that belong to the class known as hydrolases. They are thought to play an important role in the transport of sugars and phosphates in the intestine, bone, kidney and placenta. Elevated serum levels of alkaline phosphatase activity may indicate liver disease.

Alkaline soil A soil with a pH value greater than 7.0.

Alkaloids Nitrogen-containing organic compounds produced by plants; physiologically active in vertebrates; many have a bitter taste and some are poisonous.

Allele (allelomorph) Any of one or more alternative forms of a given gene; both (or all) alleles of a given gene are concerned with the same trait or characteristic, but a particular allele codes for a product qualitatively and/or quantitatively different from that coded by other alleles of that gene.

Allelopathic Substance produced by one plant which injures or suppresses the growth of another.

Allelopathy Ability of one species to inhibit or prevent the growth of another species through the production of toxic substance(s).

Allergy When a person is very sensitive to certain foods, liquids, pollen, or animals. The allergy can make the person ill, including sneezing, itching, and rashes. Severe allergies can even be fatal. Some organisms which cause uncomfortable allergies with people include Ragweed (pollen), Poison Ivy (oil), and Honey Bees (venom).

Allowable depletion The proportion of available water that can be used before irrigation is needed.

Alpha spore A fertile spore produced by the asexual state of fungi in the Diaporthaceae, especially the genus Phomopsis. Spores are fusoid to oblong and are usually produced together with beta-spores.

Alternaria sp Alternaria is a large and widespread genus, the conidia of which are easily carried by the wind, with peak concentrations in the summer and early fall. Alternaria is commonly found in house dust, carpets, textiles, on horizontal surfaces in building interiors, and window frames. It is one of the main fungal causes of allergy, being a common type I & III allergen. Outdoors, it may be isolated from samples of soil, seeds and plants, and is frequently reported in air. The large spore size suggests that this fungus will deposit in the nose, mouth and upper respiratory tract causing nasal septum infections. It has also been associated with hypersensitivity pneumonitis. It is a common cause of extrinsic asthma. Acute symptoms include edema and bronchiospasms; chronic cases may develop pulmonary emphysema. Baker's asthma is associated with inhalation of Alternaria conidia present in flour. Other diseases caused by Alternaria include: Farmer's lung, mycotic keratitis, skin infections, and osteomyelitis. Also, the species A. alternata is capable of producing tenuazonic acid and other toxic metabolites that may be associated with disease in humans or animals. Several species are pathogenic to plants and contribute to the spoilage of agricultural products. Alternaria has been isolated from substrates such as sewage, leather, stone monuments, optical instruments, cosmetics, computer disks, and jet fuel. Morphological characteristics include abundant production of conidia that are large (18-83 x 7-18 microns) and multicellular with both transverse and longitudinal septa; conidiophores are dark, mostly simple. Colonies grow fast, are suede-like to floccose, and black to olivaceous-black or grayish in color. (Aw–0.85-0.89) Culture - Potato dextrose agar or Malt extract agar, 20° - 25°C, 7 – 10 days.

Alternate Leaves arranged singly at different heights and on different sides of the stem.

Alternate host Either of the two hosts of a heteroecious pathogen, e.g. a rust, or of a heteroecious pest, e.g. an aphid. When several or many plant species are hosts of a given pathogen they may be described as alternative hosts.

One of two taxonomically different hosts required by a heteroecious rust fungus to complete its cycle; also applies to some insects.

One of two kinds of plants on which a parasitic fungus (e.g., rust) must develop to complete its life cycle.

Alternation of generations A life cycle in which there are two stages: (a) a diploid, sporophyte stage which produces haploid spores by meiosis, and (b) a haploid, gametophyte stage which produces haploid gametes of two types, usually called male and female but technically microgametes and megagametes. The female gamete is fertilized to become a diploid zygote, which matures to become a diploid sporophyte, completing the cycle. Either the gametophyte or the sporophyte stage may be, and typically is, very small and remains attached to the other stage.

Aluminum toxicity High levels of aluminum or manganese are common in acid soils and can be toxic to both the plant and rhizobia.

Alveola A small cavity or pit on a surface.

Alveolate Pitted or honeycombed on the surface.

Amanitins The cause of most fatal mushroom poisonings.

Amatoxins Cyclic octopeptides ncluding and amanitin) found in some spp. of Amanita, Galerina; extremely toxic to humans.

Ambimobile Describes systemic fungicides which can move upward in the xylem or downward in the phloem.

Amblyosporium sp Amblyosporium is a saprobe commonly found on decaying animals and feces because it grows well on ammonium and amino acids. It is considered a mitosporic (it lacks a known sexual state and belongs the Fungi Imperfecti) and ectomycorrhizal fungi (lives on the surface of the roots and forms a Hartig net). Amblyosporium has been isolated from Sitophilus oryzae L., an important insect pest of stored grain and processed foods. It has even been isolated from fabric. One species, A. botrytis forms effused orange-red tufts on decaying fungi. Culture - Potato dextrose agar or Malt extract agar, 20° - 25°C, 7 – 10 days.

Ambrosia fungi Symbiotic wood-inhabiting fungi associated with wood-wasps and wood-inhabiting beetles; the fungi feed the larvae, and are dispersed by the adults.

Amendment, physical Any substance added to the soil for the purpose of altering physical conditions (sand, calcined clay, peat, etc.).

Amerospores Amerospores are small, round, non-septate asexual spores or spore-like particles, indistinguishable from each other at 600X magnification using light microscopy. They include Trichoderma and unchained spores of Aspergillus and Penicillium. Amerospores can also include Acremonium, Verticillium, Paecilomyces, Scytalidium, Cunninghamella, Monocillium, Gliocladium, and some yeasts.

Amino acid A family of modified organic acids that serve as building blocks for the synthesis of proteins.

Ammonification The formation of ammonium ions from nitrogenous compounds.

Amoeboid Without a cell wall or a definite shape, moving and feeding by means of pseudopodia (temporary cytoplasmic protrusions).

Amorphous Shapeless, formless.

Amphibian Cold-blooded, smooth-skinned animals from the Chordate Phylum. This class of animals includes frogs, toads, salamanders, and newts. Some live on land and some in the water, but most species return to the water to mate and lay eggs.

Amphid In nematodes: A chemosensory organ, occurring laterally in pairs, located in the anterior region. Sometimes called lateral organs.

Amphidelphic In nematodes: Having two ovaries, one directed anteriorly and the other posteriorly.

Amphigynous Having an antheridium through which the oogonium grows, as in many Phytophthora species.

Amphimictic In nematodes: 1. Reproduction in which sperm and eggs come from separate individuals (cross-fertilization).

2. Capable of interbreeding freely and of producing fertile offspring.

Amphimixis Copulation of two unrelated cells and nuclei, e.g., egg and sperm; reproduction by a sexual process.

Amphimobile Able to move both upward and downward in phloem, as do some pesticides.

Amphiospore Urediniospore with thickened walls and capable of hibernating.

Amphithecium The outer, "ectodermal" cell layer of the sporophyte in basal plants.

Amphotericin B An antifungal polyene antibiotic derived from actinomycetes (Streptomyces spp.); used to treat systemic mycoses.

Amplexus The mating position of frogs and toads. The male "hugs" the female and waits for her to lay eggs so that he can fertilize them. Amplexus can last for hours, or sometimes even days. While the frogs are in amplexus, the female swims or hops away from danger, and the male gets a piggy-back ride.

Amplification Production of multiple copies of a DNA sequence, either in vivo or in vitro, starting with one or a few copies.

Ampulliform Flask-like in form.

Amygdaliform Almond-shaped.

Amylase An enzyme that breaks down complex carbohydrates such as starch.

Amyloid Of spore walls, spore ornamentation, hyphal walls, ascus tips, etc.: Staining blue or grayish to blackish-violet in Melzer's reagent, presumably because of the presence of starch or a starch-like compound.

Anaerobic Describing cells which can live without oxygen and utilize another substrate, e.g., sulphur, as an electron acceptor. The relatively recently discovered, atypical chytridiomycetes, Callimastix and allies, are unusual among fungi in being obligate anaerobes. Most fungi are aerobic.

Anaerobic digestion The breaking down of organic matter in oxygen free conditions.

Anagensis Evolutionary change along an unbranching lineage; change without speciation.

Analogous Of similar form and/or function, but of different evolutionary origin.

Anamorph The asexual form in the life cycle of a fungus, when asexual spores (such as conidia) or no spores are produced.

Anamorphosis The complete asexual, mitotic, diasporic expression of a fungus, comprising one or more anamorphs.

Anaphylactic shock An allergic reaction marked by contraction of smooth muscle and dilation of blood vessels. If not checked rapidly by an injection of epinephrine, the reaction can be lethal.

Ancestor Any organism, population, or species from which some other organism, population, or species is descended by reproduction.

Androgynous Having the antheridium and oogonium on the same hypha.

Androstenol Mammalian pheromone produced by truffles.

Anemometer An instrument used for measuring wind speed and direction.

Anesthetic A chemical which causes an organism not to feel pain. Mosquitos, ticks, and leeches have an anesthetic in their saliva so that their host cannot feel them feeding.

Aneuploid Chromosome constitution differing from the usual diploid constitution by loss or duplication of chromosomes or chromosomal segments.

Angiocarpous. Of basidium-producing organs: Hymenial surface at first exposed but later covered by an incurving pileus margin and/or excresences from the stipe.

Angular leaf spot Bacterial blight.

Anion A negatively charged ion.

Anisogamete A motile gamete dissimilar in form or physiology to the gamete with which it fuses.

Anisogamy Fusion of gametes of different size, as in Allomyces (Blastocladiales: Chytridiomycota).

Anixiella sp Anixiella is a synonym for the genus Gelasinospora, an ascomycete. Anixiella is a decomposer that thrives on feces or decaying plant material. It can be found even in artic regions. Culture - Potato dextrose agar or Malt extract agar, 20° - 25°C, 7 – 10 days.

Annellide A conidiogenous cell in annellidic conidiogenesis which, by repeated percurrent proliferations, develops an elongated tip marked by transverse bands (annellations).

Annellidic Describes blastic conidia produced in sequence by short percurrent extensions of a conidiogenous cell, which leave annular scars (annellations) on the cell.

Annual A plant that lives for only one year. It grows from a seed (usually in Spring), flowers, and then grows fruits. It dies when cold weather arrives.

Annual ring Single-year growth of xylem in a woody stem.

Annulations Grooved bands in the cuticle of some nematodes.

Annule Thickened interval between transverse striae in the cuticle of a nematode.

Annulus The skirtlike remnant of the partial veil which shields the lamellae as they produce the spores and protects the spores until they are ready for release. It surrounds the upper part of the mushroom stem below the cap and sometimes slips down. Not found on every mushroom. Also called "annular ring."

Anorexia nervosa A personality disorder manifested by extreme aversion to food, usually occurring in young women.

Antabuse (disulfiram) Substance which, for several days after it has been ingested, will interrupt the catabolism of alcohol at the acetaldehyde stage, causing headache, numbness of extremities, metallic taste, palpitation, nausea, vomiting, blurring of vision; used to treat chronic alcoholism.

Antagonism Arelationship between different organisms in which one partly or completely inhibits the growth of another or kills it. The term is usually applied to the effects of toxic metabolites of one organism on another, e.g. of Trichoderma viride on Rhizoctonia solani, but does not exclude mutual inhibition between organisms.

Antagonist Organisms that release toxins or otherwise change conditions so that activity or growth of other organisms (especially pests) is reduced.

Antagonistic symbiosis Parasitism; one organism of an association beefits at the expense of the other.

Antenatal The period between conception and birth. Same as prenatal.

Antenna Main sense organ on insects, crustac ans, and some other animals. Antennae are used to collect information from the environment, including the location of food sources and possible dangers.

Anterior In front; before; opposite of posterior.

Anterograde Moving forward.

Anthelmintic A drug, for example praziquantal, used sp:cifically against helminth or worm infections.

Anther In a flower, the terminal part of the stamen which holds the pollen; an antheridium.

Antheridium Male sexual organ (male gametangium) found in some fungi.

Anthers The pollen-producing organs of flowers.

Anthesis The period of the opening of a flower during which pollination can occur.

Anthocyanescence Having reddish-purple color in tissues that are normally green; often a symptom of plant disease appearing in the nearly dead margins around completely dead spots in green leaves.

Anthocyanosis Presence of abnormal red-purple coloration in foliage.

Anthracnose Common name of plant diseases characterised by black lesions, usually sunken, caused by certain imperfect fungi that produce conidia in acervuli, e.g. Colletotrichum, Gloeosporium and some closely-related Sphaceloma spp. We do not recommend the use of the term except as the common name for diseases caused by Colletotrichum lindemuthianum on Phaseolus, C. coffeanum on coffee, and Gloeosporium limetticola on lime, Citrus aurantifolia. The disease of raspberry and loganberry caused by Elsino veneta is better known as 'cane spot'.

Disease caused by acervuli-forming fungi (order Melanconzales) and characterized by sunken lesions and necrosis.

Anti microbials Pesticides intended to kill unwanted microbes such as fungi, bacteria, or viruses. Also called disinfectants.

Antibiosis An association between organisms, or between an organism and a metabolic product of another organism, that is harmful to one of them.

Antibiotic A substance produced by a microorganism and able to inhibit the growth of other microorganisms, or to kill them. A complex chemical substance produced by one microorganism that inhibits or kills other microorganisms (e.g., streptomycin).

Antibiotic pesticide An active ingredient that is isolated from or produced by a microorganism (e.g. a bacterium or a fungus), or a related chemical that is produced artificially.

Antibody A protein produced in the blood of vertebrates following exposure to an antigen. The antibody binds specifically to the antigen and thus stimulates its inactivation by other parts of the immune system.

The major classes of antibody are immmunoglobulin A, or IgA, found predominantly in bodily secretions like saliva; IgM and IgG which are typically produced sequentially in response to microparasitic infections; and IgE which is often elevated in the response to helminth infections. Only IgG is able to cross the placenta to provide maternal immunity.

Anticholinergic Interfering with the transmission of nerve impulses.

Anticlinal Perpendicular to the surface.

Anti-coagulant A chemical that keeps blood from clotting. In most animals blood naturally clots (stops up) when there is a wound. Some parasites, like leeches, have an anti-coagulant in their saliva so blood will flow freely for them to drink.

Antidepressants Pharmaceutical agents used to treat clinical depression.

Antidote A remedy used to counteract the effects of a poison.

Antifeedant A pesticide that prevents an insect or other pest from feeding.

Antifungal substance A substance of known or unknown chemical structure which reduces the germination of fungal spores in water or the growth rate of fungal hyphae at an EC 50 of 100 /ml or less. Complete inhibition usually occurs at a concentration of less than 500 /ml. Many substances, even nutrients, can be inhibitory at a sufficiently high concentration; hence it is necessary to stipulate that a substance should be active at a very low concentration below some arbitrary level before it can be described as an anti-fungal substance.

Antigen A substance, usually a protein or a polysaccharide which, when injected into the body of a living warm-blooded animal, induces the formation in the blood of proteins (antibodies) with which the antigen reacts specifically. The reaction may be observed in vitro by precipitation, agglutination etc.

A protein, typically foreign, that elicits a specific immune response.

Any foreign chemical (normally a protein) that induces antibody formation in animals.

Anti-inflammatory Agents that reduce inflammation without directly antagonizing the agent that caused it.

Antinuclear antibody Anti-self antibodies directed against the DNA. It is one indicator for the autoimmune disorder systemic lupus erythematosis (SLE).

Antiseptic A substance that prevents, retards, or destroys microorganisms.

Antiserum The serum from a vertebrate that has been exposed to an antigen and which contains antibodies that react specifically with the antigen.

Antiserum titre The greatest dilution of an antiserum which will give a detectable reaction with an antigen. Usually detected by a series of two-fold dilutions 1 : 2, 1 : 4, 1 : 8,.....1 : 256 and usually expressed as the reciprocal, 2, 4, 8, ... 256 etc.

Antisporulant A material preventing or decreasing spore production without killing vegetative growth of a fungus.

Antithesis The time of flower opening in a plant.

Ants (formicidae) A family of insects related to bees and wasps.

Anxiety disorders Also known as anxiety neurosis or anxiety reaction. A condition that can be caused by both psychologic and physiologic factors. It can take two general forms: 1. acute anxiety (panic disorder), marked by repeated occurrences of intense self-limited anxiety lasting usually a few minutes to an hour, or

2. chronic anxiety, characterized by less intense reactions of much longer duration (days, weeks, or months).

AONB Area of Outstanding Natural Beauty.

Apex Tip of a root or shoot, containing the apical meristem.

Aphanoplasmodium A fine plasmodial network of hypha-like strands, with finely granular protoplasm and an indefinite margin; known in a few Myxomycetes, especially Stemonitaceae.

Apheresis Fractionation of blood by centrifugation, so that specific components can be replaced.

Aphid Small, sucking insect of the family Aphididae (order Homoptera) that produces honeydew and injures plants when in large populations.

Aphids (aphididae) A family of insects related to whiteflies and planthoppers. Aphids are soft-bodied, sometimes called plant lice, and feed by sucking juices of plants.

Apiary A place where bees are kept, such as a bee hive.

Apical apparatus in an ascus, one or more areas or structures in the apex which function in the discharge of ascospores.

Apical apparatus or apical ring A specialized structure found in the tips of many unitunicate-inoperculate asci, acting as a valve or sphincter that controls the forcible expulsion of ascospores.

Apical dominance Growth of the bud at the apex of a stem or tuber while growth of all other buds on the stem or tuber is inhibited.

Apical meristem Terminal growing point.

Apiculate Having an apiculus, a small pointed projection at the tip.

Apiculture The study of bees.

Apiculus 1. A short projection at one end of an object;

2. in basidiospores, hilar appendix.

Aplerotic Of an oospore, not filling the oogonium.

Apomorphy A character state which is unique to a single, terminal taxon. Example: among primates, complex grammar is an apomorphy of human beings. It is quite diagnostic of humans, but useless in determining phylogenetic relationships because it is not a shared, derived characteristic, or synapomorphy, of any larger group.

Apophysis A swelling or swollen filament, e.g. at the end of a sporangiophore below the sporangium in Mucorales; in basidiomycetes, the swelling at the tip of a sterigma from which the basidiospore develops and which becomes the hilar appendix .

Apoplastic Describing the movement of substances via the cell walls: not entering the living cells.

Apothecia (plural) Cup-shaped, spore-bearing structures produced by certain types of fungi such as Sclerotinia.

Apothecial ascoma (apothecium) The usually cup- or saucer-like ascoma of the 'discomycetes'; the hymenium is exposed at maturity, as in cup fungi and morels.

Apothecium (mollisia dehnii)apothecium A disk-shaped or cup-shaped ascocarp (fruiting body) of a lichen or non-lichenized ascomycete. In an apothecium, the asci are born in a single, orderly layer on an open, fairly flat surface.

Appendage A process or outgrowth of any sort; in conidia, cellular (tubular) or extracellular (mucilaginous), filiform to capitate or infundibuliform ornamentation.

Appendiculate Of a basidioma, where the pileal margin is fringed with pieces of the partial veil.

Applanate Flattened.

Apple scab Disfiguring disease of apple leaves and fruit caused by the Spilocaea pomi anamorph of Venturia inaequalis (Dothideales: Ascomycetes).

Application rate The amount of pesticide formulation applied to a given area.

Appressed Closely flattened down or pressed against a surface.

Appressorium Swollen, flattened portion of a fungal filament that adheres to the surface of a higher plant, providing anchorage for invasion by a fungus.

Aquaculture The cultivation of aquatic organisms for the production of human food. (i.e. Fish farming)

Aquatic Describes an organism that lives in or near water.

Aquifer An underground formation of sand, gravel or porous rock that contains water.

Arable Able to be cultivated for agriculture.

Arachnoid Cobweb-like.

Arborescent Of tree-like growth-habit, a tree. A woody plant.

Arbovirus A virus which uses Arthropods as vectors and is transmitted in their saliva to the definitive host. For example, yellow fever.

Arbuscular mycorrhiza Symbiotic association between a nonpathogenic or weakly pathogenic fungus and the roots of plants in which fungal hyphae invade cortical cells of the root and produce vesicles and arbuscles.

Arbuscule Of vesicular-arbuscular mycorrhizae: a much-branched, microscopic haustorial structure of the fungal symbiont that forms within living cortical cells of the root. The interface of the arbuscule with the plant protoplast is a site of exchange of nutrients and growth-regulating chemicals.

Archaeasceous (of asci) Essentially bitunicate, splitting at the tip; part of the inner wall may protrude; common in lichens, esp. Lecanorales.

Archaebacteria An ancient line of bacteria, including the methanogenic, halophilic and thermoacidophilic groups found in hot springs and sea-floor vents; their ribosomes are distinctive, and they also differ from eubacteria in their ribosomal RNA, wall chemistry, and lipids.

Archegonia archegonia Reproductive structure that produces female gametes (egg).

Arcuate Curved like a bow.

Arenicolous Living or growing in sand.

Areolated Divided into small spaces or areolations; usually pertains to the cuticle of a nematode.

Areolation A condition in which the transverse body annulation of a nematode traverses the lateral field.

Arid climate A dry climate with an annual precipitation of less than 10 inches.

Ark A moveable outdoor pig shelter.

Aromatics Compounds derived from the hydrocarbon benezene (CH_6)

Arthric Of conidiogenesis, thallic conidiogenesis by which a hyphal element is fragmented into conidia after transverse septation.

Arthrinium sp Arthrinium is a widespread saprobe. It is found on plants including sugarcane and especially swamp grasses & sedges. This genus is often isolated from air near grassy places in the autumn. Only one species is considered to be allergenic. There have been no reported cases of infections or toxin related diseases in humans or animals. Conidiophores are simple, mostly hyaline except for thick dark septa. Conidia are dark, 1-celled, broadly fusoid, ovoid with an equatorial germ slit, and are attached on the side and apex of conidiophore. They usually occur in grape-like masses on white wooly colonies. Culture - Potato dextrose agar or Malt extract agar, 20° - 25°C, 7 – 10 days.

Arthroconidium A thalloconidium produced by the conversion of a pre-existing, determinate hyphal element into a conidium.

Arthrodontous Of moss sporophytes, having peristome teeth which are formed by walls growing between the rows of cells making up the mouth of the spore capsule.

Arthrographis sp Arthrographis is a widespread allergen commonly found in soils, compost, and decaying plants. Only two species have been identified: A. cuboidea and A. kalrae. A. kalrae is a known causative agent of onychomycosis and has been recovered from the skin, nails, and respiratory secretions of patients with chronic pulmonary disease. The fungi's growth is variable, but it has the ability to grow at 45°C, which is significant for its identification. Culture - Potato dextrose agar or Malt extract agar, 20 - 25°C, 7 – 10 days.

Arthropod An animal in the Arthropod Phylum. Thes animals have exoskeletons (hard part is on the outside), jointed legs, and segmented bodies (split into sections). This phylum includes insects, spiders, crustaceans, centipedes, and millipedes.

Arthropoda That branch (phylum) of the animal kingdom whose members have jointed legs and are also made up of rings or segments. For example, worms, insects, crustaceans, etc.

Arthrospore A spore resulting from the fragmentation of a hypha, as in the conidial stage of many Basidiomycetes.

Ascending Of conidiophores, curving up.

Asci Several saclike cells in which meiosis occurs and which generally contain eight spores each.

Asciascospore A meiospore borne in an ascus

Ascigerous Of fungi: having asci.

Ascocarp Sexual fruiting body of an ascomycetous fungus that produces asci and ascospores; e.g. apothecium, ascostroma, cleistothecium, perithecium, pseudothecium.

Ascoconidium A budded spore arising from an ascospore within an ascus.

Ascogenous Ascus-producing or supporting, usually restricted to hyphae.

Ascogenous hyphae The restricted dikaryophase of Ascomycetes; the dikaryotic hyphae which grow out from the fertilized ascogonium and eventually give rise to asci.

Ascogenous state The ascospore-producing state of Ascomycete.

Ascogonium In Ascomycetes: the female gametangium; it may be unicellular or multicellular, simple or complex in form.

Ascohymenial Relating to an ascocarp which forms after nuclear pairing. The ascocarps produced by ascohymenial development extend out into the medium, as in the image at apothecium, and the asci are unitunicate.

Ascolocular Relating to a mode of ascocarp growth in which a perithecium develops within a cushioning hollow of cells (the stroma) which is nested in a depression of the hymenium (a locule). The ascocarps produced by ascolocular growth do not grow into the medium, and the asci are usually bitunicate.

Ascoma 1. Same as ascocarp

2. "an ascocarp having the spore-bearing layer of cells (the hymenium) on a broad disklike receptacle." This latter definition is frequently cited but appears to be either incorrect or useless. For our purposes, ascoma and ascocarp will be treated as synonymous.

Ascomycetes A group of fungi, also called the Ascomycota, that produces sexual spores (ascospores) within a saclike structure called an ascus.

Ascomycotina A subdivision of fungi characterized by the formation of sexually derived spores (ascospores) in asci. (ascomycetes, ascomycota)

Ascospore Ascospores are a general category of spores that have been produced by means of sexual reproduction (in a sack-like structure called an ascus). These are ubiquitous saprobes and plant pathogens, many of which are easily identifiable (i.e. Chaetomium). This group contains potential opportunistic pathogens, toxin producers, and allergens depending on the genus and species. A rupture in the top portion of the ascus disperses the spores during rain or in times of high humidity. Some asexual fungi, such as Aspergillus and Penicillium can become sexual under specific conditions, these are then considered ascomycetes and are given distinct names.

Ascostroma A fruiting body containing bitunicate (double-walled) asci in locules (cavities); usually dark with multiple locules, but sometimes single.

Ascus A sac-like cell containing the ascospores cleaved from within by free cell formation after karyogamy (nuclear fusion) and meiosis. Eight ascospores typically are formed within the ascus, but this number may vary considerably.

Aseptate Having no cross walls; lacking septa; nonseptate; coenocytic.

Aseptic Free or freed from contaminating organism(s); (of microbiological technique) pertains to working under sterile conditions and using sterile techniques, e.g., working in a laminar flow bench, using flame sterilized inoculating loops and instruments.

Asexual Not forming part of a cycle which involves fertilisation and meiosis.

Asexual reproduction A form of reproduction neither preceded nor followed by meiosis; in fungi, commonly involves mitospores, such as, sporangiospores or conidia.

Asexual stage Vegetative; without sexual organs or spores.

Asexual state Either a vegetative state or a reproductive state in the life cycle of a fungus in which nuclear fusion is absent and in which reproductive spores are produced by mitosis or simple nuclear division. Synonym: imperfect state.

Asexual Vegetative; without sex organs, gametes, or sexual spores; the imperfect or anamorphic stage of a fungus.

Asperate Rough with points or projections.

Aspergilloma A 'fungus ball' composed principally of hyphae of Aspergillus, found in a pre-existing cavity (esp. in an upper lobe of the lung) or a bronchus; usually relatively benign or asymptomatic.

Aspergillosis Any disease in man or animals caused by Aspergillus (esp. A. fumigatus).

Aspergillus sp Aspergillus is a common type I & III allergen. They are frequently isolated from forest products, soils, grains, nuts, cotton, organic debris, and water damaged building materials. Spores can also be found in moist ventilation systems and house dust. There are more than 160 different species of Aspergillus, sixteen of which have been documented as etiological agents of human disease but rarely occur in individuals with normally functioning immune systems. However, due to the substantial increase in populations of individuals with HIV, chemotherapy patients and those on corticosteroid treatment, contamination of building substrates with fungi, particularly Aspergillus is of concern. Aspergillosis is now the second most common fungal infection requiring hospitalization in the United States. Many Aspergillus species produce mycotoxins that may be associated with diseases in humans and other animals. Toxin production is dependent on the species or strain within the species and on the food source for the fungus. Some of these toxins are carcinogenic including aflatoxins and ochratoxin. Aspergillus is a common cause of extrinsic asthma with symptoms including edema and bronchiospasms, and chronic cases may develop pulmonary emphysema. These fungi are frequently secondary opportunistic pathogens in patients with bronchiectasis, carcinoma, other mycosis, sarcoid, and tuberculosis. Some species can also cause onychomycosis (infection of the nail). (Aw – 0.71 – 0.94). Culture - Potato dextrose agar or Malt extract agar, 20° - 25°C, 7 – 10 days. Speciation of Aspergillus requires the culture of the fungus under different conditions of media, humidity, and temperature.

Aspergillus terreus Aspergillus terreus occurs in the tropics and subtropics with a worldwide distribution among soils, grains, straw, cotton, stored foods, and decaying vegetation, and has been isolated from air and house dust. Also found in patients with cystic fibrosis, this species has grown in the human ear causing otomycosis, and can damage human nails (onychomycosis) and skin. It can produce mycotoxins including itaconic acid, patulin, mevinolin, and citrinin, which may be associated with disease in humans and animals. Mevinolin has the ability to lower blood cholesterol in humans; the compound has been manufactured & approved for use by the USDA under the trade name Mevacor. (Aw-0.78) Culture - Potato dextrose agar or Malt extract agar, 20° - 25°C.

Aspergillus versicolor Aspergillus versicolor can be found mostly in temperate areas in air, house dust, foods, soils, hay, cotton, and dairy products. Its presence in indoor air often indicates signs of moisture problems in buildings, as it is readily found in water damaged building materials. This species produces the mycotoxin sterigmatocystin, which is reported to be carcinogenic to the liver and kidney, and it can cause such symptoms as diarrhea and upset stomach. It also produces the volatile organic compound (VOC) geosmin, this compound causes irritation of the mucus membranes of humans and pets; also causing the characteristic musty, earthy odor often connected with moldy houses. A. versicolor may be in various colors, as the name implies, and is very common and displays great variety in colony pattern and size. (Aw-0.78) Culture - Potato dextrose agar or Malt extract agar, 20° - 25°C.

Asperulate Delicately asperate.

Asporogenous Not forming spores; of yeasts, those without a known teleomorph.

Associate A plant that is known to grow with another plant. For example, Common Cattail often grows with Yellow Pond Lily, so the pond lily is associated with cattails. Sometimes associated plants compete with each other for resources (light, water, soil, space). Other times they protect each other, such as when trees block the wind from other trees or larger plants provide shade for smaller plants. Some plants even grow on top of each other, or climb up other plants.

Asterinaceous fungus Referring to fungi of the family Asterinaceae, order Hemisphaeriales, which are plant parasites with shield-shaped ascomata associated with a superficial hyphopodiate mycelium.

Asthma A disease marked by recurrent attacks of distressed breathing. It is most often due to allergic reactions to plant or animal substances or to food products.

Asymptomatic Not showing any clinical manifestations.

ATC Automatic temperature compensation.

Athlete's foot Irritation of the skin between the toes caused by dermatophytes (conidial anamorphs of the Arthrodermataceae, Onygenales: Ascomycetes).

Atomate With a finely powdered surface.

Atomize To reduce a liquid to fine droplets by passing it under pressure through a suitable nozzle, or by applying drops to a spinning disc.

Atopic Susceptible to developing allergies.

ATP (adenosine triphosphate) The major source of usable chemical energy in metabolism; on hydrolysis, ATP loses one phosphate to become adenosine diphosphate (ADP), releasing usable energy.

Atrophy Under-development of a tissue due to a decrease in size of the individual cells.

Attenuate 1. Gradually narrowed towards the end;

2. of a pathogen, having lowered virulence.

Attenuation Lessening of the capacity of a pathogen to cause disease; reduction of its virulence.

Attractant Something that is attractive. For example: the song of mole cricket males is attractive to females of the same species, and also is attractive to the gravid female fly Ormia depleta. Songs of mole cricket males also attracts some other males of the same species and, to an even more limited extent, of other species. Incandescent and fluorescent lights are attractive to some flying adult mole crickets at night and to some other insects. Traps have been designed to take advantage of these attractants. Baits have been developed as attractants for mole cricket nymphs and adults. In some insects, the scent of females is attractive to males and/or the scent of males is attractive to females, but this kind of attraction has not been demonstrated in mole crickets.

Attributable fraction The same as aetiological fraction.

AUDPC A measure of the total amount of disease over a period of time, determined from graphs of disease vs. time, which can be used to compare epidemics quantitatively.

Augmentation Release of natural enemies of pests to suppress numbers of pests.

Augmentative biological control The release of additional numbers of a natural enemy when too few are present to control a pest effectively. This relies on an ability to mass-produce large numbers of the natural enemy in a laboratory. For instance, there are native species of predatory mites which attack and kill pest mites on strawberries and other crops. Growers can buy large numbers of these predatory mites to augment those already on their crops, and so achieve biological control of the pest mites.

Aureobasidium sp Aureobasidium is a saprobe, or weak parasite, type I & III allergen, and common in a variety of soils outdoors. It is widespread in the indoor environment and is common in places that moisture accumulates like bathrooms, kitchens, shower curtains, tile grout, and windowsills. This genus has 14 species, A. pullulans being the most common. Indoors A. pullulans is often found as a black stain on damp materials in homes such as painted wood. This species has also been reported to cause chromoblastomycosis (in an immunocompromised patient), which is a chronic cutaneous infection of the skin. Morphology is characterized by producing black, shiny colonies. This fungus produces abundant spores, that are 1-celled, ovoid, and 5-7 microns in size. Culture - Potato dextrose agar or Malt extract agar, 20° - 25°C, 7 - 10 days.

Auricle Claw-like appendages occurring in pairs at the base of the leaf blade or at the apex of the leaf sheath.

Autoclaving Sterilization by steam under pressure (15 lb./in^2), for prescribed time periods; better than dry heat, as sterilization is achieved at lower temperatures.

Autoecious (of a plant parasite or pest) Completing the life cycle on one kind of host

Completing entire life cycle on one host; especially applied to the rust fungi.

The need of only one host for completing the life cycle of a rust.

Autoimmune diseases Diseases caused by the body's immune system attacking its own tissues or organs, e.g. rheumatoid arthritis, juvenile diabetes, lupus.

Autolysis Self-digestion resulting in cell breakdown and often liquefaction.

Autotroph An organism that synthesizes its nutritive substances from inorganic molecules; e.g. plants capable of photosynthesis.

Autotrophic Capable of growth independent of outside sources of nutrients or growth factors.

Autotrophic nutrition Synthesis of organic food molecules from inorganic compounds such as carbon dioxide

Auxiliary cell Cell from which the extramatrical spores of Gigaspora and Scutellospora arise.

Auxin A plant growth substance capable of inducing cell extension in vitro tests on fragments of higher plant tissue. Distinction must be made between synthetic growth-regulating compounds with these properties, such as a-naphthaleneacetic acid and 2,4,-naphthalene-acetic acid and 2,4,-dichlorophenoxyacetic acid (for which the term ' synthetic auxin ' is often used), and endogenous auxins. Endogenous auxins are all indolyl compounds or closely related substances, typical of which is the chief auxin, indol-3-yl-acetic acid (IAA).

Auxotroph A biochemical mutant deficient for one or more substances; it will grow on minimal medium only if it has been supplemented with these substances.

Auxotrophic Having nutritional requirements (e.g. for growth factors) additional to those of the 'wild type'.

Available nutrient A nutrient in soil that can be readily absorbed and used by plants; a nutrient in soluble or exchangeable form.

Available visible radiation Wavelengths between 0.4 and 0.7 μm also referred to as 'usable radiation': equal to c 45% of direct solar radiation.

Available water The amount of water held in the soil that can be extracted by plants.

Avicide The class of pesticides used to kill birds.

Avirulence (avr) gene Gene in a pathogen that usually causes a hypersensitive reaction, is associated with active plant defense reactions in a resistant plant, and causes disease in a susceptible plant.

Avirulent Descriptive of a variant of a pathogenic organism that is incapable of causing severe disease. Synonymous with non-virulent, which is the preferred term because it avoids confusion in speech with 'a virulent'.

Avoidance If a disease does occur routinely in an area, there may be ways to avoid disease development. Choice of planting site, time of planting, storage conditions, and wound avoidance are a few of these techniques. Phytophthora root rots can be avoided by not planting susceptible plants in heavy, poorly drained soils. Planting later in the year, when soils are drier and warmer, will prevent damping-off diseases that are common to many vegetables.

Wounding can provide entrances for pathogens or weaken plants so that they are less able to defend themselves. For example, the crown gall bacterium must enter raspberry, cherry, and many other host plants through wounds. Therefore, careful planting will reduce potential of this disease. Planting certified virus-free stock is a good way to avoid virus-related diseases on potatoes and many flowering plants grown in greenhouses.

Good horticultural practices, such as proper fertility, pruning, watering, and training go a long way toward minimizing disease problems.

Awn Hairlike projections, usually extending from the mid-nerve or section of a grass flower.

Axenic Describes a condition in which an organism grows alone, with no other organisms (host, symbionts or parasites) present, as in 'axenic culture'.

Axeny Inhospitality, 'passive' as opposed to 'active' resistance of a plant to a pathogen

Axil The upper angle between a twig or leaf and the stem from which it is growing.

Axillary Pertaining to or placed within an axil.

Axillary bud Bud that develops in the axil of a leaf.

Axillary lymph nodes Lymphoid organs located near the shoulder joint.

Axis Another word for stem.

Axoneme The main core of a flagellum consisting of two central microtubules surrounded by nine double microtubules.

Azygospore A parthenogenetic zygospore; formed by some vesicular-arbuscular mycorrhizal fungi (family Endogonaceae).

B

B cell A white blood cell which produces antibody.

B light A disease characterized by general and rapid killing of leaves, flowers, and branches.

B order harvesting A harvesting method that leaves a strip of uncut hay along every other border; next harvest these borders are mowed and the alternate borders are left standing.

B.T. (bacillus thurengiensis) A live microorganism (bacteria) that has insecticidal properties affecting a selective range of insects. At least 34 subspecies exist and are active against moth and butterfly caterpillars, mosquito and black fly larvae, beetle larvae, and more, respectively.

Bacillar Shaped like a short rod.

Bacilliform Rod-like in shape; of spore shape, a cylinder with length:breadth ratio greater than 3.

Bacillus 1. A member of the genus Bacillus.

2. Any rod-shaped bacterial cell, i.e., a cell whose length is ca. two or more times greater than its width.

Bacillus thuringiensis A bacterium that causes disease in many insects, especially caterpillars; formulations of the bacteria are used as insecticides.

Backcross To cross (mate) an offspring with one of its parents.

Bacteraemia The presence of bacteria in the blood.

Bacteria Bacteria are single cells surrounded by a cell wall. Bacteria lack a true nucleus in that they do not have a membrane surrounding the DNA (genetic material). Plant parasitic bacteria lack chlorophyll and obtain nutrients from living or nonliving sources. None of the plant parasitic bacteria now known are obligate parasites. Plant parasitic bacteria are rod-shaped and reproduce by fission. During fission, one cell splits into two cells after the DNA has replicated.

Bacteria enter plants through natural openings in the plant surface (stomates, hydathodes, lenticels, and nectaries) or though wounds made by insect feeding, mechanical injuries, pruning, grafting, or other means.

Bacteria cause many different types of symptoms on plants. These include leaf spots, blights, soft rots of fruit, roots, and storage organs, wilts, overgrowths, scabs, and cankers. Fire blight on apple, pear, cotoneaster, mountain ash and related plants is one of the most serious bacterial diseases in Montana.

Bacterial diseases account for about 11% of the biotic diseases in Montana. A wide variey of tests are used to identify bacteria. These include microscopic, cultural, immunoassay, and bioassay tests.

Bacterial insecticide An insecticide in which the active ingredient is a living bacterium that infects an insect pest and then kills or inhibits it. A type of microbial pesticide. Bacterial insecticides are not included in the Compendium.

Bacterial streaming Large populations of bacteria that exude from the cut surface of infected plant tissue when observed with a microscope.

Bactericide A pesticide that is used to kill or inhibit bacteria in plants or soil.

(Chemicals that kill bacteria for medical or veterinary applications are termed "antibacterial agents"; they are classed as pharmaceuticals and so are not included in the Compendium.)

Bacteriocin A highly specific antibacterial protein produced by certain strains of bacteria and active mainly against other strains of the same species.

Bacteriocinogenic Having the capacity to produce a bacteriocin.

Bacteriophage A virus causing lysis of bacterial cells.

Bacteriosis Infection by bacteria, or a disease so caused.

Bacteriostatic Able to inhibit the growth and reproduction of at least some types of bacteria.

Bacteristat Material that inhibits bacterial growth, but does not kill the bacteria.

Bacterium A living microorganism characterized by cell membranes and cell walls. About 100 species of bacteria cause diseases in plants (e.g., leaf spots and blights caused by Pseudomonas species), and in animals (e.g., anthrax).

Bagasse The fine organic material removed from vacuum drums during the processing of sugar cane. Because it is of high organic matter and sugar content bagasse makes a good inoculant carrier, but bagasse must first be sterilized to stop the growth of fungi and other contaminant organisms.

Bait A food or other substance used to attract a pest to a pesticide or trap.

Bait box A small structure in which anticoagulant bait is placed to attract an animal.

Balansioid With characteristics similar to those of species of Balansia (Ascomycota, Clavicipitales).

Ballistoconidium A forcibly discharged conidium.

Ballistospore A spore that when mature is actively projected.

Ballistospory The ability to launch spores into the air. The mechanism of ballistospory is not fully understood. This ability is common in Basiodiomycota and is also known in some Ascomycota.

Band application An application in which a material such as fertilizer or herbicide is applied in strips, usually to the bed or seed row.

Banding A foliar symptom characterized by a limited zone of necrotic or discolored tissue traversing the leaf, e.g., the band of tissue on a pine needle injured by SO_2 or 0_3.

Barbate Bearded.

Barbel A whisker-like sense organ of certain fishes, including catfish and carp. These fish use their barbels to "feel" along the bottom for food.

Bark All tissues outside the vascular cambium of a woody plant, generally including the cork layers, cork cambium (phellogen), and phloem.

Barrage The space between two mycelia which have an aversion for one another; aversion zone.

Basal group The earliest diverging group within a clade; for instance, to hypothesize that sponges are basal animals is to suggest that the lineage(s) leading to sponges diverged from the lineage that gave rise to all other animals.

Basal knob Structure at the base of a nematode stylet.

Basal treatment Applied to encircle the stem of a plant above and at the ground such that foliage contact is minimal. A term used mostly to describe treatment of woody plants.

Basal whorl Groups of leaves attached at the same point at the base of a plant.

Basauxic Growing from the base, rather than at the apex.

Base Having a pH of greater than 7.

Base map A map of a field or other management unit showing the coordinates (latitude and longitude or other coordinate system) and boundaries.

Basidial stage (basidium) A spore stage of the rust fungi; a specialized structure in the Basidiomycetes bearing basidiospores.

Basidiocarp A sporocarp produced by a member of the Basidiomycotina and which bears basidiospores.

Basidiole A structure in the hymenium of a member of the Basidiomycotina that is morphologically similar to a basidium without sterigmata. It may be an immature basidium or a permanently sterile structure in the hymenium.

Basidioma Any multihyphal structure producing basidia; formed by most basidiomycetes, e.g., mushroom, puffball.

Basidiomycetes A fungus group, also called the Basidiomycota, characterized by the formation of external basidiospores on basidia.

Basidiomycotina Subphylum of Dikaryomycota forming exogenous meiospores on basidia, and have an extended dikaryon; generally called Basidiomycetes.

Basidiospore Basidiospores are a general category of sexual spores that have been released from the basidium of a fungus. A ubiquitous type I & III allergen, saprobe and plant pathogen, mainly found in gardens, forests, and woodlands. Spores disseminate during rain or in times of high humidity. Rarely opportunistic pathogens, Basidiospores may produce toxins, including amanitins, monomethyl-hydrazine, muscarine, ibotenic acid, and psilocybin. Basidiospores are an agent of dry wood rot, which may destroy the structure wood of buildings.

Basidium Specialized cell or organ, often club-shaped, in which karyogamy and meiosis occur, followed by production of externally-borne basidiospores (generally four) that are haploid. There are several types of basidia.

Basin A portion of a rice field bounded by levees.

Basipetal Downward from the apex toward the base of a shoot; referring to development in the direction of the base so that the apical part is oldest.

Basipetal Successive from apex to base.

Basipetospora Basipetospora is the anamorphic state of the genus Monascus. It is a widespread saprobe found on substrates with high water tension, such as dried foods. Monascus is a yeast known as the "red yeast rice". It contains Mavinolins, which includes

Monacolin-K, this is found on M. ruber or B. ruber. This species contains the highest natural Monacolin-K, therefore known as Ruby Monascus. (Aw - 0.75 - 0.78) Culture – Potato dextrose agar or Malt extract agar, 20° - 25°C, 7 – 10 days

Bask What snakes, turtles, lizards, and other reptiles do to maintain their body temperature. Reptiles need sunlight to help them digest their food, so they often lay on logs or rocks to soak up sunlight.

Bastien treatment An apparently successful treatment for amanitin poisoning, involving:
1. twice-daily injection of 1 g vitamin C,

2. 2 capsules of nifuroxazide, three times a day,

3. two tablets of dihydrostreptomycin, three times a day,

4. penicillin therapy,

5. maintenance of fluid and electrolyte balance.

Batch culture The growth of microorganisms in a fixed volume of medium.

Bean rust A disease of beans caused by the fungus Uromyces appendiculatus.

Beaumont period A period of two days during which the temperature does not fall below 1OC and the relative humidity exceeds 75%, such conditions being favourable for infection of potato leaves by Phytophthora infestans; potato blight can be expected to appear ten days after a Beaumont period.

Beauveria sp Beauveria is a type I allergen and saprobe reported as mainly an aggressive parasite of insects (may attack at larval or adult stages), and was first recognized as the agent of muscardine disease of the silkworm. B. bassiana is the best known member of this genus and is under research for biocontrol of insects. It is also a rare human pathogen, associated with keratitis and pneumonia in the immunocompromised. B. alba commonly occurs in indoor environments and appears to be less strongly associated with insects. Beauveria is commonly found in plant debris, soils, dung, and foods. Mycelium is white or slightly colored with a white fluffy to powdery appearance; conidiophores are single or irregularly grouped; conidia are hyaline, rounded to ovoid, 1-celled, dry and borne along a thin filament. The conidia are produced on short spikes or denticles, giving the conidiogenous cells a spiny appearance. Culture - Potato dextrose agar or Malt extract agar, 20° - 25°C, 7 – 10 days.

Bed A raised area of soil formed in which to plant crops, or an area in which seedlings or transplants are grown for planting in the field later.

Fig. Bed

Bees (apoidea) A superfamily of insects related to ants and wasps. While there are several familes of bees, it is the carpenter, bumble and honey bees that are most commonly known. Many of the other familes of bees are solitary.

Beetles Any species in the order of insects called Coleoptera that has four wings of which the outer pair are modified into stiff covers (elytra) that protect the inner pair when at rest.

Beltrania Beltrania is a mitosporic fungus that is very widespread and commonly found in dead leaves and plant debris in subtropical to tropical areas. It is known as an ascomycete, which is one of the major classes of fungal organisms. This class contains the "sac fungi" and the yeasts. Many are reported to be allergenic.

Bench setting The height at which the bed knife of a mower is set above a firm, level surface.

Beneficial Organisms that provide a benefit to crop production, applied especially to natural enemies of pests and to pollinators such as bees.

Beneficial insects Insects which are natural predators or parasitoids of harmful insect pests,or are important plant pollinators, etc. Examples include: lady beetles, lacewings, spined soldier bugs, social or parasitic wasps, bees, and ants.

Benomyl (benlate) The first successful systemic fungicide: one of the Benzimidazoles, has very low toxicity to plants and animals, and controls ascomycetous fungal pathogens by interfering with spindle formation during nuclear division.

Benzimidazoles Systemic fungicides, including Benomyl.

Beta spore An infertile spore form produced by certain fungi in the family Diaporthaceae, especially in the genus Phomopsis. Spores are slender and usually curved or bent and are produced together with alpha-spores.

Bicellular Two-celled.

Biconic Conical at each end.

BIDE (birth, immigration, death, emigration) The four demographic processes which might act on a population compartment in a typical compartmental model.

Biennial A plant that lives for two years. In the first year, the plant grows a rosette. In the second year, it sends up stems with flowers and fruits. A biennial plant dies at the end of the second year.

Bifacial necrosis Death of plant tissues, extending from the upper to the lower leaf surface.

Biflagellate Having two flagella, as in zoospores of the Oomycota .

Bifurcate Divided into two branches of equal or unequal length.

Bile duct Tubular structures responsible for conducting bile (a substance that aids in digestion) from the liver to the intestine.

Biliary obstruction Blockage or clogging of a bile duct.

Bilirubin A red pigment formed from hemoglobin during normal and abnormal destruction of red blood cells in the body.

Bill The mouthpart of a bird; also called a "beak."

Binary fission A type of asexual reproduction in which two cells, usually of similar size and shape, are formed by the growth and division of one cell.

Binding hyphae Thick-walled, much-branched, aseptate, interwoven, narrow, often coralloid hyphae which bind skeletal and generative hyphae together in a dimitic or trimitic hyphal system; ligative hyphae.

Binomial, latin The scientific name of an organism, composed of two names, the first designating the genus and second the specific epithet, together making the species name.

Binomial sampling A sampling method that involves recording only the presence or absence of members of the population being sampled (such as an insect pest) on a sample unit (such as a leaf), rather than counting the numbers of individuals; presence/absence sampling.

Binucleate Having two nuclei.

Bioaccumulation The uptake (via breathing, eating, drinking, direct absorption) and storage of a chemical in animal tissue. Over a period of time, a higher concentration of chemical may be found in the organism than in the environment.

Bioassay Any quantitative procedure in which a given organism is used for assay purposes.

Biochemical A chemical that occurs naturally in an organism, or an identical substance that has been made artificially.

Biochemical oxygen demand (BOD) The amount of oxygen required by micro organisms, usually in polluted water, slurry or industrial effluent for oxidation processes. Measured as milligrams of O_2 taken up by 1 l of the sample when incubated at a standard temperature of 20^{0}C for 5 days

Biochemical pesticide A naturally-occurring substance that controls pests by a mechanism other than toxicity. Examples include sex pheromones that are used as mating disrupters for insect pests, and plant extracts that are used as attractants to lure insect pests to traps or that are used as insect repellents.

Biochemistry The branch of organic chemistry that deals with the chemical processes that occur in living organisms.

Biocide A wide-spectrum poison that kills a great number and variety of organisms.

Bioconcentration The increase in concentration of a chemical in the tissues of an organism relative to the levels in the surrounding environment (e.g., air, water).

Bioconcentration factor The (unitless) measure of the tendency for a chemical to accumulate in (especially aquatic) organisms. The ratio of the concentration of a substance in a living organisms (mg/kg) to the concentration of that substance in the surrounding environment (mg/l for aquatic systems.) The BCF value is highest for hydrophobic (or lipophilic) chemicals. concentration in organism (mg/kg) BCF = concentration in environment (mg/l)

Biocontrol (syn. biological control) Exploitation by humans of the natural competition, parasitism and/or antagonism of organisms for management of pests and pathogens.

Bioconversion The enzyme-mediated conversion of organic substrates, such as cellulose, to other more valuable substances, such as protein, by other organisms.

Biodegradable Capable of being broken down by microorganisms (bacteria and/or fungi).

Biodegradation The breaking down of a chemical by organisms in the environment.

Biodiversity The Term Biodiversity was coined in 1985, and has been defined in many different ways. Holdgate regarded it as the 'Total sum of life's variety on Earth, expressed at the genetic, species and ecosystem level'

Biofix An identifiable event that signals when to begin degree-day accumulation.

Biofuel Fuels devised from biological materials including crops (especially trees) and animal wastes

Bioindicator species Species, varieties, or cultivars sufficiently sensitive to a specific pollutant to be useful as indicators for the presence of that pollutant.

Biointensive pest management Accurate diagnosis of pest problems and use of biological and preventive tactics to keep pests from causing unacceptable losses, based on understanding pest ecology.

Biological accumulation The acccumulation of toxic substances, e.g., DDT, in higher levels of the food web by consumption of organisms of the same or lower levels which have acquired the substance but which have not been affected by it.

Biological control Any biological agent (parasite or parasitoid, natural predator, bacteria, virus, etc.) that adversely affects a pest species. Once introduced, parasitoids and predator populations can often become self-sustaining in the natural environment, providing permanent pest control. Examples of biological control agents lady beetles (vs. aphids), cinnabar moth (vs. tansy ragwort plant), birds (vs. mosquitoes and many other insect pests), B.t. (vs. moths and butterflies)

Biological control by habitat manipulation A method of promoting populations of natural enemies by making conditions more favorable for it. For example, Larra bicolor (a wasp parasitoid of mole crickets) adults require nectar as an energy source, and obtains nectar only from a small number of plant species. The deliberate planting of such plants is biological control by habitat manipulation.

Biological efficiency The efficiency of an organism or biological system

Biomagnification Process by which substances such as pesticides or heavy metals move up the food chain, becoming more concentrated with each succeeding step up the chain.

Biomass The total mass (amount) of living organism(s) in a particular area or volume.

Biopesticide (also called biorationals) Pesticides composed of very large numbers of a biological control agent, generally a pathogen. For example, Bacillus thuringiensis (often called B.t. for short) is an bacterium that kills insects and can be produced on an industrial scale. When sprayed (much like a chemical) this pathogen kills insects and does not persist long in the environment. Its advantage over many chemicals is that it is not toxic to birds, mammals and fish. The fact that it is not a specialist natural enemy allows commercial companies to sell it for use against various pest insects and thus obtain a large market for the product. The disadvantage is that it may kill non-target insects in the same family as the pest, or even of many other insect families which are more distantly related.

Biotechnology Use of genetically modified plants or other organisms; production of goods or substances using biological agents such as microorganisms, animal or plant cells.

Biotic Of or pertaining to a living organism.

Biotic disease Disease caused by a pathogen, such as a bacterium, fungus, mycoplasma, or virus.

Biotic pathogen A living organism capable of inciting disease, e.g., fungi, bacteria, viruses, etc.

Biotroph An organism entirely dependent upon another living organism as a source of nutrients; an obligate parasite.

Biotrophic Entirely dependent upon another living organism as a source of nutrients. The term is applicable to such organisms as mycorrhizal fungi, obligate parasites etc.

Biotype 1. A group of organisms occurring in nature and assumed to be genetically almost identical; a population of genetically identical individuals.

2. A sub- division of a species, subspecies or serotype which may be distinguished from other subdivisions of the species by the possession of some special or usefully diagnostic physiological character .

3. a sub- division of a physiologic race; a sub-race.

Bipolar 1. Describes a system of heterothallism in which sexual compatibility is controlled by many different alleles which occur at a single locus

2. describes yeasts in which budding occurs at opposite ends of the long axis of the cell, by percurrent extension; they are blastic-annellidic.

Bipolar affective disorder A mood disorder that commonly begins with depression and is characterized by at least one period of elation sometime during the course of the illness.

Bipolaris sp Bipolaris is a plant parasite mostly on subtropical and tropical plants, particularly to Graminae. Some species are pathogenic to grasses and animals such as the dog, and may cause nasal mycotic granuloma in cattle. This fungus can grow in semi-dry environments. Bipolaris is commonly found in dead or dying plant debris, soils, and grasses. It has been reported to produce the mycotoxin, sterigmatocystin that has been shown to cause liver and kidney damage when ingested by laboratory animals. This fungus is associated with phaeohyphomycosis, a disease consisting of a group of mycotic infections characterized by the presence of demataceous septate hyphae. Infections of the eyes and skin by black fungi could also be classified as phaeohyphomycosis. This fungus causes allergic fungal sinusitis, characterized by the presence of Bipolaris in the sinuses. In certain people with severe allergies, the large spores of this fungus can travel to the sinuses or upper respiratory tract, where they attach to the mucus and grow; producing an unrelenting allergic reaction that progressively and permanently damages the sinuses. Morphological characteristics of this genus are the production of brown conidia that are multi-celled, elliptical, straight or curved. Culture - Potato dextrose agar or Malt extract agar, 20 - 25°C, 7 – 10 days.

Bird repellent A chemical that deters birds from approaching or feeding on crops or stored products.

Bird Warm-blooded, feathered animals from the Chordate Phylum. Birds usually move by flying, and they have hollow bones which make them light enough to do so. Birds build nests and lay eggs.

Bird's nest fungi Members of the order Nidulariales (Gasteromycetes): the basidiospores are in peridioles ('eggs') and the 'nest' is a splash-cup dispersal mechanism.

Bisaccate pine pollen grainbisaccate A pollen grain with two sacci. The two "sacs" or bladders are used to promote wind dispersal. Pollen of this type is known from as early as the Devonian.

Biseptate Having two septa.

Biseriate Of spores in an ascus, in more or less two rows.

Bispora Bispora is a widespread mitosporic fungus. It has been isolated from dead wood in temperate areas in the northern hemisphere. Culture – Potato dextrose agar or Malt extract agar, 20 - 25°C, 7 – 10 days.

Bitunicate Describes a type of ascus with two distinct, persistent wall layers: the 'jack-in-a-box' ascus that is diagnostic of the Bitunicatae and is found in pseudothecial ascomata.

Blackarm Bacterial blight lesions on stems.

Blacklight trap A monitoring device for certain types of flying insects that are attracted to the ultraviolet light source built into the trap.

Blade The flattened portion of a grass leaf located above the sheath; the portion of the leaf that grows away from the stem.

Blank Nut with no kernel-consists of only the collapsed pellicle (skin).

Blanking Producing no grain or seed (used to describe individual florets of the rice panicle).

Blastic Of conidiogenesis, one of the two basic types of conidiogenesis, characterised by a marked enlargement of a recognisable conidial initial before the initial is delimited by a septum. The conidium differentiates from part of a cell. Types include enteroblastic and holoblastic conidia.

Blasting A symptom of plant disease characterized by shedding of unopened buds; classically, the failure to produce fruit or seed.

Blastoconidium A conidium formed by blastic conidiogenesis.

Blastomyces Blastomyces is a thermally dimorphic fungus, and is rarely isolated from the environment but has been found in acidic, organic soil around waterways and beaver dams. The species B. dermatitidis can cause blastomycosis. Blastomycosis is generally acquired by inhalation and initially presents with a respiratory infection, which may spread and cause disease to other organs and systems such as the central nervous system (CNS), eyes, skin, sinuses, tongue, reproductive tract, gastrointestinal tract, liver, spleen, and bones. This is a very serious disease that can be fatal, and a compromised immune system is the primary risk factor. This disease is most prevalent in males' ages 40-60 years and in children. Blastomycosis can also infect dogs, which are infected by inhaling the infectious particles. There is no evidence of animal to human transmission. B. dermatitidis is found predominantly in the Missouri, Ohio, Tennessee, and Mississippi river valleys. When grown at 25°C, Blastomyces is a filamentous fungus, and when grown at 37°C it has a yeast-like form. Culture – Potato dextrose agar or Sabouraud glucose agar, 20 - 25°C, 7 – 10 days.

Blastomycetous Of a yeast, budding.

Blastomycosis A disease in man caused by Blastomyces dermatitidis (teleomorph in Ajellomyces dermatitidis).

Blastoschizomyces Blastoschizomyces has a single species, B. capitatus, which is a yeast commonly found in soils, beach sand, poultry feces, and wood pulp. As well as being a saprobe to the environment, it is found in the normal microbial flora of human skin, and digestive and respiratory tracts. An opportunistic fungi that is potentially pathogenic in cases of human immuno-supression. Disseminated infections of the lungs, kidneys, liver, spleen, and brain may occur and are likely in neutropenic patients with leukemia or those undergoing bone marrow transplantation. Development of endocarditis, osteomyelitis, meningitis, encephalitis, urinary tract infection, mycetoma, and pneumothorax may result. Colonies grow rapidly and mature in about 5 days; the colony is yeast-like and becomes mold-like with age. B. capitatus may be wrongly identified as Trichosporon, Geotrichum candidum, or Candida krusei; but because of its ability to grow at 45°C, its resistance to cycloheximide and biochemical characteristics helps to accurately identify the species. Culture – Potato dextrose agar or Malt extract agar, 20° - 25°C, 7 – 10 days.

Blastospore A spore that arises by budding, as in yeasts.

Bleeding Sap flow from a wound.

Blend A combination of two or more cultivars of a single turf-grass species.

Blepharoplast In a zoospore, the basal body or granule (kinetosome) from which arise the longitudinal fibres constituting the axoneme (main core) of a flagellum; joined to the nucleus by a rhizoplast.

Blight A term used from the 17th century to denote any obscure malignant influence; later, a disease of unknown or 'atmospheric' origin affecting plants. In popular use the term generally refers to any agency causing widespread blackening of leaves and shoots, including intensive infestation by aphids. The only valid use in plant pathology is in the common names of certain diseases characterised by potentially catastrophic epidemic spread, e.g. potato blight (Phytophthora infestans), fireblight (Erwinia amylovora) of pear, etc.

A common term for several different diseases; usually applied to those where leaf damage is sudden and severe.

Any sudden, severe, and extensive spotting, discoloration, or destruction of leaves, flowers, stems, or entire plants, usually attacking young, growing tissues. (In disease names, often coupled with the name of the affected part of the host; e.g., leaf blight, blossom blight, shoot blight).

Blind node The first node formed on a strawberry runner that usually does not form a daughter plant.

Blister rusts Serious diseases of pines, caused by species of Cronartium (Uredinales: Teliomycetes).

Blotch A large area of discolouration of a leaf, fruit etc. On leaves its outline is unrelated to the distribution of main or minor veins. A blot or spot, usually superficial and irregular in shape and size, on leaves, shoots and fruit.

Blotting Following electrophoresis: the transfer of nucleic acids and/or proteins from a gel strip to a specialized, chemically reactive matrix on which the nucleic acids, etc. may become covalently bound in a pattern similar to that present in the original gel.

Blue baby syndrome An illness that begins when large amounts of nitrates in water are ingested by an infant and converted to nitrite by the digestive system. The nitrite then reacts with oxyhemoglobin (the oxygen-carrying blood protein) to form metheglobin, which cannot carry oxygen. If a large enough amount of metheglobin is formed in the blood, body tissues may be deprived of oxygen, causing the infant to develop a blue coloration of their mucous membranes and possibly digestive and respiratory problems. This condition is also known as methemoglobinemia.

Blue cheeses Cheeses ripened and flavoured by Penicillium roquefortii (Hyphomycetes), e.g., Roquefort, Stilton, Gorgonzola, Danish Blue.

Blue mould A downy mildew of tobacco, caused by Peronospora tabacina (Peronosporales: Oomycota).

Blue stain A deep-seated discolouration of wood resulting from fungus attack; predominantly bluish but sometimes having a grey or brown tint; confined almost exclusively to sapwood.

Blueing reaction A colouration of bruised or broken tissues in some agarics (esp. Boletaceae), caused by an oxidative reaction.

Bolete Common name for a member of the Boletales.

Boletinoid Of a hymenophore, with a structure intermediate between lamellae and pores.

Bolt To initiate the growth of flower structures.

Boot A bulge in the upper leaf sheath caused by the expansion of the developing panicle.

Bordeaux mixture The first practical fungicide used by Millardet in 1883-1885 against downy mildew of grape vine (Plasmopara viticola), and still in general use for controlling many plant diseases; it contains 4 parts copper sulphate, 4 parts quicklime, 500 parts water.

Borer Insect or insect larva that forms tunnels or cavities in the bark or within the wood of trees.

Borers The name given to immature stages of beetles and moths that live and feed inside plants by boring through the living tissue.

Borrelia A genus of bacteria with numerous species that cause disease in humans. The diseases associated with these organisms are typically relapsing fevers.

Borrow pits Depressions on either side of a levee created when soil is removed from the field to build the levee.

Botanical Derived from plants or plant parts.

Botryoderma Botryoderma is a mitosporic hyphomycete fungus, which has been isolated from roots of fumigated and nonfumigated Douglas-fir stumps in Oregon. It has also been seen in South Africa and Brazil. Culture - Potato dextrose agar or Malt extract agar, 20° - 25°C, 7 - 10 days.

Botryosporium Botryosporium is a plant pathogen and resembles gray mold, and can often be found on greenhouse tomatoes, and has also been found in hydrilla and/or soil. Botryosporium has been detected in moldy buildings along with Acremonium, Oedocephalum, Cladosporium, Aspergillus, and Stachybotrys. A laboratory study shows hygrosporic growth in the human respiratory system. Culture - Potato dextrose agar or Malt extract agar, 20° - 25°C, 7 - 10 days.

Botrytis sp Botrytis sp. is a Type I & III allergen, not a known toxin producer or opportunistic pathogen. Mostly reported to be found in tropical and temperate areas. This fungus is a parasite of plants, soft fruits and vegetables. The cause of leaf/root rot on fruits and vegetables such as, strawberries and onions. It is also known as "gray mold" or "noble rot" on wine grapes. Botrytis is known to cause allergies and induce asthma attacks, and is also a rare agent of keratomycosis. In some types of agricultural settings, such as greenhouses, the concentration of aerosolized spores may be greatly enhanced. Botrytis is also used in some types of wine production. Conidia are hyaline or gray in mass, 1-celled, ovoid, and 7-14 x 5-9 microns in size. (Aw - 0.93-0.95) Culture - Potato dextrose agar or Malt extract agar, 20° - 25°C, 7 – 10 days.

Boundary layer The calm layer of air on surfaces below the layer of more turbulent air.

Bovine spongiform encephalopathy (BSE) A disease of cattle (often called 'mad cow disease') caused by an agent that is neither a bacterium or a virus. First recognized in 1986 it has a long incubation period.

Bracket fungi Corky or woody, often perennial, basidiomata of the polypores (Aphyllophorales: Basidiomycetes). Also known as shelf fungi.

Bract A reduced leaf associated with a flower or inflorescence; modified leaf from the axil of which a flower arises.

Brand name A name used by a pesticide manufacturer or supplier for a formulation, and the most prominent name on product labels and in advertisements. Often a registered trade mark.

Brassicas Plants of the brassica classification e.g. turnip.

Break crop A change in crop in an arable rotation in order to help control weeds and plant diseases, i.e Oilseed rape grown in a sequence of cereal crops would be a break crop.

Breakdown 1. disintegration of plant tissue; rotting.

2. (of fungicides, insecticides etc.) failure to control (disease, pest etc.) by dosage or concentration of the material that has previously been effective, e.g. as a result of occurrence of resistant strains of the organism.

3. the occurrence of severe disease in a formerly resistant host variety. A general term describing a condition which may have several alternative causes. It commonly refers to the advent of a race of the pathogen having a gene for virulence corresponding with a particular major gene for resistance in the host, i.e. the change occurs in the pathogen and not in the host. In this sense the word is misleading and should be abandoned.

The term may also be applied in circumstances in which resistance (usually field resistance) becomes ineffective, e.g. under adverse environmental conditions, with no change in the virulence of the pathogen.

Breakdown, low temperature Disintegration of plant tissues (especially fruits etc.) associated with storage at low temperatures.

Breaking Disease symptom, usually caused by a virus, involving addition or loss of flower color to create a variegated pattern.

Breeding The identification and selective combination of desirable genes, with the goal of enhanced yield, reduced susceptibility to plant disease, improved seed quality or better nitrogen-fixing ability.

Breeding line plant strain used in a plant breeding program and usually containing one or more desirable agronomic or breeding characteristics.

Broad spectrum pesticide A pesticide which kills a wide range of pest species, as opposed to a pesticide which kills a single or limited number of species.

Broadcast application The application of a material such as fertilizer or herbicide to the entire surface of a field.

Broadcast treatment Application of a herbicide over an entire field area.

Broadleaf One of the major plant groups known as dicots, with net veined leaves usually broader than grasses. Seedlings have two leaves (cotyledons).

Bromatia Enlarged hyphal tips induced by Attine (leaf-cutting) ants in their domesticated fungus, and eaten by them.

Bronzing A golden brown discoloration that usually appears on the lower surface of leaves and is often an advanced stage of the silvering or glazing typical of injury by PAN and other oxidants; brown coloration on needles due to spider mite infestation.

Brood All the individuals that hatch at about the same time from eggs laid by one series of parents and that normally mature at about the same time.

Broom An abnormally dense mass of host branches and foliage in which the typical growth pattern is replaced by a disordered cluster of foliage at the branch tips.

Broth In microbiology: Any of a variety of liquid media, especially nutrient broth or any liquid medium based on nutrient broth and/or hydrolysed protein.

Brown rot (of wood) Wood decay resulting from selective removal of cellulose and hemicellulose, leaving a brown amorphous residue that usually cracks into cubical blocks and consists largely of slightly modified lignin.

Browse The name for plants that deer, rabbits, beavers, or other mammals munch on. Browse is usually small plants, very young trees, or leaves and twigs on larger trees.

Bt The bacterium, Bacillus thuringiensis.

Buccal capsule In nematodes: Structure connecting the oral opening with the anterior portion of the esophagus. The buccal capsule (also called the stoma) is subject to great variation among different nematodes.

Bud A terminal or axillary structure on a stem consisting of a small mass of meristematic tissue, generally covered wholly or in part by modified scale leaves.

Bud scale A specialized protective leaf of a bud.

Bud wood Wood consisting of strong, young shoots bearing buds suitable for use in budding.

Budding Reproduction by binary fission, a characteristic form of propagation in yeasts. "The onset of the cellular events is accompanied by the nuclear events of mitosis. ... The initial events of budding can be seen as the development of a ring of chitin around the point where the bud is about to appear. This reinforces and stabilizes the cell wall. Enzymatic activity and turgor pressure the act to weaken and extrude the cell wall. New cell wall material is incorporated during this phase. Cell contents are forced into the progeny cell, and as the final phase of mitosis ends a cell plate, the point at which a new cell wall will grow inwards from, forms."

Buffer An additive that lowers the pH of a solution and, depending on its concentration, can maintain the pH within a narrow range even if acidic or alkaline materials are added to the solution.

Bug In the entomological sense many families of 4-winged insects (including tens of thousands of species worldwide) in the order Hempitera in which the second pair of wings is used for flying. The first pair is partially toughened, partially membranous, and helps to protect the second pair.

Bulb A short, flattened, usually globose or disc-shaped, underground, perennial, storage organ composed of concentric layers of overlapping fleshy scale leaves attached to a stem plate at the base; essentially a subterranean bud.

Bulbil A small sclerotium-like structure consisting of a small number of cells, homogeneous throughout and pseudoparenchymatous at maturity.

Bulbous Shaped like a bulb, especially at the base of a stem.

Bulimia nervosa A disorder marked by morbidly increased appetite, often alternating with periods of anorexia.

Bullae In nematodes: Blisterlike prominences near the vulval fenestra of some Heteroderidae.

Buller drop The tiny droplet of liquid which appears at the base of a basidiospore just before it is discharged.

Bunch type growth Plant development at or near the soil surface without production of rhizomes or stolons.

Bunt A disease caused by a member of the order Tilletiales, especially species of Tilletia.

Bunt ball Smut sorus filled, with teliospores, that replaces a cereal or grass kernel but is covered by plant tissue at maturity.

Burn The condition in which the cells of the host become reddish or dark brown and collapse.

Burr knot A rough outgrowth, often present on the trunks or roots of certain trees.

Burrow An underground excavation used by some rodents and related pests for food storage and shelter.

Bursa Extension or flap of cuticle at the side of the male nematode sex organ, used for orienting during mating.

Bush type Plant with determinate growth.

Button The rounded, immature stage of a mushroom with a cap, before the cap has expanded.

Byssochlamys Byssochlamys is a widespread ascomycete fungus, and is also the anamorph of Paecilomyces. Byssochlamys is commonly found in soil, dust, canned or bottled fruit, barley grain, silage, and wood in temperate areas. This is an extremely thermo-tolerant (heat-resistant) fungus, and may cause spoilage in acidic foods, such as fruits. The optimal growth for this fungus is 30°-37°C, but may grow up to 98° - 100°C. Byssochlamys may also produce the mycotoxins, patulin and mannitol. The teleomorph form, Paecilomyces is the most commonly found in viable cultures, and also has ascospores that can survive in temperatures up to 80°C. (Aw - 0.84 - 0.92). Culture – Potato dextrose agar or Malt extract agar, 20° - 25°C, 7 – 10 days.

C

Cache A pile of food hidden away for Winter. Some squirrels and birds cache nuts and seeds to get them through cold weather.

Caeoma An aecium in the Uredinales that is not surrounded by a peridium; from the form genus Caeoma.

Caespitose Growing in tufts.

Calcarate With a spur-like process, a calcar.

Calcareous soil Doil containing high levels of calcium carbonate.

Calibrate To standardize or correct the measuring devices on instruments; to adjust nozzles on a spray rig properly.

Calibration The process of adjusting the output of pesticide application equipment so that the proper amount of pesticide can be applied to a given area.

Callose Amorphous, hardened carbohydrate constituent of plant cell walls, commonly developing upon injury.

Callus Parenchymatous tissue of cambial origin that forms in response to wounding, infection by pathogens, etc.

Parenchyma tissue that grows over a wound or graft and protects it against drying or other injury.

Callus tissue A protective tissue of thin-walled cells developed on wound surfaces, often beginning at the edges of a wound.

Calyptra Cambium in mosses, a derivative of the distal archegonium forms a hard, protective layer around the developing sporophyte called the epigonium. Later the epigonium breaks up, but the sporophyte retains a conical "dunce cap" over its distal end, and sometimes a ring around the base. The cap is the calyptra. The piece retained around the base is the vaginula.

Calyx The sepals of a flower; they enclose the unopened flower bud.

Cambium A layer of actively dividing cells (lateral meristem), found within stems and roots, that gives rise to secondary growth in perennial plants, causing an increase in girth. There are two main types of cambium: vascular cambium, which gives rise to secondary xylem and phloem tissues, and cork cambium (or phellogen), which gives rise to secondary cortex and cork tissues, as in bark. The vascular cambium surrounds the roots, trunk, branches, and shoots, extending throughout a tree. Each year, the vascular cambium produces a new layer of phloem toward the outside of the tree and a new layer of xylem, or wood, toward the inside.

Camouflage The ability of an animal to blend in with its surroundings. Color, patterns, and body shape all help the animal do this.

Canaliculate With a longitudinal groove or channel.

Cancellate Anastomosing and forming a network; reticulate; latticed.

Cancer The unregulated overgrowth of cells.

Candida albicans A common saprophyte of the digestive tract and female urogenital tract. It does not ordinarily cause disease, but may do so following a disruption of bacterial flora of the body, or in patients with depressed immune systems.

Candidiasis A disease of man and animals caused by species of Candida, esp. C. albicans; also called moniliasis and candidosis.

Canker A sunken necrotic lesion of main root, stem or branch arising from disintegration of tissues outside the xylem cylinder, but sometimes limited in extent by host reactions which can result in more or less massive overgrowth of surrounding tissues; concentric zonation may indicate successive host responses to advancing infection.

In a perennial canker infection and host responses continue for more than one season. Because no new wood is formed under such cankers the lesions appear to penetrate deep into the wood: this is because continued growth outside the limits of the lesion tends to bury it. Infection normally continues to be restricted mainly to the bark tissues.

A hold-over canker is one in which the pathogen survives the winter and from which it may re-infect healthy tissues in the spring. A term particularly used of fireblight cankers caused by Erwinia amylovora.

A plant disease symptom characterized by a sharply defined necrosis of cortical tissue, often sunken below bark surface.

A dead area on a stem surrounded by living cortical tissues.

Canopy The expanded leafy top of a plant or plants.

Cap The top "umbrella-like" part of a fungus; espcially a mushroom. Underneath the cap, most mushrooms have gills which release spores.

Capillaceous, capilliform Hair-like.

Capillitium In gasteromycetes, a mass of sterile thick-walled, brown, infrequently septate hyphae among the spores; in Myxomycota, the thread-like structures in the fruit-bodies, ranging from few to a dense network and, in order Physarales, containing lime.

Capitate Having a well-formed head, usually more or less globose.

Capnodiaceous fungus A member of the Capnodiales (Ascomycota); a sooty mould.

Capsid The protective layer of protein surrounding the nucleic acid core of a virus; the protein molecules which make up this layer.

Capsule In bacteria: A layer of material external to but contiguous with the cell wall.

Carbamate insecticide A synthetic compound derived from carbamic acid. Carbamate insecticides are contact killers with relatively short-lived effects. Examples are carbaryl and bendiocarb.

Carbamate pesticide One of a group of pesticides formed primarily as esters of carbamic acid with the general formula (HO-$CONH_2$). Certain of these compounds have insecticidal or herbicidal properties. The insecticides are cholinesterase inhibitors. Examples carbaryl (Sevin), aldicarb (Temik), bendiocarb.

Carbohydrate Organic compound consisting of a chain of carbon atoms to which hydrogen and oxygen are attached in a 2:1 ratio, e.g., sugars, starch, glycogen, cellulose.

Carbohydrate utilization Plants metabolize carbohydrates during growth; use them in the synthesis of cell components such as cellulose; or store them as starch against future need. Carbohydrate utilization patterns refer to the balance between these activities.

Carbon cycle The continuous circulation of carbon atoms from inorganic carbon dioxide (CO_2) to organic molecules and back to CO_2.

Carbonaceous Dark coloured and readily broken, brittle.

Carboxamides (carboxin, vitavax) Systemic fungicides used as seed dressing to control smut on barley and wheat.

Carcinogen A substance or agent that produces or incites cancerous growth.

Cardia In nematodes: Valvular apparatus connecting the esophagus and intestine. Sometimes called the cardiac valve or esophago-intestinal valve.

Cardinal temperatures The minimum, maximum and optimum temperatures of growth of an organism.

Carlavirus (siglum of carnation latent virus.) Member of a group of plant viruses with slightly flexuous, rod-shaped particles containing a single molecule of linear RNA, most of which are transmitted by aphids in a noncirculative manner.

Carnivore An animal that eats mostly other animals. A predator. A few plants are also carnivorous.

Carotenoids Fat-soluble pigments including carotenes (yellow and orange) and xanthophylls (yellow).

Carpel The female reproductive organ of a flower, consisting of stigma, style and ovary.

Carrier 1. A material serving as diluent and vehicle for the active ingredient (e.g. of fungicide), usually in a dust.

2. an organism harbouring a parasite without itself showing disease (BMS).

3. a bacterial strain in which a population equilibrium between phage-resistant and phage-sensitive cells exists, the latter being constantly infected by free phage particles. Such cultures can be freed from phage by successive colony reisolation.

An individual who is infected but has no symptoms of disease.

There are two types of carrier state: silent carriers retain their infectiousness, while latent carriers are not infectious.

For example, many of those infected with tuberculosis are silent carriers, while infection with herpes virus may create latent carriers.

A plant or animal that carries a virus or other infective agent without showing symptoms.

Carrion Dead animals. Some animals, such as crows and vultures, eat carrion. This is important since they help "clean up" this dead material.

Carrying capacity The maximum number of individuals a habitat can sustainably support. Typically defined so as not to include losses from predation or disease.

Cartilaginous Firm and tough, but readily bent.

Case control study A study in which the risk factors of people with a disease are compared with those without a disease.

Case definition In the example of CFS, a combination of symptoms, signs, and physiologic characteristics that serve to distinguish a case of chronic fatigue syndrome from other disease states.

Case-control study An epidemiologic study that examines selected patients who have a defined disease (cases) with persons without the disease (controls).

Cash flow Movement of funds through a business.

Casting Premature loss of abscised leaves or twigs.

Castings, earthworm (wormcasts) Soil and plant remains excreted by earthworms that are deposited on the turf surface or in the burrow; forms a relatively stable soil granule that can be objectionable on closely mowed turf.

Casual agent An organism, such as fungus, bacterium, or virus, that produces a disease.

Catabolic Describes metabolic chemical reactions resulting in the breakdown of complex materials and the release of energy in the metabolic process.

Catalytic model A type of compartmental model in which the force of infection is treated as a parameter to be estimated.

Catch crop Crop utilizing land between two longer term crops

Catenate In chains, e.g. of a row or chain of cells or spores.

Catenulate In short chains of two or three.

Caterpillar Usually applies to the larvae of the Lepidoptera, but the larvae of sawflies in the Hymenoptera are also caterpillarlike.

Catfacing Disfigurement or malformation of fruit; in the case of strawberries, usually the result of injury to developing achenes by Lygus bugs or low temperature.

Catkin A spikelike cluster of unisexual flowers, e.g., the male flowers of walnut.

Caudal In nematodes: Pertaining to or located near the posterior region or tail.

Cauloneme A lateral process from the protoneme (initial, threadlike process) which develops into the gemetophore in mosses.

Causal agent Organism or agent that incites and governs disease or injury.

Causal organism An organism causing disease. This implies that the organism is the sole cause but it may be used in the sense of being part of a causal complex.

Cavity A hole or hollow area, especially inside a tree. Many animals, such as woodpeckers and raccoons, live in them.

Cell The unit of plant and animal structure; it consists of a small bit of protoplasm which, in plants, is usually surrounded by a cell wall.

Cell cycle The period from one cell division to the next.

Cell membrane Structure that bounds a cell and helps control the movement of substances into and out of the cell.

Cell wall Protective, resistant, but permeable structure secreted externally to the cell membrane in plants, bacteria, fungi, and certain other organisms.

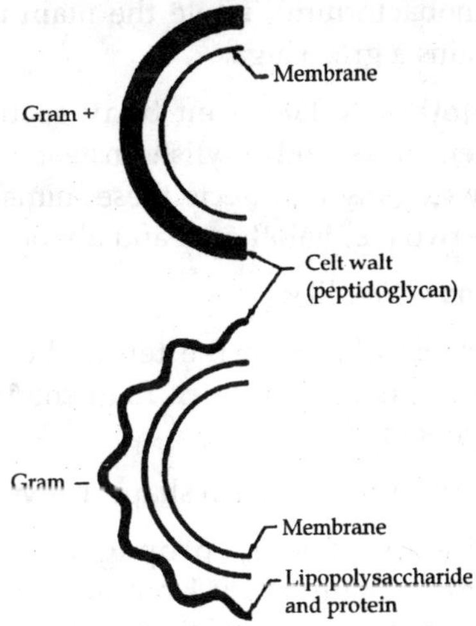

Fig. Cell walls of Gram positive and Gram negative bacteria (schematic representation)

Cellulose Principal polysaccharide of plant cell walls; a polymer of glucose; walls of Oomycetes are partly composed of a similar substance called 'fungal cellulose'.

Center of origin Geographical area where a plant originated.

Centriole Cytoplasmic organelle identical to a kinetosome; found in flagellated cells of fungi; the centriole divides and organizes spindle fibres during mitosis and meiosis.

Centripetal Directed, or developing from the outside towards the centre or axis.

Centromere The constricted portion of a chromosome to which, in mitosis, the chromosomal fiber is attached.

Centrum The structures inside an ascoma, which are concerned with development of the asci.

Cephalids In nematodes: Two structures (posterior and anterior) situated in the cephalic region and extending in a complete circle around the body; possibly part of the nervous system. Sometimes called hypodermal commisures.

Cephalodia Delimited parts of a lichen thallus containing a blue-green alga (cyanobacterium), while the main thallus contains a green alga.

Cephalothorax The front body section on spiders, ticks, and crayfish. Instead of three body sections like insects, these animals only have two (cephalothorax and abdomen).

Ceraceous Wax-like.

Cerci Appendages of the tenth abdominal segment, usually slender, filamentous, and segmented.

Cerebriform Brain-like in shape; convoluted.

Certification Seeds, propagative plant material, or nursery stock produced and sold under inspection to maintain genetic identity and purity, freedom from harmful pathogens, insect pests, and weed seeds. It is approved and certified by an official certifying agency.

Certified seed or planting stock Seeds, tubers or young plants certified by a recognized authority to be free of or to contain less than a minimum number of specified pests or pathogens.

Certified transplants Strawberry plants that have received a certification tag from the California Department of Food and Agriculture, production practices must meet standards for freedom from pest problems and plant samples from production fields must test free of viruses.

Cervical lymph nodes Lymphoid organs located in the neck.

CFU The number of colonies formed per unit of volume or weight of a cell or spore suspension.

Chaetomium sp Chaetomium sp. is found on a variety of substrates containing cellulose including paper and plant compost. Several species have been reported to play a major role in decomposition of cellulose-made materials. These fungi are able to dissolve the cellulose fibers in cotton and paper and thus cause the materials to disintegrate. The process is especially rapid under moist conditions. During the Second World War countries lost a great deal of equipment to these species. It is reported to be allergenic, although recent research has indicted that it may be more toxic then Stachybotrus. It is an ascomycete, in most species, the spores are lemon-shaped, with a single germ pore. The spore column results from the breakdown of the asci within the body of the perithecium. The perithecia of Chaetomium are superficial and barrel-shaped, and they are clothed with dark, stiff hairs. It can produce an Acremonium-like state (imperfect stage) on fungal media. Cultivation - Potato dextrose agar, 24° to 26°C.

Chaetothyriaceous fungus A member of the Chaetothyriaceae (Ascomycota, order uncertain); a sooty mould.

Channel In a phialide, zone at the phialide apex in which the area occupied by the protoplast is usually of considerably reduced diameter and surrounded by periclinal thickening; the conidial primordium emerges through the channel to produce a new conidium.

Chanterelle Edible species of Cantharellus (Aphyllophorales: Holobasidiomycetes).

Character Heritable trait possessed by an organism; characters are usually described in terms of their states, for example: "hair present" vs. "hair absent," where "hair" is the character, and "present" and "absent" are its states.

Check The part of a rice field between two levees.

Chemoprophylaxis Drug treatment designed to prevent future occurrences of disease. Treatment may be chemotherapy as far as an individual is concerned but chemopropylactic for the population as a whole.

Chemosterilant A chemical that renders an insect infertile and thus prevents it from reproducing. Some insects that mate only once can be controlled or eradicated by releasing huge numbers of sterilised insects.

Chemotaxis Movement or growth of an organism in response to changing concentration of a chemical stimulus, often in relation to food or for mating.

Chemotherapy Treatment of disease by chemicals. Drug treatment of a diseased individual.

Treatment of disease by chemicals (chemothera-peutants) working internally. Chemical agent has toxic effect directly or indirectly on the pathogens without injury to the host plant.

Chemotrophic Obtaining energy for growth etc. from inorganic nutrients without utilizing light.

Chemotropism Growth of an organism up a chemical concentration gradient.

Chestnut blight A devastating disease caused by Cryphonectria (Endothia) parasitica (Diaporthales: Ascomycotina) which has almost wiped out the edible cheṣtnut (Castanea dentata) in eastern North America.

Chewing insects Insects that consume all tissues of leaves or portions of leaves, using robust mandibles for chewing.

Chiastobasidium A holobasidium that is clavate with nuclear spindles transverse across the basidium and located near the top of the basidium.

Chilling In strawberries, exposure to temperatures low enough to induce the production of food reserves needed to support vigorous vegetative growth.

Chilopoda This class consists of the centipedes which are elongate, flattened animals with 15 or more pairs of legs. Each segment contains a single pair of legs.

Chimaera A plant or plant part consisting of tissues of diverse genetic constitution, arising as a result of grafting, somatic mutation etc. In such a way that the components are represented either as separate germinal layers of the apical meristem, giving rise to periclinal and mericlinal chimaeras in which one tissue is completely or partially enveloped by a ' skin' of the other, or as discrete segments involving all layers of the apical meristem, forming sectorial chimaeras.

A plant with several tissue sectors or layers differing in genetic or chromosomal constitution from the original plant.

Chitin The dense substance forming the indigestible outer skeleton of insects, and the material from which the walls of the mycelia are made.

Chitinchitin A polymer of repeating sugar molecules (a slightly modified glucose, poly-N-acetyl-D-glucosamine). Chitin is the material which makes up the exoskeleton of insects and, in more or less modified form, in almost all arthropods. In arthropods, chitin occurs in a cross-linked form, chitin. Significantly, it is also found in the radular "teeth" of mollusks, the setae (bristles) and jaws of annelid worms, and the cell walls of

Fungi. So, this is exceedingly ancient stuff, possibly predating the split between bacteria and metazoans. [What may be of sociological interest is that the 1,5 aldose linkage was missing from the middle glucose in this diagram and no one spotted it for over a year ....]

Chitosan Chitosan (poly-D-glucosamine) is one of the most common polymers found in nature. Structurally, it is related to cellulose, which consists of long chains of glucose molecules linked to each other. Chitosan is chitin with acetic acid amide linkage hydrolyzed from the 2-amino group. In fact, this is exactly how chitosan is commercially prepared.

Chlamydospore An asexual spore, primarily for surviving adverse conditions, originating by modification of (a) hyphal segment(s) and possessing an inner wall usually impregnated with hydrophobic material.

Chloranemia The necrotic symptom of yellowing; a loss of chlorophyll.

Chlorinated hydrocarbon A group of pesticides that contain chlorine, carbon, and hydrogen.

Chlorophyll The green pigment of plants that captures the energy from sunlight necessary for photosynthesis.

Chloroplasts Discrete photosynthetic organelles within plant cells, contain chlorophyll

Chlorosis Partial or complete absence of normal green colour, affected organs becoming yellow-green, yellow or white. The condition may result from reduction in size or number of chloroplasts, destruction of chlorophyll, or from a failure of synthesis of chlorophyll as in iron deficiency.

Failure to produce chlorophyll can be the result of genetic factors (albinism) or chimaeral growth or growth of plants in darkness (etiolation).

Yellowing of plant tissue due to failure of chlorophyll synthesis or to chlorophyll destruction.

The abnormal plant color of yellowish-white or gray condition of plant parts resulting from the incomplete destruction of the chlorophyll.

Chlorotic Abnormally yellow.

Chlorotic dwarf An abiotic disease of Pinus strobus characterized by reduced growth, chlorosis and mottling of the needles, and premature abscission of all but current needles.

Choke A disease of grasses caused by Epichloë (Clavicipitales); the stroma of the teleomorph encircles the main axis of the grass and prevents it from flowering.

Cholangitis Inflammation of the bile duct.

Cholinergic Stimulating or augmenting the transmission of nerve impulses (acetylcholine is an important agent of neurotransmission across synapses).

Chord In nematodes: A longitudinal internal thickening of the hypodermis.

Chorion The outer membrane of an insect egg.

Chromatid One of the two daughter strands of a duplicated chromosome which are joined by a single centromere.

Chromatography A separation technique which uses the differential rates of diffusion of different sizes of molecules in gas (GC), liquid (HPLC), on paper, and in thin layers of silicate on glass, aluminum or plastic plates (TLC). Now much used in lichenology.

Chromista One of the seven Kingdoms. It is mentioned here because the Oomycota and the Hyphochytriomycota are atypical member groups, placed in it along with many photosynthesizing organisms including the diatoms (Bacillariophyta), the Xanthophyta, the Silicoflagellates, the

Chrysophyta and the Phaeophyta (brown algae), which all have chlorophyll c, and do not store food as starch (but usually as oils). Most also produce cells that have two dissimilar flagella.

Chromoblastomycosis Fungal skin disease of humans caused by species of Phialophora (Hyphomycetes).

Chromosome The structure that contains the genes of an organism; in eukaryotes, chromosomes are in the nucleus and can be visualized with an optical microscope as threads or rods during meiosis and mitosis; in bacteria, the chromosome is usually a single circle of DNA that cannot be visualized with an optical microscope.

Chronic fatigue and immune dysfunction syndrome (CFIDS) A synonym for chronic fatigue syndrome used by some patients and physicians. It should be stressed, however, that no immune dysfunction or aberration has been persuasively linked to chronic fatigue syndrome.

Chronic health effects Describes symptoms or a disease that is of slow progression and long continuance. Opposite of acute.

Chronic injury Injury which develops after long-term or repeated low dose exposure to an air pollutant expressed as chlorosis, bronzing, premature senescence, reduced growth, etc.

Chronic Of long duration, denoting a disease of slow progress and long continuance.

Chronic toxicity The potential of a substance to cause injury or illness following repeated exposures over a period of time or long exposure. Learning About Pesticides at School 25

Chrysalis The pupa stage of a butterfly.

Chrysosporium sp Chrysosporium sp. - This fungus is a saprophyte and is commonly found in soil, dung, plant debris, and seeds. Conidiophores poorly differentiated, mostly erect and branching irregularly, hyaline. Conidia are 1-celled, globose to pyriform, single or in short chains, and usually with a broad basal scar. Cultivation - Malt extract agar, 24° to 26°C.

Chytrid A member of the Chytridiomycota.

Chytridiomycetes A fungus group, also called the Chytridiomycota, characterized by the production of motile zoospores and resting sporangia; the plant-pathogenic species are all biotrophs that are restricted to the cells of their host.

Chytridiomycota Phylum of simple eumycotan fungi mostly with posteriorly uniflagellate zoospores; hence, chytrid.

Cilium A hair-like organelle of motility that protrudes from the cell and consists of an axoneme covered by the cell membrane.

Circadian rhythms Regular rhythms of growth and activity, which occur in an approximately 24-hour cycle.

Circulative propagative transmission Pathogen transmission characterized by a long period of acquisition of the pathogen (usually a mollicute, e.g. phytoplasma or spiroplasma, and sometimes a virus) by a vector (typically an insect), a latent period before the vector is able to transmit the pathogen, and retention of the pathogen by the vector for a long period because the pathogen reproduces or replicates in the vector.

Circulative transmission Virus transmission characterized by a long period of acquisition of the virus by a vector (typically an insect), a latent period of several hours before the vector is able to transmit the virus, and retention of the virus by the vector for a long period, usually several days; the virus circulates in the body of the vector.

Circulative virus A virus that systemically infects its insect vector and usually is transmitted for the remainder of the vector's life; persistent virus.

Circumscissile Opening along a circular or equatorial line.

Cirrhosis Progressive disease of the liver, characterized by liver cell damage, scarring of the liver, and abnormal liver architecture and function.

Cirrhus 1. A mass of spores in the form of a ribbon or tendril, forced from the fruiting body of a fungus.

2. A discrete group of somatic cilia (several to over 100) that act primarily as a unified locomotive organelle; the typical cirrus is conical.

Clade A "natural group" or phylogenetically defined group of organisms. A clade is usually defined as a particular organism "and all of its descendants." This apparently innocuous concept, introduced to systematics by Willi Hennig, has completely changed the way systematics is done and has fundamentally altered the way we view evolutionary change.

Cladogenesis The development of a new clade; the splitting of a single lineage into two distinct lineages; speciation.

Cladogram A diagram, resulting from a cladistic analysis, which depicts a hypothetical branching sequence of lineages leading to the taxa under consideration. The points of branching within a cladogram are called nodes. All taxa occur at the endpoints of the cladogram.

Cladosporium sp Cladosporium sp. - (Aw – 0.84 – 0.88). C. herbarum is the most frequently found species in outdoor air in temperate climates. It is often found indoors, usually in lesser numbers than outdoors. The dry conidia become easily airborne and are transported over long distances. This fungus is often encountered in dirty refrigerators, especially in reservoirs where condensation is collected. On moist window frames, it can easily be seen covering the whole painted area with a velvety olive-green layer. Cladosporium often discolors interior paint, paper, or textiles stored under humid conditions. Houses with poor ventilation, houses with thatched straw roofs and houses situated in low damp environments may have heavy concentrations of Cladosporium, which will be easily expressed when domestic mold is analyzed. It is commonly found on the surface of fiberglass duct liner in the interior of supply ducts. It is also found on dead plants, woody plants, food, straw, soil, paint, and textiles. The ability to sporulate heavily, ease of dispersal, and buoyant spores makes this fungus the most important fungal airway allergen; and together with Alternaria, it commonly causes asthma and hay fever in the Western hemisphere. A few species of this genus cause disease, which range from phaeohyphomycosis, a group of mycotic infections characterized by the presence of demataceous septate hyphae. Infections of the eyes and skin by black fungi (also classified as phaeohyphomycosis), and chromoblastomycosis, chronic localized infection of the skin and subcutaneous tissue that follows the traumatic implantation of the etiologic agent are also caused by this fungus. Chromoblastomycosis lesions are verrucoid, ulcerated, and crusted. Skin abscesses, mycotic keratitis and pulmonary fungus ball have been recorded in immuno-compromised patients. It may also cause corneal infections and mycetoma, characterized by localized infections that involve cutaneous and subcutaneous tissue, fascia, and bone consisting of abscesses, granulomata, and draining sinuses, usually in immuno-compromised hosts. Fungal colonies are powdery or velvety olive-green to olive-brown. Other characteristics include dark conidia 1- or 2-celled, variable in shape and size, ovoid to cylindrical and irregular, typically lemon-shaped. Cultivation - Potato dextrose agar, 24°C.

Clamp connection A recurving outgrowth of a cell that, at cell division, acts as a bridge to allow passage of one of the products of nuclear division into the penultimate cell, thereby assuring maintenance of the dikaryotic condition (of members of the Basidiomycotina).

Clamp connectionclamp connection A bridge-like hyphal connection involved in maintaining the dikaryotic condition. a. Terminal cell of hypha. Growth only takes place at hyphal tips; b. Hyphal tip elongating. c. Synchronous division of nuclei and the beginning of hyphal branch that will become the clamp connection. One nucleus (b) migrates into the new clamp. d. Septum forms at base of the clamp trapping nucleus b. Nuclei a′ and b′ migrate to the hyphal tip, while nucleus a migrates away from the tip. e. Septum forms below clamp forming new cell at hyphal tip. Fusion of the clamp to the adjacent cell releases nucleus b to the adjacent cell. Now both the terminal and subterminal are binucleate, each with a compatible pair of nuclei."

Class Part of the system of biological classification. Insects, mammals, nematodes, millipedes, centipedes, etc. each have their own classes. A class is below a phylum, but above an order.

Classical biological control The introduction of specialist natural enemies from the homeland of a pest of foreign origin. The objective is to establish populations of these natural enemies to attack the pest and to reduce its numbers. Sometimes quite small numbers of the natural enemy succeed in establishing a population when they are released.

Classical epidemiology Our term for the varieties of epidemiology primarily concerned with the statistical relationships between disease agents, both infectious and non-infectious; for example a study to establish the relative risk of lung cancer associated with smoking.

Classification the systematic arrangement of organisms, based on everything we know about them.

Clathrate, clathroid Latticed.

Clavate Clublike in shape, narrowing toward its base.

Claw Sharp, hollow appendages on the end of the last tarsal segment of insects, used for clinging to objects.

Claypan A dense, compact layer in the subsoil having a much higher clay content than the overlying material from which it is separated by a sharp boundary.

Cleistothecium Closed, usually spherical, ascus-containing structure of powdery mildew fungi. A sexual fruiting structure.

Cleistothecial Of an ascocarp (especially Aspergillaceae and Erysiphaceae), have a closed spore-bearing structure without a pore (ostiole) for spore release, and from which spores are released only by decay or disintegration.

Cleistothecial ascoma An ascoma, which is closed at maturity; the unitunicate but frequently evanescent asci are often spherical and are not arranged in an hymenium; characteristic of Eurotiales.

Climber A soft- or non-woody-stemmed vine that clasps the stems or branches of trees and shrubs to raise its foliage and flowers above the ground; a type of growth form.

Cloaca In nematodes: A common duct or cavity in which the digestive and reproductive systems terminate in males.

Clonal lines A group of plants originating from buds or cuttings from the same individual.

Clone 1. A group of organisms descended by mitosis from a common ancestor. A genetically uniform assemblage of individuals (which may be chimaeral in nature), derived originally from a single individual by vegetative propagation ′ and which constitute one kind of cultivar.

2. by extension, isolates with the same genotype, obtained by vegetative propagation; which are distinguishable from

others on the basis of morphological and/or physiological traits including pathogenicity; in this sense used by R. H. Stover for a division of a cultivar.

Cloning An in vitro procedure in which a particular sequence of DNA (e.g., a gene) is reproduced in large amounts by inserting ("splicing") it into a suitable replicon, introducing the resultant recombinant (hybrid) molecule into a cell in which it can replicate, and finally growing the cells in culture.

Close contact infection An infection which requires close contact, other than sexual contact, between susceptible and infectious individuals, for transmission.

Closterovirus Member of a group of plant viruses with very long, flexuous, rod-shaped particles containing a single molecule of linear RNA, some members of which are transmitted by whiteflies.

Club fungi Members of family Clavariaceae (Aphyllophorales: Basidiomycetes) which have club-shaped basidiomata.

Club mosses Short plants that are more like ferns than real mosses. Examples include Running Cedar and Ground Pine.

Cluster cup Aecidium.

Cluster investigation An epidemiologic investigation mounted to determine if there has been an unexpected increase in the number or prevalence of cases of illness. The increase can be with respect to a particular interval in time, a particular location, or both.

Cluster Large group of flowers or fruits on a plant. Some plants have many flowers growing close together in a cluster. Clusters may be long or ball-shaped. Later, fruits replace the flowers.

Clutch A group of eggs (especially reptiles).

Clypeus A shield-like stromatic growth, with or without host tissue, over one or more perithecia or pycnidial conidiomata.

Co evolution The co-existence of legume hosts and rhizobia in a particular location or ecosystem over considerable periods of time can lead to modification of their interaction in nodulation and nitrogen fixation. Thus the pea landrace Afghanistan will nodulate with pea rhizobia from Middle Eastern countries, but not with those from Europe. This difference is regulated by genes in both host and bacteria.

Coacervate Massed or heaped together.

Coadnate United, cohering.

Coalesce To merge or grow together into a similar but larger structure.

Coalescent Joined together.

Coarctate Crushed together, crowded.

Coat protein The protective layer of protein surrounding the nucleic acid core of a virus; the protein molecules which make up this layer.

Coccidioidomycosis Deep-seated mycosis caused by Coccidioides immitis (Hyphomycetes); common in dry S.W. of U.S.A.

Coccus A spherical (or near-spherical) bacterial cell.

Cochleate Shaped like a shell, spiralled.

Cocoon The pupa stage of a moth or other insect. Usually "cocoon" refers to the outer layer that contains the pupa inside. Cocoons may be made of silk spun from a caterpillar, or dead leaves or wood, cemented together with saliva. Earthworms (which are not insects) lay their eggs in cocoons; a different kind of cocoon, which protects the egg.

Codon A particular sequence of three nucleotides in a nucleic acid that codes for a specific amino acid or termination of a polypeptide chain.

Coelomycete Common name for a member of the Coeloanamorphoses, fungi with acervular, pycnidial, pycnothyrial or stromatic condiomata.

Coenocytic Referring to the multinucleate vegetative structures (for example, mycelium) in which the cytoplasm is not separated by cross walls. Synonym: nonseptate.

Coenozygote A cell containing more than one zygote.

Coenzyme An organic molecule, which plays an accessory role in enzyme-catalyzed processes, often by acting as a donor or acceptor of a substance involved in the reaction; ATP and NAD are common coenzymes.

Coevolution The interdependent evolution of two or more species having an obvious ecological relationship.

Cofactor Non-protein component(s) needed by an enzyme to be functional; some cofactors are metal ions, others are coenzymes.

Cohort A subsection of a population with a common feature, usually age. For example, all those individuals in the UK born in 1964 form a birth cohort.

Cole crops Any of the group of crucifer family crops that are varieties of the species Brassica oleracea, including cabbage, broccoli, cauliflower, and brussel spouts.

Coleoptile In a plant, the first leaf of a monocotyledon forming a protective sheath around the plumule.

Collabent Collapsed in the middle.

Collar The portion of the seedling or plant near the surface of the soil; in grafted woody plants, the scion portion of the plant near the soil surface.

Collar region In grasses, the region found where the leaf blade and sheath meet; used in identification of species; in trees, the trunk area at the soil line.

Collar rot Rotting of the stem at or near ground or soil level.

Collarette A frill.or collar (often cup-shaped) of outer wall material remaining at the apex of a phialidic conidiogenous cell, following dehiscence of the first conidium.

Collenchyma Supporting tissue in soft stems and other plant parts, composed of elongated living parenchyma cells with unevenly thickened primary walls, often bordering veins in dicot leaves; the "strings" in celery.

Colonization Establishment and ramification of a pathogen within a host plant.

Colonize To infect and ramify through plant tissue with the growth of a pathogen.

Colony A large group of a single species of organism, living close together. Some animals that live in colonies include ants and termites. Some plants that live in colonies include Yellow Pond Lily, cattails, and crabgrass.

Colony forming unit The number of colonies formed per unit of volume or weight of a cell or spore suspension.

Colorant A paintlike material, usually a dye or pigment, applied to (a) brown warm-season turf grasses that are in winter dormancy; or (b) brown cool-season turf grasses that are in summer dormancy; or (c) turfs that have been discolored by environmental stress, turf-grass pests, or the abuse of human beings. Its purpose is to maintain favorable green appearance.

Columella This is another one of those anatomical terms of almost unbelievable generality. It refers to any small rod-like structure (and some that are not rod-like in the slightest), such as the ear ossicle of reptiles, the basal column of univalves in gastropods, the sterile support of the fruiting body in fungi, and, for our purposes, either (a) a structure closely analogous to the fungal columella in the sporophyte of basal plants; or (b) the central portion of the root cap in higher plants.

Columellacolumella The swollen tip of the sporangiophore (the stalk) on which mitospores develop. The image shows an intact columella, with spores attaches. When the spores have dispersed, the columella collapses into a structure that looks remarkably like a microscopic mushroom.

Coma A tuft of hairs.

Commensal One of two partners living in permanent close association, which gains a slight benefit from the association without causing serious disadvantage to the other.

Commensalism A relationship between two organisms where one species is helped, but the other is unaffected.

Comminute To reduce to small fragments, usually by mechanical means.

Commisure In nematodes: Connecting bands of nerve tissue.

Common Often seen or known to be in an area.

Common bunt of wheat Common bunt or stinking smut of wheat is so important in the history of plant pathology in Washington that it is given detailed coverage. Common bunt, hereafter referred to as bunt, is caused by T. caries and T. foetida, two species so closely related that they hybridize readily. The spore wall of T. caries has reticulate ornamentation, the wall of T. foetida is smooth. Their life and disease cycles are similar. Both invade wheat seedlings shortly after the wheat germinates, are systemic within the host, and they sporulate within the ovary, completely replacing normal ovary contents with black, spherical spores. The ovary wall remains intact, firmly enclosing the spore mass to form bunt balls. The bunt balls are easily crushed, releasing the spores and the foul odor typical of 'stinking' smut, attributed primarily to trimethylamine. When wheat is threshed most of the bunt balls rupture and loose spores contaminate healthy kernels and some are blown from the rear of the machine. Some bunt balls escape rupture and remain within the threshed grain. Their size and shape make bunt balls difficult to remove with grain-cleaning equipment. Bunt balls are slightly lighter than kernels and can be removed from seed by flotation in salt brine. The best description of bunt that I have seen is that of Heald, and readers are referred to it. In Heald's time dwarf bunt was not distinguished from common bunt. The spores of T. controversa are reticulate like those of T. caries, and Heald stated that T. foetida causes a 'tall' bunt, T. caries a 'short' bunt. This statement should be discounted as both species of common bunt cause some degree of stem shortening not comparable to shortening caused by dwarf bunt. Dwarf bunt has a different life cycle and will be treated separate from common bunt.

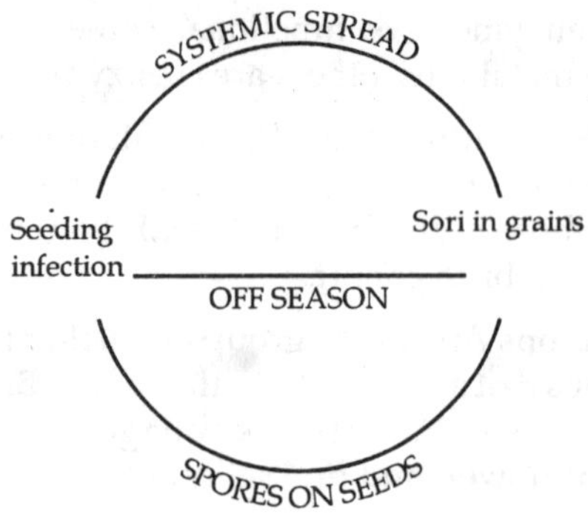

Fig. Disease cycle of hill bunt of wheat caused by Tilletia caries and T. foetida.

Common name Unlike higher plants and animals few pathogenic fungi, bacteria or other pathogenic agents have common names. On the other hand,many plant diseases have vernacular names and some of these, such as smut and rust, have been extended to cover the pathogens.

Names of specific diseases, e.g. choke of grasses caused by Epichlo typhina, are excluded but a number of general disease names such as blight, damping-off and mildew have been listed.

Plant virus names, which are not yet subject to an international agreement, are frequently covered by adding ' virus ' to the common name of the species in which the infection was first recorded. It is often convenient to derive common names of pathogens from their generic names. When this is done the name is printed in roman (not italic) type and without the initial capital. Thus, an aspergillus for a species of Aspergillus: fusaria, phytophthoras, for species of Fusarium and Phytophthora, respectively.

Co-morbid Two or more disease conditions that occur simultaneously within the same person. A diagnosis of CFS requires that certain other conditions that may also cause fatigue, such as cancer, are not present.

Comovirus Member of a group of multicomponent plant viruses with small, isometric particles containing two linear RNA species, readily transmitted mechanically and by beetles.

Compaction A compression of soil that results in poor water drainage, air movement and root growth.

Companion cell Specialized cell adjacent to sieve tubes in phloem.

Companion planting The practice of planting certain plant species-often herbs-in close association with crop plants to repel pests.

Compartmental model A mathematical model which divides hosts into different compartments according to their infectious state. A typical model for microparasites might be an SEIR model. Sometimes referred to as a prevalence model.

Compartmentalization In trees: the processes that result in isolation of wounded or diseased xylem from normal xylem by the formation of chemically and anatomically specialized tissue around the damaged zone.

Compatible 1. Of a relation between host and pathogen in which disease can develop. When host and pathogen are incompatible disease does not develop, commonly because of hypersensitivity of the host.

2. Of a relation between root-stock and scion in which a complete and functional union is formed between the two. Structural or functional anomalies causing breakage of the union or defective growth of the plant indicate an incompatible relation between stock and scion.

Competition Because each plant only forms a finite number of nodules, the ability of indigenous rhizobia to form nodules will limit the number produced by inoculant strains. This can limit nitrogen fixation, and in the American Midwest can mean that the plant derives less than 50% of its nitrogen from symbiosis. While this is generally seen as a competition between inoculant and indigenous rhizobia, various factors play a role. These include the great numerical superiority of the indigenous rhizobia in the bulk soil, and the limited mobility of the inoculant rhizobia. It is also assumed that indigenous rhizobia are also better adapted to soil conditions.

Competitive exclusion agent Organism capable of outcompeting other organisms, thus excluding them from the environment.

Complanate Flat.

Complement A thermolabile substance present in the blood serum of higher animals and able to combine with antigen-antibody complexes. The complement on combining with antigen-antibody complexes is removed from the serum. This removal of the complement, or complement fixation, is the basis of a sensitive test for detecting antigen-antibody reactions.

Complete flower One possessing all floral parts sepals, petals, stamens and pistils.

Complete resistance A resistant reaction in which some aspect of disease development, usually symptom expression or pathogen reproduction, is completely stopped.

Complex 1. A collective term for all the strains of a virus, or for viruses shown to be related.

2. A collective term for two or more virus strains which,together in a host, cause a disease with symptoms that are not predictable from knowledge of the symptoms caused by the components separately. Welsh recommends that the term composite disease should be applicable in such cases, and that the use of complex in this sense should be discontinued; we support this recommendation.

Complex disease A disease caused by the interaction of two or more pathogens.

Complex tissue Tissue that consists of more than one cell type, e.g. phloem.

Complicate Folded upon itself.

Composite disease A disease caused by two or more unrelated viruses.

Compost A mixture of organic residues and soil that has been piled, moistened, and allowed to decompose.

Compound ascoma A teleomorphic fructification incorporating several to many distinct ascomata on or in a single structure, as in the Clavicipitales and Xylariales.

Compound leaf One divided into two or more distinct leaflets.

Compressed Flattened in one plane either dorsally or laterally.

Concatenate Linked together as in a chain.

Concentration The amount of active material in a given volume. Specifications for concentration of herbicides should be on the basis of pounds per unit volume of diluent.

Concentric One circle within another with a common center.

Conceptacle A hollow structure producing spores or spermatia.

Conchate Shaped like a bivalve shell.

Concolorous Having the same colour.

Concrescent Coalescent; becoming joined.

Conditioning The process by which fungi must enzymically soften up substrates like dead leaves before the detritivorous animals can eat them.

Conditions for disease In order for a biotic plant disease to occur, three conditions must be met: 1. the host plant must be susceptible;

2. a pathogen must be present; and

3. the environment must be favorable for infection by the pathogen. All three of these factors must occur simultaneously. If one or more is absent, the disease does not occur.

The genetics of the plant determine its susceptibility or resistance to a particular disease. Specific pathogens vary in their ability to infect different plant species. For example, the late blight fungus can infect certain solanaceous plants such as tomato, potato, and eggplant. However, cucumbers, beans, and other types of plants are not susceptible to late blight.

Various physical and biochemical factors of a given plant species influence susceptibility. These factors include plant defense mechanisms, carbohydrates and protein types, cuticle thickness, and stomatal shape. The developmental stage of the plant also can influence disease development. For example, damping-off diseases occur in younger plants, but not in older plants.

Pathogens differ in their ability to survive, spread, and reproduce. For instance, a certain viral disease may occur only when a specific insect vector transports virus particles to the susceptible plant. Or, a fungus that cannot survive cold winters may be problematic only in the growing season following the shipment of the infected plant material from a warmer state.

Certain temperature, light, or moisture conditions generally are necessary for a disease to occur. Cool, moist conditions are ideal for many fungal and bacterial pathogens. For example, the bacterial disease fire blight and the fungal disease peach leaf curl both are more severe following wet springs.

Cone A cylinder or egg-shaped structure of scaly plates which contains seeds. Certain plants, especially pine trees, have cones instead of regular fruits.

Confluent Running into one another.

Conglobate Formed into a ball; of the bases of stipes, together making a fleshy mass.

Conidial fungi Those anamorphic expressions of Ascomycetes or Basidiomycetes which form conidia . Most moulds belong here. Some have no teleomorph and are effectively holomorphs.

Conidiation The process of producing conidia.

Conidiogenesis The process by which individual conidia develop.

Conidiogenous cell A conidium-producing cell. .

Conidiogenous Producing and bearing conidia.

Conidiole A small conidium, especially one on another; a secondary conidium.

Conidioma A specialised, multihyphal, conidium-producing structure; includes acervulus, pycnidium, pycnothyrium, sporodochium, synnema.

Conidiophore A simple or branched, fertile hypha bearing conidiogenous cells from which conidia are produced.

Conidiosporangium A conidium-like or deciduous zoosporangium of certain of the Peronosporales; it contains only one spore and hence resembles a unicellular conidium.

Conidium A non-motile, usually deciduous, asexual spore which is not formed by cytoplasmic cleavage or free cell formation.

Conjugate 1. Joined in twos;

2. copulation, especially isogamic copulation.

Conjugation In general, any of various sexual processes in microorganisms in which gene transfer follows the establishment of direct contact between two (or more) cells which typically show little or no morphological differentiation from vegetative cells. In bacterial conjugation, one bacterium (the "male" or donor cell) transfers DNA to another (the "female" or recipient cell) while the cells are in physical contact; a recipient that has received DNA from a donor is called a transconjugant.

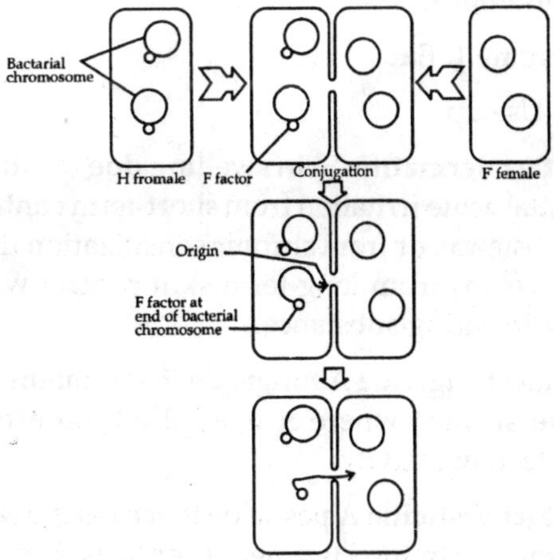

Fig. Conjugation in bacteria: chromosome from an Hfr male is transferred to F^- female

Conk A shelf-like, typically hardened basidiocarp of a wood decaying fungus, usually a polypore.

Connate Joined by growth.

Connective tissue disorder A variety of inflammatory diseases of connective tissue, the most common of which is rheumatoid arthritis. Much, if not all, of this disease is now attributed to autoimmune processes.

Connective tissue The supporting tissues of the body, such as tendons, ligaments, bone, and cartilage.

Connivent Coming into contact, but not organically joined.

Conservation Any biological control practice designed to protect and maintain populations of existing natural enemies.

Conservation headland A sterile strip, minimum of 1m wide, between the crop and the field boundry which prevents weeds such as sterile brome and couch encroaching on the crop. A band of about 6m wide along the outside of the crop is left unsprayed and receives no pesticide treatments in the spring.

Constitutive Present or produced under all conditions.

Consume 1. Eat.

2. Use up.

Contact dermatitis Skin swelling due to either inital acute irritation from short-term contact substance, or from chronic sensitization that develops from long-term skin contact with an irritating substance.

Contact fungicide A fungicide that remains on the surface where it is applied; no after-infection activity.

Contact pesticide A pesticide that causes death or injury to insects when it contacts them. It does not have to be ingested. Often used in reference to a spray applied directly to a pest.

Contact rate The rate at which susceptibles meet infecteds. Usually measured as individuals per unit time.

Contagious distribution Same as an aggregated distribution.

Contaminant A substance, microorganism or virus accidentally introduced into a culture, substrate or medium; generally, any ' unwanted ' material in an experimental system.

Context The inner or body tissue of a fruit body which supports the hymenophore in the larger and especially the pileate species of Hymenomycetes.

Contiguous Touching.

Continuous cropping Growing the same crop in the same location repeatedly.

Continuous Of spores, hyphae, etc., without septa; of a stipe, one with the tissue of the pileus or peridium.

Continuous time model A model in which the system changes continuously over time. Derivatives (e.g. dY/dt) are the mathematical formalism for describing such continuous change. The differential equation which embodies a model provides the values of these derivatives at any particular time point; calculus or a computer can then be used to move the state of the model forwards in time.

Continuous models have the advantage over discrete time models in that they are more amenable to algebraic manipulation, although they are slightly harder to implement on a computer.

The same as a differential equation model.

Contract growing Crops grown under contract to supply a specific outlet or to a factory for further process. Often grown to a detailed specification.

Control A device used to verify or regulate a scientific experiment or study. A case-control study serves as a useful example. Since patients with a specific illness are examined for various characteristics, a group of healthy individuals who otherwise have as much in common with the patients as possible must be examined in parallel for the same characteristics.

Control action guideline A guideline used to determine if pest control action is needed.

Control action threshold Pest population level at which treatment is necessary to prevent economic loss, also called economic threshold.

Controlled Atmosphere Regulation of O_2 and CO_2 levels as well as temperature to improve storage of fruit and vegetables

Convergence Similarities which have arisen independently in two or more organisms that are not closely related.

Cool-season turf grass Turf-grass species adapted to favorable or optimal growth during cool portions (60 to 75 degrees F.) of the growing season.

Coordinates A pair of numbers used in a grid system to specify location, usually latitude and longitude.

Coppice Regrowth of woody plants by sprouts from stumps or roots.

Coppice woodland Consists of shoots arising from stumps of trees which are cut on a regular cycle. Used to supply small diameter poles for building, fencing and fuel. Creates a variable habitat.

Coprine A chemical found in the Coprinus genus, especially prominent in C. atramentarius. Responsible for Antabuselike reactions if ingested within forty-eight hours of drinking any form of alcohol.

Coprophilous fungi Fungi living on dung; include many zygomycetes, ascomycetes and basidiomycetes.

Coprophilous Growing on dung.

Coral fungi Hymenomycetes with highly branched, upright basidiomata, usually arising from the ground; some members of the family Clavariaceae (Aphyllophorales, Holobasidiomycetes).

Coralloid Having the form of branching coral, especially basidiomata of Ramaria.

Core aerification A method of turf cultivation in which soil cores are removed by hollow tines or spoons to control soil compaction and to aid in the penetration and distribution of pesticides and water.

Coremium Compact or fused, generally upright conidiophores, with branches and spores forming a headlike cluster.

Coriaceous Leathery.

Cork cambium A narrow cylindrical sheath of meristematic cells that produces cork cells to replace the epidermis during secondary growth (growth in width). A type of cambium, also called phellogen.

Cork External protective tissue of a stem or root, impermeable to water and gasses; the primary component of bark.

Corneal opacity Clouding of the transparent front portion of the eyeball.

Corneous Consisting of a horn-like substance.

Cornicles The posterior dorsal erect or semi-erect tubules of aphids which secrete a waxy defensive liquid to protect the insect against enemies; short, blunt horns or rounded projections occurring on the abdomen.

Cornute bearing horns; horn-like in form.

Corolla Petals, collectively.

Coronate Crowned.

Corpus In nematodes: The anterior cylindrical part of the esophagus. The basal region of the corpus at times may be swollen to form a bulb.

Cortex Region of parenchyma tissue between the epidermis and the phloem in stems and roots; region beneath the rind of a sclerotium.

Cortical Of or pertaining to the cortex.

Corticolous Living or growing on bark.

Cortina (of agarics) A filamentous or web-like partial veil covering the mature gills; typical of the rusty-brown-spored mycorrhizal genus Cortinarius.

Coryneform 1. Essentially rod-shaped with one end thickened or bulbous.

2. A name applied, loosely, to any Gram-positive, asporogenous, pleomorphic rod-shaped bacterium; as such it covers bacteria from a range of genera.

Cosmid A plasmid into which has been inserted the cos site of bacteriophage.

Cost/benefit analysis A quantitative evaluation of the economic, social, environmental and human health costs which would be incurred versus the economic, social, environmental and human health benefits of a proposed action such as the establishment of an acceptable dose of a toxic chemical.

Costa The midline vein, nerve or midrib of a bryophyte leaf .

Costate Bearing ribs.

Cotyledon leaves The first leaf, or pair of leaves, of the embryo of seed plants.

Cotyledon Seed leaf, one in moncots and two in dicots; primary embryonic leaf within the seed in which nutrients for the new plant are stored.

Fig. Cotyledon

Cotyledonscotyledon The embryo leaf or leaves of seed plants that absorbs nutrients packaged in the seed, until the seedling is able to produce its first true leaves and begin photosynthesis. "The number of cotyledons present in an embryo is an important character in the classification of flowering plants (angiosperms). Monocotyledons (such as grasses, palms, and lilies) have a single cotyledon, whereas dicotyledons (the majority of plant species) have two. In seeds that also contain endosperm (nutritive tissue), the cotyledons are thin, but where they are the primary food-storing tissue, as in peas and beans, they may be quite large. After germination the cotyledons either remain below ground (hypogeal) or, more commonly, spread out above soil level (epigeal) and become the first green leaves. In gymnosperms there may be up to a dozen cotyledons within each seed." Tiscali.

Courtship The way two animals attract each other to mate.

Cover crops Cultivation of a second type of crop primarily to improve the production system for a primary crop; examples include grasses or legumes maintained in orchards or vineyards and legume or other crops grown during the winter season to improve soil condition.

Cover Protection for an organism, usually a plant. Animals use plants to hide and protect themselves.

Cover, coverage The proportion of the surface area of plants or plant parts on which an applied material is retained following dusting, spraying etc.

Cover, overall An application of a material to the whole area of a crop rather than concentrated in rows, bands etc.

Covey A family or small flock of birds. Especially refers to game birds, such as turkeys, doves, or bobwhites.

Crawler This is the term used to describe the active first instar in a few families of whiteflies and scales. This instar moves to find a suitable feeding site and then embeds its piercing-sucking mouthparts into a plant. Later instars usually do not move from this site so the term crawler is is often used in place of "first instar."

Creatinine A component of urine, and the final breakdown product of creatine, which is an important molecule for building energy reserves, for example, in muscle cells.

Crenate Having the margin scalloped so as to form rounded teeth.

Crenulate Minutely crenate.

Crepuscular Times in the morning and evening when it is between daylight and dark.

Crepuscular times are when some animal species are most active, including Eastern Gray Squirrels and White-tailed Deer.

Crinkling Bending or twisting of foliage without breaking; wrinkling.

Crista 1. Of mitochondria, folds in the internal membrane of the mitochondrion which gives the organelle its characteristic appearance. This is the site of the electron transport chain in oxidative metabolism. The cristae, therefore, serve as the physical link between the tricarboxylic acid cycle and oxidative phosphorylation.

2. more generally, a crest or ridge.

Cristate Crested.

Crochets Tiny hooks on the prolegs of caterpillars.

Crop rotation The practice of growing a sequence of different crops on the same land in successive years or seasons; done to replenish the soil, curb pests, etc.

Cross absorption test A test of serological affinities of two antigens, e.g. viruses. This involves testing dilutions of an antiserum against the antigen used in its preparation after previously incubating the antiserum with an excess of dissimilar antigen.

Cross fertilization A process in which sexual reproduction occurs as a result of the fusion of sex cells from different individuals.

Cross inoculation groups Groups of legumes, any one of which will form nodules when inoculated with rhizobia isolated from another legume in the group.

Cross pollination Transfer of pollen from the anthers to the stigma of another flower.

Cross protection The process whereby a normally susceptible host is infected with a less virulent pathogen (usually a virus) and thereby becomes resistant to infection by a second, usually related, more virulent pathogen.

Cross resistance In pest management, resistance of a pest population to a pesticide to which it has not been exposed that accompanies the development of resistance to a pesticide to which it has been exposed.

Cross sectional study Same as a horizontal study.

Cross walls Septa.

Crossing over (crossover) Exchange of genetic material between two homologous chromosomes by the breakage and joining of sister chromatids; occurs during meiosis (reduction division).

Cross-sectional study In epidemiology, a study in which participants are examined at only a single time for characteristics of a disease.

Crown The part of the alfalfa plant from which new shoots are produced, emerging at soil surface just above taproot; the point at or just below the soil surface where the main stem (trunk) and roots join. Also used to refer to the topmost limbs on a tree or shrub; the shortened stem of a strawberry plant, from which roots, leaves, and fruit trusses arise.

Crown gall disease A disease occurring on many species of plant and caused by the bacteria Agrobacterium tumefaciens. In this disease, in which large galls develop, the bacteria transfers part of its genetic information to the host plant.

Crown group All the taxa descended from a major cladogenesis event, recognized by possessing the clade's synapomorphy.

Crozier formation Process of ascus development from coiled tips of ascigerous hyphae.

Crozier In ascogenous hyphae, a terminal hook in which conjugate nuclear division takes place just prior to nuclear fusion, meiosis, and ascus formation.

CRS Congenital rubella syndrome.

Cruciate In the form of a cross; of basidial septa, with a vertical septum and one at right angles to it.

Cruciately-septate Describes the basidia of order Tremellales (Phragmobasidiomycetes) which are divided into four more or less equal parts by vertical cross-walls.

Crucifer A plant in the mustard or cabbage family, such as broccoli and cauliflower.

Crude birth rate The number of live births in a year divided by the population size.

Crude death rate The number of deaths in a year divided by the population size.

Crustacea Most species in this group (Class Crustacea) are marine or fresh water animals, although there are a few terrestrial species. While the familiar larger types, lobsters, crayfish, crabs and shrimp, are in this group, it also contains many small to minute aquatic forms that are very important in aquatic food chains.

Crustaformeria In nematodes: Glandular region of the distal part of uterus that may play a role in the formation of the egg envelope; sometimes called the quadricolumella.

Crustose describes a lichen thallus that adheres closely to the substrate (often rock).

Cryptic Inconspicuous or hidden.

Cryptobiosis The ability to enter a state of suspended metabolic activity during unfavorable environmental conditions (drying, cold); survival strategy of some plant pathogenic nematodes.

Cryptococcosis A systemic mycosis caused by a basidiomycetous yeast, Cryptococcus neoformans (teleomorph in Filobasidiella neoformans).

Cryptogram A descriptive code summarizing some of the main properties of a virus.

Cryptograms aid identification and add precision to the vernacular names of viruses. They also show how much is known about a given virus and they help to overcome the difficulties of synonymy, homonymy, and translation into other languages.

According to A. J. Gibbs, B. D. Harrison, D. H. Watson & P. Wildy, each cryptogram consists of four pairs of symbols (for example, R/1:2/5:E/E:S/ for tobacco mosaic virus) with the following meanings:-1st pair. Type of nucleic acid/strandedness of nucleic acid. Symbols for type of nucleic acid: R = RNA; D = DNA. Symbols for strandedness: 1 = single-stranded; 2 = double-stranded.

2nd pair. Molecular weight of nucleic acid (in millions)/percentage of nucleic acid in infective particles.

3rd pair. Outline of particle/outline of 'nucleocapsid ' (the nucleic acid plus the protein most closely in contact with it). Symbols for both properties: S = essentially spherical; E = elongated with parallel sides, ends not rounded; U = elongated with parallel sides, end(s) rounded; X = complex or none of above.

4th pair. Kinds of host infected/kinds of vector. Symbols for kinds of host: A =actinomycete; B =bacterium; F = fungus; I = invertebrate; S = Symbols for kinds of vector: Ac = mite and tick (Acarina, Arachnida); Al = white-fly (Aleyrodidae, Hemiptera, Insecta); Ap = aphid (Aphididae, Hemiptera, Insecta); Au = leaf-, plant- or tree-hopper (Auchenotthyncha, Hemiptera); Cc = mealy-bug (Coccidae, Hemiptera); Cl = beetle (Coleoptera, Insecta); Di = fly and mosquito (Diptera, Insecta); Fu = fungus (Chytridiales and Plasmodiophorales, Fungi); Gy = mirid, piesmid, or tingid bug (Gymnocerata, Hemiptera); Ne = nematode (Nematoda); Ps = psyllid (Psyllidae, Hemiptera); Si = flea (Siphonaptera, Insecta); Th= thrips (Thysanoptera, Insecta); Ve = vectors known but none of above. Symbols for all pairs: = property of the virus is not known; () enclosed information is doubtful or unconfirmed.

Cucumovirus Member of a group of multicomponent plant viruses with isometric (icosahedral) particles consisting of three linear RNA species (RNAs 1, 2, and

3), transmitted by sap and in nature by aphids in a noncirculative manner.

Cucurbit A plant in the gourd or cucumber family, such as melon, squash and pumpkin.

Cuitlacoche The spore mass of the corn smut fungus, Ustilago maydis, as eaten by Mexicans.

Cull A seedling that is rejected because it does not meet certain specifications.

Culm Stem of grasses, cereals and bamboos.

Cultivar 1. An assemblage of cultivated individuals which is distinguished by any characters (morphological, physiological, cytological, chemical or others) significant for the purposes of agriculture, forestry or horticulture and which, when reproduced (sexually or asexually), retains its distinguishing features.

2. variety;

3. strain, e.g. barley ' Balder '. A cultivar may be a clone (Solanum tuberosum ' King Edward VII ') or a line (Triticum aestivum ' Marquis '). Plant pathologists should use this useful term in the sense of the Code, e.g. for cultivars (varieties) of the cultivated mushroom. Use of the term in the sense of R. H. Stover ' a distinct group of isolates that have some common morphological or physiological characteristics but which may consist of one or more clones or races ', e.g. Fusarium oxysporum (cultivar) ' Inodoratum ' (clone) B, is to be deprecated.

A cultivated plant variety or cultural selection. Used synonymously with variety.

Cultivation A tillage operation used in preparing land for seeding or transplanting, or later, for weed control and loosening the soil.

Cultural practices The manner in which plants are grown, such as: application of nutrients, irrigation practices, type of cultivation; may be used for disease management.

Culture 1. A general term for a growth of a microorganism in vitro or in vivo its use in the sense of isolate or strain should be avoided.

2. to cultivate a microorganism, usually in vitro in a prepared medium and in the absence of other microorganisms, but the term can be applied also to the maintenance of an obligate parasite on its host plant.

Culture collection A repository of cultures of characterized viruses, bacteria, and other organisms. Used for reference and comparison with new isolates.

Culture medium A prepared substrate for the growth of a microorganism or of other living cells in vitro.

Cumulate Heaped up.

Cumulative effects The additive effects of exposures to a given pesticide from various sources (e.g., dietary, environmental), or to several pesticides simultaneously or over time.

Cuneate Wedge-shaped, thinner at one end than the other.

Cunninghamella sp Cunninghamella sp. Commonly found as saprophyte in soil. Can cause disseminated and pulmonary infections in immune compromised hosts. It is a zygomycete with zygospores being of the Mucor type. Asexually it produces extensive white mycelium, nonseptate, conidiophores (sporangiosphores) simple or branched, with enlarged tips bearing heads of conidia; conidia are colorless, 1-celled, globose, conidia may be spiny or smooth. Cultivation - Potato dextrose agar, 24°C.

Cup fungus A 'discomycete' any ascomycete with an open, shallow, cupulate apothecial ascoma; a hetergeneous grouping because the asci may be unitunicate operculate, unitunicate inoperculate, or in many lichens, bitunicate.

Cup The remnants of the universal veil surrounding the base of certain mushrooms, such as Amanita andVolvariella. Also called "volva."

Cupulate Cup-shaped.

Cupule Structures that surround one or more ovules or seeds, e.g., the "cap" of an acorn. The cupule lobes may be free or united.

Curculio Beetles of the suborder Rhynchophora which have slender snouts curving downward and backward, with chewing mouthparts at the tip.

Curing Holding potato tubers under warm, humid conditions that favor wound healing.

Curl Distoration, puffing, and crinkling of a leaf resulting from unequal growth.

Curvularia sp Curvularia sp. - Reported to be allergenic. It may cause corneal infections, mycetoma and infections in immune compromised hosts. This fungus can be parasitic or saprophytic. Conidiophores brown, mostly simple, bearing conidia apically; dark conidia, end cells lighter, 3- to 5-celled, more or less fusiform, typically bent, with one of the central cells enlarged. Cultivation - Potato dextrose agar, 24°C.

Cuticle Noncellular outer layer of an insect or a nematode; water-repellent, waxy layer of epidermal cells of plant parts, such as leaves, stems and fruit.

Cutin An insoluble polymer that, embedded in waxes, forms the cuticle covering the epidermal cell walls in the aerial parts of higher plants.

Cutin monomers V SativaCutin: the waxy substance of which a cuticle is composed.

Cutinolytic Of certain enzymes: able to digest cutin.

Cutis Of basidiocarps of certain wood-decaying fungi: the outer layer consisting of compressed hyphae parallel to the surface, sometimes with varnishlike incrustation.

Cutout A period of reduced growth and square production following a fruiting cycle.

CV A plant type within a species, resulting from deliberate manipulation, which has recognizable characteristics (color, shape of flowers, fruits, seeds and height or form).

Cyanescent Becoming blue.

Cyanobacteria Prokaryotic chlorophyllous organisms often capable of fixing nitrogen; also known as 'blue-green algae.'

Cyathiform Like a cup, slightly wider at the top than at the bottom, and sometimes stalked.

Cycloheximide (actidione) An antibacterial and antifungal antibiotic derived from Streptomyces, which is used in isolating fungi pathogenic to humans.

Cyclopeptides (cyclic oligopeptides) Substances formed of amino acids joined in a ring.

Cyclosporine A cyclic polypeptide, produced by the Hyphomycete, Tolypocladium niveum; used as an effective immunosuppressant after organ transplantation.

Cylindrical Having the form of a cylinder; of spore shape, with a length:breadth ratio of 2-3.

Cymbiform Broadly reniform.

Cyst In fungi, a resting structure in a protective membrane or shell-like enclosure;in nematode females, the egg-laden carcass of a female nematode; in bacteria, a specialized type of bacterial cell enclosed in a thick wall, often dormant and resistant to environmental conditions.

Cystidia Sterile accessory cells, frequently of distinctive shape, occurring at any surface of a basidioma, particularly the hymenium, from which they frequently project. The shaggy mane, Coprinus comatus, for example, has spacing cystidia to hold its crowded gills apart.

Cystidiole In hymenomycetes, a simple hymenial cell of about the same diameter as a basidium, but remaining sterile and extending beyond the hymenial surface.

Cystidium A sterile cell occurring among basidia and often projecting beyond the hymenium, differing morphologically from the basidium.

Cystosorus An aggregation of united cysts or resting spores, the presence and morphology of which are of taxonomic significance, e.g. in the Plasmodiophorales.

Cytokine Proteins manufactured by cells of various lineages that, when secreted, drive specific responses (e.g., proliferation, growth, or maturation) in other susceptible cells.

Cytokinins Chemicals which, regardless of their other activities, promote cell division in cells of various plant origins usually when grown as call us in aseptic culture. For the most part endogenous cytokinins are all N6 substituted adenine derivatives, for example zeatin from Zea mays and N6 purine from cultures of Corynebacterium fascians. A number of non-purine synthetic compounds such as azakinetin, N, N1 diphenyl urea and benzimidazole are known to possess cytokinin-like activity.

Cytomegalovirus (CMV) One of the eight known types of human herpesviruses, also known as human herpesvirus 5 (HHV-5). It belongs to the beta subfamily of herpesviruses. CMV can cause severe disease in patients with immune deficiency and in newborns when the virus is transmitted in utero.

Cytopathology The study of changes induced by disease at the cellular level.

Cytoplasm All the living contents of the cell, except the nucleus.

Cytoplasmic inheritance Inheritance of genes not located in the nucleus, i.e. those in mitochondria and chloroplasts.

Cytospore A type of spore released from an encysted amoeba phase in the Amoebidiales (Trichomycetes).

D

Damage threshold level The lowest pest population density at which damage occurs.

Damping off Collapse and death of seedling plants resulting from the development of a stem lesion at soil level. This is a growers term, apparently deriving from the damp conditions among seedlings crowded in a box, which favour the attack and spread of such pathogens as Pythium spp. Pre-emergence damping-off is the death of seedlings, as a result of disease, before they have emerged above the surface of the soil or other medium.

Decay of seeds in the soil or young seedlings before or after emergence.

Data gap Missing information; used to refer to those health or environmental effects studies which are required by the Environmental Protection Agency for registration of pesticides (but not yet completed by the pesticide manufacturer.).

Daughter plants Vegetative progeny of strawberry plants; plants that develop along the runners produced by another strawberry plant called the mother plant.

Day neutral The term applied to strawberry cultivars that produce flower buds more or less independently of day length; everbearing.

Days to harvest Term for the prescribed minimum number of days required by U.S. government regulations between a pesticide application and the harvest of the crop.

Debris Scattered, old remains of plants and animals. Underwater, debris found on the bottom may include leaf and stem parts, old mussel shells, sticks, etc. Debris on a forest floor may include dead leaves, pieces of bark, twigs, and nut shells.

Decant To pour off gently from the top of a liquid so as not to allow sediment or solid material to leave the bottom of the container.

Decay To break down into smaller parts. When plants or animals die, their bodies decay. The nutrients that made up their bodies then go back into the soil and are re-used. Decomposers are organisms that help decay.

Deciduous Falling off, as leaves that are shed in the autumn; used to describe the sporangia of downy mildew fungi (peronosporales: oomycota).

Decline Poor growth of plants; leaves small, brittle, yellowish, or red; some defoliation and dieback commonly is present.

Decomposer An organism that breaks down the bodies or parts of dead plant or animal matter into smaller pieces (decay).

Decomposers, such as mushrooms, bacteria, and earthworms, are very important in food webs.

Decorticate The cortex having been eroded, removed, etc.

Decoy crop Crop that stimulates germination of seeds of a parasitic plant such as witchweed (Striga spp.), but is not susceptible to infection by the parasitic plant; helps reduce seed populations of the parasite in soil so a susceptible crop can be planted.

Decumbent Resting on the substratum but with the ends turned upwards.

Decurrent Extending downwards beyond the point of insertion, e.g. of lamellae extending down the stipe.

Decurved Bent downwards.

Dedicated fuel cropping Use of resources solely towards the production of a crop destined for use as a fuel

Deep planting Tree death often arises from planting trees too deeply. If the root collar area of newly-planted trees is several inches below the landscape grade, the tree has been planted too deeply! This is more of a problem when trees are planted in clay soils, and/or compacted soils, but surface mulches, exposed sites, or turf established to the stem of the tree can aggravate the problem.

Deep planting results in reduced oxygen and water to the fine root systems. This in turn causes poor root regeneration of transplanted trees, gradual death of existing roots, lower tree vitality, and increased vulnerability to biotic problems and environmental extremes. Sometimes deep planting causes a slow and progressive decline in vitality that may progress for 5 to 10 years.

Deer truffle Elaphomyces (elaphomycetales, ascomycetes).

Deficiency payments Payment to producers to make up the difference between the average market price for a commodity and the guarantee price when the market falls below the guaranteed price.

Definitive host The host in which a parasite reproduces sexually.

Defoliant A plant growth regulator that causes the leaves or other foliage to drop from a plant, usually to facilitate harvest.

Defoliate To strip or become stripped of leaves.

Defoliation Loss of leaves from a plant, whether normal or premature.

Defoliator Any chewing insect that consumes the leaves or needles of plants.

Degenerate bacteria that do not have cell walls. Mycoplasmas are smaller than bacteria but larger than viruses. They cause animal and human diseases.

DEGO (dorsal esophageal gland outlet) In nematodes: The point at which the dorsal gland empties into the lumen of the esophagus.

Degree days The departure of the average daily temperature from a defined base (e.g. the minimum recognized temperature for the growth of a plant species). The number of day degrees may be totaled to assess the accumulated warmth of a particular year's growing season.

Dehiscence The process of releasing spores. That is actually the botanical definition, although it is widely used in mycology. Dehiscence is also used in a more restricted, mycological sense to mean a process for realsing all of the spores in an ascocarp simultaneously, or in some coordinated manner.

Dehiscent Of an ascus or fruit-body, opening when mature, by a pore or by rupturing or fragmentation; of conidia and other spores, falling off.

Deirid In nematodes: Paired, porelike organs located in the lateral fields, in the vicinity of the nerve ring of many of the Tylenchoidea; believed by some workers to be sensory in nature.

Delayed dormant Refers to the treatment period in fruit tree crops, beginning when buds begin to swell until the beginning of green tip development.

Deliquesce To liquefy; to autolyse, as in the gills of the agaric, Coprinus, or 'Prototunicate' asci.

Deliquescent To ramify into fine divisions, such as abnormal numbers of buds, twigs, branches, or leaves, e.g., witches' broom development.

Deltoid Triangular in shape.

Delusional disorders A psychiatric disorder characterized by states of heightened self-awareness and a tendency toward paranoia.

Dematiaceous Of spores, mycelium, etc., more or less darkly pigmented.

Deme The ' -deme ' terminology was introduced for natural populations within a specific taxon of flowering plants by J. S. L. Gilmour & J. W. Gregor, but it is a concept of general application. According to Heslop-Harrison ' the essence of the deme system is the construction of a series of category- terms by the addition of one or more virtually self-explanatory prefixes to the neutral suffix " -deme ". The suffix itself, always to be used in terminology with a prefix, denotes any group of individuals of a specific taxon. The prefixes establish on what grounds they are individuals considered together; they thus carry information concerning the kind of criterion applied to forming the grouping, unlike any categories of nomenclature taxonomy '. Examples are: topodeme (a -deme occurring in a specified geographical area), gamodeme (a -deme composed of individuals which are so situated spatially and temporally that, within the limits of the breeding system, they can interbreed), agamodeme (a -deme composed of predominantly apomictic individuals).

Some biologists have wrongly used the suffix alone in the restricted sense of topodeme or gamodeme and Robinson has not lessened confusion by, in plant pathological usage, restricting ' -deme ' for parasites. Robinson thus defines pathodeme as ' A population of a pathogen in which all individuals have a particular character of pathogenicity in common '. Further, Robinson states that pathodeme and pathotype should always be qualified by an adjective such as ' vertical ' or ' horizontal '.

Plant pathologists who feel a need to use the -deme terminology would probably do bast to use ' pathodeme ' for pathogens and ' resistodeme ' for groups of resistant hosts.

Dementias A physically caused permanent or progressive decline in intellectual function that interferes with the patient's normal social or economic activity.

Demicyclic A rust fungus that lacks the urediniospore (repeating) stage (e.g. many species of Gymnosporangium) .

Dendroid Tree-like, arborescent, branching like a tree.

Denitrification Chemical reduction of nitrogen compounds in the soil; normally due to the escape of nitrogen into the air.

Dense Tightly packed together.

Density dependent 1. Effects whose intensity changes with increasing population density.

2. Effects whose intensity increases with increasing population density, for example a mortality which becomes very high near a carrying capacity. Effects which become less pronounced at higher population sizes (perhaps the reduction in fecundity caused by the difficulty of mating) are sometimes called inverse density dependent.

Density dependent effects are often a cause of nonlinearity in a model.

Dentate Toothed.

Denticle In nematodes: Minute tooth or "prickle".

Denticulate Finely toothed.

Deoxyribonucleicacid The double-stranded, helical molecule that contains genetic code information; each repeating unit, or nucleotide, is composed of deoxyribose (a sugar), a phosphate group, and a purine (adenine or guanine) or a pyrimidine (thymine or cytosine) base.

Dependent transmission Ttransmission of a virus by a vector only in the presence of a second (helper) virus. The virus that is not transmissible alone is referred to as the dependent virus.

Deposit Amount and pattern of spray or dust deposited per unit area of plant surface.

Depressed With the centre lower than the margin.

Depression A neurotic or psychotic condition marked by an inability to concentrate, insomnia, and feelings of dejection and guilt.

Derived Describes a character state that is present in one or more subclades, but not all, of a clade under consideration. A derived character state is inferred to be a modified version of the primitive condition of that character, and to have arisen later in the evolution of the clade. For example, "presence of hair" is a primitive character state for all mammals, whereas the "hairlessness" of whales is a derived state for one subclade within the Mammalia.

Dermal Pertaining to the skin; absorption through the skin is one of the main ways in which pesticides can enter the body.

Dermatitis Non-contagious but often chronic skin inflammation; sometimes caused by fungal infections.

Dermatophytes Anamorphs of some Onygenales (Ascomycetes), which live on keratin and can cause skin disease in humans.

Desalination The removal of salts from sea water (generally)

Desiccant An insecticide that kills insect pests by damaging their cuticle thus causing them to dehydrate, or a herbicide that promotes the drying of living plant tissues, such as unwanted plant tops.

Desiccate To dry out.

Desiccation The drying out of a plant.

Destroying angel The pure white agaric, Amanita virosa: ingestion of 1 mL of this fungus can prove fatal; its toxins are cyclic polypeptides, esp. amanitins.

Detassel To remove the immature tassel, as done in production of hybrid seed corn, to prevent the plant from forming mature pollen.

Determinate Having stems and branches that stop growing at a certain point, usually after producing flowers. In cotton, this term is applied to varieties with a distinct interruption in growth following fruit set.

Deterministic model A mathematical model in which the parameters and variables are not subject to random fluctuations, so that the system is at any time entirely defined by the initial conditions chosen.

Detersile Of a villose surface, removable so that the surface becomes bare.

Dethatching Removing excessive thatch from a turf-grass area by hand-raking or with various types of machinery equipped with vertical knives or tines.

Detritivore Animal which eats organic 'garbage,' dead remains of other life forms.

Detritus Tiny bits of dead plant and animal material in water.

Deuteromycetes Group of fungi without a sexual stage; the asexual stage of ascomycetes and basidiomycetes.

Deuteromycotina (also fungi imperfecti) A non-phylogenetic category originally created for fungi with no known sexual stage; the category still includes fungi with no known sexual stage, but it also includes the asexual stages of various fungi now known to have a sexual stage in the Ascomycotina or Basidiomycotina.

Developmental threshold The lowest temperature at which growth occurs in a given species.

Dewlap A fold of loose skin under the throat of an animal. Male frogs and toads fill thier dewlaps with air to make a sound to attract females during breeding season.

Dewpond Small pond which forms on high chalky soils, filled by rainwater. Often man made for livestock drinking, kept full by rain.

Dextrinoid Staining yellowish or reddish brown in iodine-containing solutions such as Melzer's reagent; pseudoamyloid.

Diabetes mellitus A metabolic disease in which carbohydrate utilization is reduced and fat and protein utilization is enhanced. It is caused by insulin deficiency.

Diabetic acidosis A condition in diabetic patients in which the levels of alkali are reduced relative to the level of acids.

Diabetic ketoacidosis A form of acidosis in diabetic patients caused by the enhanced production of ketone bodies.

Diabetic nephrosclerosis A condition in diabetic patients marked by hardening of the kidney from overgrowth and contraction of the connective tissue of the organ.

Diageotropism The tendency to grow in a direction horizontal to the surface of the earth.

Diagnostic A distinguishing characteristic important for the identification of a disease or other condition.

Diagnostic distinctive A distinguishing characteristic serving to identify or determine the presence of a disease or other condition.

Diapause A period of spontaneous dormancy, independent of environmental conditions, interrupting developmental activity in an embryo, larva or pupa.

Diaphragm In gasteromycetes, a homogeneous wall of hyphae that separates the gleba from the sterile base, e.g. in some species of Calvatia.

Diaspore Any disseminated propagule, either sexual or asexual.

Dichotomous branching A type of branching in which the tip divides into two more-or-less equal apices; by repetition of this type of branching in various planes distinctive shoot systems may be produced.

Dichotomous Branching, often successively, into two more or less equal arms.

Diclinous Having the antheridium and oogonium on different hyphae.

Dicot Plant with two cotyledons or seed leaves.

Dictionary A device for starting futile arguments over definitions.

Dictyoseptate Having cross-walls running horizontally and vertically, like the layers of cement between bricks (also described as muriformly septate).

Dictyospore An asexual spore with intersecting septa in more than one plane; with a length:breadth ratio not exceeding 15:1; if elongated, with only a single axis, and that axis not curved through more than 180 degrees; any protuberances, other than setulae, not more than 1/4 the length of the spore body; a muriform spore.

Dictyosporous Having spores with cross and longitudinal walls.

Didelphic Of nematodes: Possessing two complete genital tubes or ovaries.

Didymospore an asexual spore with one septum across the body; with a length:breadth ratio not exceeding 15:1; if elongated, with only a single axis, and that axis not curved through more than 180 degrees; any protuberances, other than setulae, not more than 1/4 the length of the spore body.

Didymosporous Having two-celled spores.

Dieback Necrosis of a shoot starting at the apex and spreading towards the older tissue, death of the stem may occur; a disease caused by Phytophthora cinnamomi Rands and a number of other species of Phytophthora.

Diet What an animal eats for food.

Differential equation The mathematical formulation corresponding to a continuous model; an equation involving derivatives.

The mathematical formulation corresponding to a discrete time model.

Differential host 1. A host that gives reactions which distinguish between race specific isolates of a pathogen, e.g. a rust.

2. A plant that gives reactions distinctive for a specific virus, permitting its differentiation from other viruses commonly present. Distinguish from indicator plant.

Differential medium Culture medium that is used to distinguish between organisms that can grow on it.

Differentiation The physiological and morphological changes that occur in a cell, tissue, or organ during development from a juvenile state to a mature state.

Digitate Shaped like the fingers of a hand; with finger-like divisions.

Dikaryomycota Phylum of eumycotan fungi comprising the Subphyla Ascomycotina and Basidiomycotina and their anamorphs; characterized by hyphae with chitinous walls and perforate septa, and (usually) the occurrence of a dikaryotic phase.

Dikaryon A nuclear phenomenon unique to phylum Dikaryomycota, in which compatible nuclei pair off and cohabit without copulating; this phenomenon is restricted to ascogenous hyphae in ascomycetes; but is the dominant phase in basidiomycetes; the paired nuclei always divide synchronously.

Dikaryotic Having two separate but compatible nuclei in each cell or compartment; a nuclear condition unique to the fungi.

Dikaryotisation Where a homokaryon is converted into a dikaryon, typically by two compatible homokaryons fusing.

Diluent An inert material added to a plant protectant to reduce its concentration.

Dilution end point The stage of a serial dilution of cells or virus preparations at which growth or infection from a standard sample of the suspension no longer occurs.

Dilution plating A method to obtain pure colonies of bacteria and fungi in which infected plant material or infested soil is diluted in sterilized water. Small samples of the water are spread on the medium surface of several petri plates to find which dilution will produce pure colonies.

Dilution streaking Repeated streaking of bacteria on the surface of a nutrient medium with a sterile metal loop to allow pure colonies to grow.

Dimidiate Appearing to lack one half, or having one half very much smaller compared with the other half; of a pileus, without a stalk and semicircular; of a lamella, stretching only halfway to the stipe; of a perithecial wall, having the outer wall covering only the top part .

Dimitic Having two types of hyphae: generative and skeletal or generative and binding.

Dimorphic Having two distinct forms, as in some fungal pathogens of humans, which are yeast-like in the host, but mycelial in culture; also sometimes used to describe sexual dimorphism; having an anamorph and teleomorph.

Dimorphism Existence of two morphologically different forms in one organism.

Dioecious Having male and female reproduction structures on separate individuals.

Diorchic In nematodes: Possessing two testes.

Diploid A nucleus is diploid if it contains two copies of each non-redundant gene. In Fungi, it is necessary to distinguish between diploid nuclei and diploid cells. A hypha may contain several haploid nuclei (either identical or from different individuals). Technically, the cell is diploid or polyploid. However, there may be no diploid nuclei.

Diploidization The process which occurs when compatible haploid hyphae from different individuals exchange nuclei. This does not necessarily result in the formation of diploid nuclei, meiosis, or genetic exchange. In fact, some fungi can maintain populations of nuclei from up to six independent individuals in a single hypha.

Diplolepidous opposite Of moss sporophytes, having an inner (endostome) and outer (exostome) row of plates forming the peristome teeth as radial elements between them.

Diplopoda This class consists of the millipedes which are elongate, wormlike animals with many pairs of legs. Most millipedes have 30 or more pairs of legs and most body segments bear two pairs.

Diptera An order of insects that includes flies and mosquitoes.

Direct life cycle A life cycle in which a parasite is transmitted directly from one host to the next without an intermediate host or vector of another species.

Direct penetration Penetration of plant tissues by a pathogen through barriers such as leaf cuticle by chemical and physical means (e.g. penetration peg).

Directed application A pesticide directed onto weeds or soil in a way that avoids contact with the turf grass.

Disc Of Valsa and related fungi: a more or less flat apical part of a stroma that protrudes above the bark surface; also, of Discomycetes: the exposed fertile portion of an apothecium.

Discolichens Lichens in which the mycobiont (fungal partner) forms apothecial ascomata.

Discomycetes Cup fungi; an informal grouping of ascomycetes which mostly form unitunicate asci in apothecial ascomata, though many lichenized taxa have bitunicate or archaeasceous asci.

Discrete time model A model in which the system jumps from one state to the next at fixed intervals or timesteps. These difference models are simple to understand but often difficult to analyse. The parameters in such a model refer to the amount of change over the finite timestep; they are sometimes referred to as finite rates.

In a (rather precise) sense, a differential equation is what you eventually get from a difference equation when you let the timestep get smaller and smaller and smaller.

Disease Harmful deviation from normal functioning of physiological processes (BMS). Deviation from normal functioning of physiological processes, of sufficient duration to cause disturbance or cessation of vital activity (APS). These two definitions are broad enough to allow malfunctions caused by nutritional deficiencies or excesses, toxic chemicals, adverse environmental factors, genetic anomalies etc., to be classed as diseases in addition to those caused by infective agents. The two classes of malfunction are sometimes classified as non-infectious and infectious diseases respectively; we recommend that the term ' disease ' should apply only to malfunctions caused by pathogenic organisms or viruses, and that those caused by other factors should be termed disorders.

The APS definition implies a minimum degree of malfunction is necessary before the term ' disease ' can be applied; ' harmful ' in the BMS definition perhaps carries the same implication. This interpretation is consistent with the common practice of describing a plant as diseased only when sufficient symptoms have developed to enable an observer to distinguish it with the naked eye from a normal one. However, some workers consider that any plant infected by a parasite is diseased, because any infection must cause some malfunction of the host, e.g. acceleration of respiration rate commonly associated with the earliest stages of pathogenesis; we consider that, in general, it is helpful to distinguish between plants that are infected and those that are both infected and showing symptoms, by describing the former as ' infected ' and the latter as ' diseased '. Nevertheless, when plants become heavily infected without showing obvious symptoms, as with many viruses, there may be epidemiological justification for regarding such plants as ' diseased ', because they may be at least as important as sources of inoculum as plants which show symptoms. If the term ' disease ' is applied in such cases it is desirable that the usage should be defined.

The debilitating effects on a host of infection by a parasite.

Any malfunctioning of host cells and tissues that results from continuous irritation by a pathogenic agent or environmental factor and leads to development of symptoms.

The sum of the deviations of the vital functions beyond the latitude of health.

Disease cycle Of a disease caused by a biotic agent: the cyclical sequence of host and parasite development and interaction that result in disease, in reproduction or replication of the pathogen, and in the readying of a new generation of the parasite for infection.

Disease diagnosis Disease diagnosis is both an art and a science. Each disease problem requires a thorough search for obvious or subtle clues to reveal the factors and their interactions that possibly are involved. To accomplish this, a plant disease diagnostician must be both a detective and a plant ecologist.

There is no one key set of questions or techniques for diagnosing plant diseases. Experience, practice, and study are the best teachers. It is very important to know what diseases and common problems occur on the crop being inspected and whether these problem are likely to occur in Montana. Disease lists, written descriptions, photographs, and people experienced in working with the crop and plant diseases are the most useful resources.

One way to start is by looking at textbook photographs of common problems and attempting to match the problem with the picture. Read disease descriptions carefully to decide whether they correspond to the problem you are diagnosing. Since about 60% of the diseases in Montana are caused by abiotic factors, consider the possibility of environmental, physiological, cultural, and chemical causes first.

It is easier to diagnose plant problems by making a personal, on-site inspection. An inspection can reveal subtle influences of the site, the environment, and possible management practices overlooked by the grower. Ask the grower a series of in-depth questions about the nature of the problem.

Difficulties arise when you are given only a portion of a plant that may or may not indicate the real problem. The worst situation is a diagnosis by phone, because misunderstandings and an inaccurate diagnosis can occur easily.

Because more than one disease may be involved or the real cause may not be obvious, refined detective skills often are needed. The check-list method of using a series of questions to build a case history of the problem is helpful. Always first identify the plant and determine what it should look like at the current time of year. Familiarity with the plant in question is extremely important. This includes cultivar characteristics, methods of culture, and the appearance of a healthy crop A client may mistake a normal stage of development for a diseased plant. For example, certain juniper and spirea cultivars normally have very yellow needles. Therefore, this characteristic is not a disease symptom for these particular cultivars.

Symptoms and signs are very useful in diagnosis. Symptoms are the visible effects of disease expressed by the plant. Describe which plant parts or growth stages are showing symptoms. Are symptoms showing on roots, tubers, bulbs, corms, seedling, foliage, stem, branches, trunks, flowers, fruit, or the entire plant.If leaves are affected, are only older or younger leaves affected.Are only leaf margins affected, etc. If leaf spots are present, observe the color, size, shape (angular or circular) of the spots and note the spot margins are definite or indefinite. If you suspect the problem is in the roots, dig up a plant when practical and wash soil off the roots to view them. Also, determine when symptoms first began occurring. In some cases, there has been a gradual decline in plant health over a number of years. In others, the problem occurred in a matter of days.

Signs are physical evidence of the pathogen causing the disease. Signs can include fungal fruiting bodies (such as mushrooms or pycnidia), mycelia, bacterial slime, or the presence of nematode cysts on roots. Often, signs are not visible to the eye. In these cases, microscopic, cultural, or other laboratory techniques are required to view signs of a pathogen.

Generally, the presence of signs is more concrete evidence of a particular disease than are symptoms. This is because many different factors can cause confusing look-alike symptoms. For example, drought, pesticides, root rot pathogens, and vascular wilt pathogens all can cause plants to wilt. However, a poinsettia may have wilted leaves and stem lesions with visible fungal growth on them. You can then view the spores under a microscope and identify them as a fungus known to cause stem blight and wilted leaves.

The symptom patterns on individual plants, on multiple same-species plants, and on different plant species provide useful clues about the cause of plant injury.

Uniform damage is indicative of abiotic (non-living) factors, such as weather or nutrient deficiencies. If damage does not spread, or there is a clear line between damaged and non-damaged tissues, abiotic factors likely are involved. Examples include damage on:

many different plant species in the same area, all plants in a row, field, or area,

all leaves or shoots on one side of the plant, or

the same aged portion of each leaf (i.e., only needle tips turning brown or only older leaves turning yellow).

Spread of the damage from plant to plant or to other plant parts over time may indicate damage by living organisms, such as pathogens or insects. Biotic problems may result in non-uniform patterns such damage to:

scattered same-species plants,

scattered leaves or shoots on a single plant, or

scattered spots on a single leaf.

Question whether the plant species is suited to the environment in which it is planted. Adverse environment alone may be the cause of the problem, such as in the case of winter injury. In other cases, environmental factors often render plants more susceptible to disease organisms. For example, powdery mildew is more severe under humid conditions; damping-off generally affects seedings emerging in cold, wet soils.

Plant species and varieties vary widely in their ability to tolerate differences in temperature extremes, sun, shade, moisture, low humidity, road salt, wind, frequent chinooks, late spring freezes, lack of snow protection, soil problems, etc. Common soil problems include pH, poor drainage, high water table, rocky, sandy, or heavy clay soil, or a hardpan layer that does not allow roots to penetrate deeply into the soil.

Consider whether the plant was planted correctly in the first place. Common planting errors include planting in a small or shallow hole, allowing bare root tranplants to dry out before planting, and planting trees too deeply. Other planting errors include burning the roots with fertilizer and watering too little or too much while the transplants are establishing themselves in the new location. Poor quality seed, not hilling potatoes correctly, and planting tomatoes in cold soil are other examples of planting problems.

Consider both recent weather and weather in the last few years (in all seasons). Inquire about pesticide and fertilizer use as well as watering regimes. Also, ask about other cultural practices such as mowing frequency, use of tree wraps, pruning methods and timing, insect problems, construction near trees, etc.

The information and checklist above are guidelines to help you narrow down the potential causes of problems. Some problems can be identified readily while others require considerable sleuthing. Asking thorough questions about the situation and knowing the range of possible causes are extremely helpful.

Laboratory tests are required to prove the identity of many pathogens.

Disease escape The failure of a host to become diseased because of separation, in space or time, of susceptible host tissues and the infective units of the pathogen, e.g. an air-borne pathogen may not arrive in contact with the host until after the susceptible period has passed; climatic conditions may not permit the germination of a pathogen in contact with the host at a time of low resistance.

Disease gradient 1. The change in incidence of disease with distance from a source of infection.

2. a mathematical expression of the change in incidence of a disease from a (point- or line-) source. This is usually inferred from the amount of disease present in successive samples along a traverse.

Disease incidence Number of plants affected by a disease within a population.

Disease potential A term coined by J. Grainger to indicate the physiological condition of a host plant that determines its liability to contract disease and which is represented by the ratio Cp/Rs, where Cp is the total carbohydrate of the whole plant, and Rs the residual dry weight of the shoot. Originally devised in studies on the infection of potato by Phytophthora infestans, in which disease failed to occur if the ratio was less than unity at the time of a Beaumont period, the concept was subsequently applied to other diseases.

Disease progress curve Graph of some measure of disease (i.e. severity) over time.

Disease pyramid A memory aid similar to the disease triangle but including, in addition, the factor of time in the development of a disease.

Disease severity The measure of damage done by a disease.

Disease symptoms root symptoms Injury to the root system often includes yellowing, stunting, or wilting of above-ground parts. Wash off the roots when possible and look for the following:

Small discolored or dead areas (fungi).

General death of the feeder roots or the entire root system (fungi). Discoloration of the vascular tissue in the crown and lower stem. Galls on roots.

Disease triangle A memory aid that diagrams the three important components necessary for disease: susceptible plant, virulent pathogen and favorable environment.

Disease, 'simple interest A disease which develops from a constant source of inoculum (i.e. the 'capital' is constant), and in which often there is only one ' generation ' of infection in a season, e.g. fusarium wilt of cotton. The transformation, log [l/(l-x)] where x is the proportion of the population that is diseased, is recommended for plotting the progress of the disease with time. The term (l-x) allows for a diminishing proportion of plants remaining to be infected.

Although diseases such as cotton fusarium wilt show only a 'simple interest' epidemiology in one season, their development from year to year may be of the 'compound interest' type. A 'compound interest disease' (from van der Plank, I963) is one in which the inoculum increases in successive 'generations' of infection in a season (i.e. the capital is increased by the amount of 'interest'). The recommended transformation in this case is log [x/(l-x)].

Disease, incidence of The frequency of occurrence of a disease; commonly the proportion of plants affected in a given population.

Disinfect To eliminate a pathogen from infected plant tissues.

Disinfectant An agent that frees from infection by destroying parasites established in plants. This term is best avoided in plant pathology, to prevent confusion with its popular use relating to destruction of bacteria etc. on objects or animals, not necessarily in situations where infection is present.

An agent that kills organisms present on the surface of plants or plant parts, or in the immediate environment, e.g. in soil.

Any agent for destroying the causal agent of disease after infection.

Disinfest To kill pathogens that have not yet initiated disease, or other contaminating microoganisms, that occur in or on inanimate objects as such soil or tools, or that occur on the surface of plant parts such as seed.

Disinfestant An agent that kills or inactivates pathogens in the environment or on the surface of a plant or plant organ before infection takes place.

Disjunctor A structure between conidia to promote their disarticulation .

Disk A type of cultivator made up of many circular blades used for weed control and soil preparation.

Disorder A harmful deviation from normal functioning of physiological processes, arising from causes other than pathogenic organisms or viruses, e.g. mineral deficiency or toxicity, genetic anomaly, low-temperature injury etc.

Dispersal To disperse: to dispel or scatter; in plants dispersal refers to mechanism of dispersing reproductive propagules like seeds or pollen.

Dispersal spore A spore disseminated by wind, water or other agent; diaspore.

Disperse To spread seeds or spores. Plants (seeds) and fungi (spores) need to spread to new areas. They rely on wind, water, or animals to carry their seeds (or spores) to new places.

Dispore In a basidiomycete, One of the spores from a two-spored basidium.

Dissemination The spread of infectious material (inoculum) from a diseased to a healthy plant by wind, water, humans, insects, animal, machinery, or other means.

Dissepiment A partition, for example, that between the pores of a bolete or polypore.

Distal Far from the point of attachment or origin; in a direction away from main body.

Distribution The pattern by which parasite numbers are partitioned amongst available hosts. If the fraction of the population who have i parasites is p(i) then the distribution is the set of numbers p(0),p(1), p(2), ... Summary statistics of this distribution include prevalence (1-p(0)) and mean intensity (p(1)+ 2p(2)+ 3p(3)+...). Useful distributions include the Poisson and negative binomial.

Dithiocarbamates Organic fungicides; dimethyl- (DMDC): Thiram, ferbam, ziram; ethylene-bis- (EBDC): nabam, maneb, mancozeb, zineb.

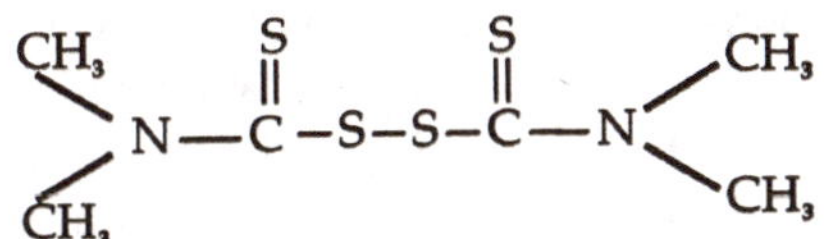

THIRAM

Fig. Thiram

Diuretics Agents that promote the excretion of and/or increase the amount of urine.

Diurnal Active during the day.

Diversity Term used to describe numbers of taxa, or variation in morphology.

DNA The double-stranded, helical molecule that contains genetic code information; each repeating unit, or nucleotide, is composed of deoxyribose (a sugar), a phosphate group, and a purine (adenine or guanine) or a pyrimidine (thymine or cytosine) base.

DNA fingerprinting A laboratory technique in which the banding patterns of DNA fragments from two different individuals are compared.

DNA ligase An enzyme which repairs breaks in the phosphate backbone of DNA.

DNA polymerase An enzyme that synthesizes a daughter strand(s) of DNA (under direction from a DNA template). May be involved in repair or replication.

DNA replicase A DNA-synthesizing enzyme required specifically for replication.

DNA sequencing Determining the actual sequence of bases in parts of DNA or RNA molecules. The first fungus to be fully sequenced was Saccharomyces cerevisiae.

Dodder Parasitic seed plant without leaves; a yellow filamentous vine.

Doliiform Barrel-shaped.

Dolipore septum A septum of a dikaryotic basidiomycete hypha which flares out in the middle portion forming a barrel-shaped structure with open ends; a septal pore cap, the parenthosome, is developed at each side of the pore.

Dolipore The complex barrel-shaped septal pore apparatus found in basidiomycetes other than Uredinales and Ustilaginales.

Domestic Animals that are tamed and raised by humans. Dogs, cats, cows, horses, goats, and sheep are all examples of domestic animals.

Domestication The collection from the wild state and the use in agriculture of plants having desirable traits.

Dominant A plant, usually a tree, that becomes the largest and most common in an area over time. Trees such as oaks, hickories, and elms will eventually take over a forest in a particular area.

Dormancy A condition of suspended growth and reduced metabolism of an organism, generally induced by internal factors or environmental conditions as a mechanism of survival.

Dormant sprays Sprays used when plants are dormant, to control overwintering pathogens.

Dormant To become inactive during winter or periods of cold weather.

Dorsal Of or belonging to the upper surface; top.

Dorsal The upper part of a given area.

Dorsoventral From top to bottom; from the upper surface to the lower surface.

Dorsum The upper surface; top.

Dose A measure of exposure. It is often expressed in milligrams per kilogram (mg/kg) or parts per million (PPM).

Dose response relationship A relationship such that with increasing levels of exposure to a factor, a corresponding rise in occurrence of disease or adverse effect is found.

Dose, dosage A measured concentration of a toxicant for a known duration of time (concentration per unit time). 1. The quantity of fungicide etc., applied per individual or per unit area, volume or weight.

2. The quantity of a pathogen used to inoculate living hosts particularly in quantitative response studies.

Double fertilization In flowering plants: the more or less simultaneous union of one sperm and one egg to form a zygote (N=2) and another sperm with two polar nuclei to form triploid (N=3) endosperm in the ovule; in the Gnetales: the fusion of two sperm with two eggs to produce two zygotes, only one of which will mature into an embryo.

Double integument The two outermost layers of an ovule in angiosperms, one of which will differentiate into the seed coat.

Dough stage A stage in grain development when the grain turns from a liquid to a soft doughy consistency before hardening.

Downy mildew Plant disease caused by certain members of the Peronosporales. Downy mildews are characterized by the formation of superficial hyphal growth in which, typically, individual spore-bearing structures can be distinguished.

Draft Ewes Female sheep which are sold out of a flock, usually upland to a lowland farm.

Drag off The practice of removing soil from the tops of potato hills before sprout emergence.

Drift (of pesticides) Movement of airborne particles of a spray, dust, or vapor away from the target area during or shortly after an application.

Drop spectrum The distribution (by number or volume of drops) of spray into intervals of drop size.

Drought damage Individual trees vary widely in their resistance to drought.However, in years of low rainfall, many tree species can be substantially weakened or killed by drought. Deciduous trees appear to die from the top down, have small, off-colored leaves, and narrow growth rings. Conifers generally die from the bottom up when subjected to drought. Drought and any factor that weakens a tree may allow invasion by many secondary fungi and insects.

Drought Insufficient soil water for normal plant growth.

Drupe A fleshy fruit with a single seed. Cherries, peaches, and plums are all drupes.

Dry rot Wood-rot caused by certain basidiomycetes, e.g., Serpula lacrymans (Meruliaceae: Aphyllophorales). A misnomer, because if the wood were really dry, it would not rot.

DTPA A chemical solution used to determine available zinc in the soil: [[(carboxymethyl) imino] bis (ethyleneitrilo)l tetra-acetic acid.

Dual culture (pot culture) Fungus plus host plant or phytobiont; the only way yet known of growing certain obligately biotrophic fungi (e.g. VAM fungi) in reasonably controlled conditions.

Dual organisms Organisms which invariably consist of two interdependent symbionts, e.g. lichens. In fact, most plants are also dual or multiple organisms because of their intimate association with endo- or ecto-mycorrhizal fungi. The eukaryotic cell is widely believed to have arisen as a multiple symbiosis.

Durable resistance Resistance that remains effective during prolonged and widespread use in an environment favorable to disease.

Dust A finely ground, dry pesticide formulation containing a small amount of active ingredient and a large amount of inert carrier or diluent such as clay or talc.

Duster An apparatus for applying pesticides in dry form.

Dutch elm disease A usually fatal disease of the American elm tree, Ulmus americana, caused by Ophiostoma ulmi (Ophiostomatales: Ascomycetes); the fungus is spread by a bark-beetle vector.

Duxelles A method of preparing minced mushrooms for later use.

Dwarfing A decrease in overall size without alteration of the proportions between parts of the plant. Synonymous with nanism.

E

Ear emergence Main heading date of a crop. i.e for a grass sward, the date at which 50% of the inflorescences have emerged.

Ear fungi The gelatinous, edible basidiomata of the Auriculariales.

Earth A geologic deposit of fine, grayish material composed chiefly or wholly of the remains of diatoms; it may occur as a powder or as a porous, rigid material; used in insecticide.

Earth balls The non-ostiolate basidiomata of the Sclerodermatales.

Earth stars Specialized soil-inhabiting puffballs (Lycoperdales: Gasteromycetes) in which the outer peridium splits and curls back in segments, raising the basidiospore mass (gleba) above the dead leaves, etc.

Eccentric Not in the center.

Echinate Having sharply pointed spines.

Echinulate Having small spines projecting from cell walls.

Ecological epidemiology A branch of epidemiology which views disease as a result of the ecological interactions between populations of hosts and parasites.

Ecological niche The environment favourable to an ecotype.

Ecology (Applied to plant pathology) The study which deals with the effect of environmental factors such as soil, climate and culture on the occurrence, severity and distribution of plant diseases.

Economic damage Damage caused by pests which results in loss of income or a reduction of value.

Economic injury level The lowest population density of pests that will cause economic damage.

Economic threshold Levels of pest populations at which damage from pests is considered more expensive than pest control. In an urban setting, this is complicated by the fact that a single visible insect in a residence or business can be considered offensive.

Economics of biological control Classical biological control typically is performed by department of agriculture. Once a classical biological control agent is imported and populations of it are established, the benefits of its actions are free; the natural enemy is not for sale, nor is there any point in selling it. In contrast, although basic research toward the use of a biopesticide may be performed by university personnel, the later development is performed by commercial companies which expect to profit by repeated sales. Therefore, classical biological control depends only upon funds for research and dissemination, and may yield an extremely good cost-benefit ratio. Biopesticides are marketable products which recoup their developmental costs through sales, and may not be less expensive than chemical pesticides.

Economies of scale Unit cost reductions which result from increasing total output.

Ecorticate Without a cortex.

Ecosystems Systems which include both living and non-living substances interacting to produce an exchange of materials between the living and the non living units

Ecotype (ecological type) A form arising by selection in a particular habitat.

Ectal Outer, outermost; used most often as ectal exciple.

Ectomycorrhiza or ectotrophic mycorrhiza Mycorrhiza in which a dikaryomycotan mycelium ramifies through the soil, forms a mantle around individual rootlets, and grows between cells of the root cortex, forming a Hartig net (the interface between the symbionts). The fungus exchanges phosphorus for photosynthates from the root. Many forest trees, esp. Pinaceae, Fagaceae, have ectomycorrhizal associations with agarics or boletes.

Ectomycorrhizae A type of mycorrhizal association in which the fungal component grows between and/or external to the cortical cells of the plant root.

Ectoparasite A parasite that lives on the outside of its host.

Ectoparasitic A parasite that lives on the outside of its host.

Ectoplasm The outermost, relatively rigid and transparent, granule-free layer of cytoplasm in many cells (e.g. in amoebae).

Ectoplasmic net An extracellular matrix; a branching and anastomosing, hyaline, membrane-bounded network of ectoplasmic filaments without cytoplasmic organelles which functions as an attachment and absorbing structure and is produced by specialised organelles, called sagenogens, on the cell surface of labyrinthulomycetes. In Labyrinthulales, the ectoplasmic net completely surrounds the cells and joins them in a common network through which the cells move by a gliding motion. In Thraustochytriales, the ectoplasmic net arises from one side of each cell and does not surround it.

Ectosporium The hardly visible spore wall layer external to the perisporium.

Ectosymbiosis Symbiosis in which one member (microsymbiote) develops on the outside of the other member.

Ectotrophic Refers to a mycorrhiza in which the mycelium forms an external covering on the root.

Ectotunica The outer wall of a bitunicate ascus.

ECU (European currency unit) A standard unit of currency used throughout the EC for calculating grants, commodity prices, subsidies etc. Related to the currencies of member states at variable rates.

Edaphosphere The soil away from the influence of plant roots, i.e. beyond the rhizosphere.

Edema Blisters produced on leaves and other plant parts under conditions of high moisture and restricted transpiration.

Edible Able to use as a food by people for nutrients; not poisonous. Inedible means something that people can not gain nutrients from, or which may cause sickness when eaten.

EEC European Economic Community.

Eelworms Nematodes.

Effective dissemination Synonymous with inoculation.

Effective inoculum dose The amount of inoculum which is available for penetration of the plant tissues. This may be equivalent to the inoculum concentration used or only a proportion of it, e.g. in leaf scar inoculations the amount of inoculum which enters is dependent upon vascular tension in the plant and rate of evaporation etc.

Effectiveness The ability of nodules once formed to actively fix nitrogen.

Efficacy An index of the potency of a drug or disease treatment.

For an anthelmintic, usually estimated as the proportion of the worm burden in a host killed by a single dose or short-term course of the treatment. For a vaccine, efficacy is the percentage of persons who are protected by the vaccine. Specifically, if Iv is the incidence in vaccinated individuals and Iu then vaccine efficacy is v=1-Iv/Iu.

Effluent Drainage from a cattle yard, manure heap or silage clamp.

Effuse Spread out flat especially as a film-like growth.

Egg The first stage of life for most animals. Most insects have four stages in their life cycle, and the first stage is the egg. Frogs change from eggs to tadpoles to adults. Birds hatch from eggs and grow into adulthood. So do reptiles. So do spiders.

Egg gallery A long, narrow tunnel along the sides of which eggs are deposited in small niches; the pattern of construction is often diagnostic of a particular species of insect.

Egg mass Cluster of eggs, usually in a matrix of body hairs or wing scales from the female adult and/or a mucilaginous cementing secretion.

Eguttulate Of a spore, Lacking guttules (oil-like globules).

Elater A free capillitium thread; a capillitial element in the gleba of Battarraea and Battarraeoides that tapers at both ends and has spiral thickenings when mature.

Electrolytes Substances that dissociate in water to form a cation (positively charged ion) and and anion (negatively charged ion). Charged ions are central to a variety of important processes in the body, including muscle contraction and nerve impulse conduction.

Electron microscope Microscope that uses a focused beam of electrons to produce a greatly enlarged image of minute objects, such as a virus, in the same way that light is used in a compound microscope.

Electrophoresis A procedure by means of which the members of a heterogenous population of charged particles can be separated by virtue of their dissimilar migration characteristics in an electric field.

Electroporation A method by which nucleic acids or virus particles can be introduced into protoplasts or cells by creating transient pores in the plasma membrane using an electric pulse.

Elicitor A molecule produced by the host (or pathogen) that induces a response by the pathogen (or host).

ELISA A serological test in which the sensitivity of the reaction is increased by attaching an enzyme that produces a colored product to one of the reactants.

Elisa A technique using the antigen binding properties of antibodies to detect specific antigens or antibodies. Visualisation is typically made possible by enzyme induced colour formation.

Ellipsoid Having every plane section an ellipse or a circle.

Ellipsoidal Oval in outline and three-dimensional.

Elytra The anterior leathery or chitinous wings of beetles and leafhoppers; serve as coverings to the hind wings and commonly meet at rest in a straight line down the middle of the dorsum.

Emarginate Of a lamella, attached by only a portion of its width and the resulting shape is as if a small triangular piece had been removed from that portion of the lamella where it meets the stipe; notched.

Embryo An organism in the early stages of development, such as a young plant in the seed, or a nematode before hatching from the egg.

Embryophyte The group of all organisms that retain the zygote on the parent to form an embryo, includes Bryophyta and vascular plants.

Emerged The emergence of the adult from the cocoon, pupal case or nymph.

Emergence Appearance of the first part of the crop plant through the ground.

Emesis Vomiting (sometimes caused by emetics).

Emulsifying agent A material which facilitates the suspending of one liquid in another.

Emulsion A mixture in which one liquid is suspended in tiny globules in another liquid; for example, oil in water.

EN English Nature.

Enation An outgrowth (in plant pathology an abnormal outgrowth) from the surface of an organ such as a stem or leaf.

Epidermal outgrowth.

Enations on psilotumenation A non-vascularized, epidermal outgrowth found in some early land plants.

Encapsidate To cover virus nucleic acid with a protein coat.

Encyst To form a cyst or protective covering.

Encysted Surrounded by a hard shell (cyst).

Endemic A term to describe levels of infection which do not exhibit wide fluctuations through time in a defined place.

For microparasites like measles, the term is used slightly differently to indicate an infection which can persist in a population in the long term without needing to be reintroduced from outside.

Endemic disease A disease permanently established in moderate or severe form in a defined area, commonly a country or part of a country.

Endemic fadeout Parasite extinction occurring because endemic levels are so low that it is possible for small stochastic fluctuations to remove all parasites.

Endobiotic Growing inside a living organism.

Endocarp Inner layer of a fruit wall.

Endocarpous Of Basidiomycota, With the mature hymenium covered; angiocarpous.

Endoconidiophore A conidiophore that produces conidia within itself.

Endoconidium A conidium produced endogenously in a hypha or conidiophore.

Endocrine disruptor A chemical capable of disrupting the development or function of the reproductive system by mimicking or blocking the action of hormones.

Endocrine glands Ductless glands which secrete hormones; include the pituitary, adrenal, thyroid.

Endocrine system The system in humans and other animals comprised of the internally-secreting glands (e.g., the pituitary and thyroid glands), their (hormonal) secretions, and their target organs (e.g., prostate glands, testicles, ovaries, uterus, breasts, and others).

Endodermis The differentiated, innermost layer of the cortex of roots, rhizomes and certain non-seedbearing vascular plant stems; surrounds the stele.

Endogenous Formed within another structure, as are the meiospores of Ascomycetes.

Endogenous inhibitor A substance produced by an organism which at low concentration will inhibit the growth of itself or another organism, or some physiological process associated with growth, in a relatively specific manner. A characteristic of higher plant growth inhibitors such as abscisic acid (ABA) is that their effects are often antagonised by specific growth promoters especially gibberellins and cytokinins.

Endomycorrhiza or endotrophic mycorrhiza An ancient symbiosis of fungi with green plants; hyphae gathering nutrients from the soil, esp. phosphorus, are continuous with others that grow between and within root cells and produce arbuscules; found in 90% of angiosperms and conifers, except Pinaceae; also called vesicular-arbuscular mycorrhizae.

Endonuclease A nuclease which cleaves phosphodiester bonds within a nucleic acid strand.

Endoparasite Parasitic organism that lives and feeds from inside its host.

Endoperidium The inner layer of a multilayered peridium.

Endophyte An organism which completes its life cycle in a plant which shows no external sign of the infection; one of a group of fungi which are parasitic on grasses and are toxic to grazing animals.

Endophytic fungi Fungi (such as some anamorphs of Clavicipitales) which grow systemically within plants without causing symptoms; now known to produce neurotoxins that discourage herbivores, and therefore to be participating in a mutualistic symbiosis.

Endophytic Living within another plant.

Endoplasmic reticulum System of interconnected cytoplasmic membranes that transports materials within the cell.

Endosperm The triploid (N=3) product of double fertilization in angiosperms; during seed maturation the endosperm will develop into a storage tissue that will provide nutrients to the seedling as it emerges (in monocots) or that will be digested and stored by the cotyledons before germination (in dicots)

Endospore Resistant, thick-walled asexual spore formed within a bacterial or fungal cell.

Endosporium The innermost layer of the spore wall which is usually thin and is the last to develop during sporogenesis.

Endospory A condition in which the gametophyte develops within the spore wall, rather than externally.

Endostome Of moss sporophytes, the inner row of walls forming the peristome teeth.

Endosymbiont An organism which lives in a mutualistic symbiosis within the cells of another organism.

Endosymbiosis When one organism takes up permanent residence within another, such that the two become a single functional organism. Mitochondria and plastids are believed to have resulted from endosymbiosis.

Endothecium The inner cell mass of the sporpphyte in basal plants. It usually produces both the central columella and the spores, which accumulate around the surface of the columella.

Endotrophic Refers to a mycorrhiza in which the mycelium grows within the cortical cells of the root (e.g., in orchids).

Endotunica The often extensible inner wall of a bitunicate ascus.

Energy flow The rate of energy transfer between elements of an ecological system

English walnut The walnut species used for the selection of commercial scion cultivars; origin believed in Persia.

Engraver Beetle which feeds in the phloem-cambium region of woody plants, often scoring or engraving adjacent sapwood tissues.

Enphytotic Plant disease that causes about the same amount of injury each year.

Enteroblastic Of conidiogenesis, Mode of blastic production of cell walls in which, following completion of any developmental stage, the fungus in a new stage does not lay down a wall layer or layers continuous with the outer wall layer(s) of the previous stage.

Enterovirus A genus of RNA viruses with over 70 types identified in humans. They reproduce in the intestinal tract, and various members can cause a variety of human diseases, including poliomyelitis, aseptic meningitis, hepatitis, inflammatory heart disease, and rhinitis.

Entire Having a smooth margin, not dissected or toothed.

Entire margin In leaves, the condition in which the margin forms a smooth line or arc without noticeable serrations; note that lobed leaves can also have entire margins; compare to lobed margin and toothed margin.

Entomogenous Living in or on insects, especially as pathogens.

Entomologist A person who studies insects. Nowadays, entomology is such a large subject that entomologists specialize in some aspect of it. They may work only on identification of some group of insects, or on biochemistry or genetics of insects, or on control of insects in a small range of crops, or on mathematical modelling of insect populations.

Entomology The science which includes all phases of studies on insects.

Entomopathogen An organism (generally a bacterium, virus, protozoan or fungus) causing disease in insects.

Entomopathogenic Causing disease in insects.

Entomophagous nematodes Nematodes that eat insects.

Entomophilous Having insects involved with dispersal of spores.

Entropy The degree of randomness of or disorder of a system.

Environment All of the living organisms and non-living features of a defined area.

Environmental fate What happens to a pesticide in the environment, including how long it persists in soil, water or foliage, how it moves (leaches) through the soil, how it degrades in water or light, how it is metabolized aerobically or anaerobically by organisms, how it accumulates in animals or plants.

Enzyme A protein that, even in low concentration, speeds up (catalyzes) specific chemical reactions; usually becomes inactivated or unstable at high temperatures; name usually derived from substrate or function, with suffix -ase. Constitutive enzymes are produced even in the absence of the substrate; adaptive enzymes are not produced until the presence of the substrate has been recognized.

Enzyme linked immunosorbent assay A serological test in which the sensitivity of the reaction is increased by attaching an enzyme that produces a colored product to one of the reactants.

Eosinophil A white cell of the category known as granulocytes. These cells contain numerous dense granules in their cytoplasm that comprise a battery of highly active digestive chemicals and toxins. Their chief role is thought to be in combatting large parasites, although occasionally their activity may be triggered by other agents, potentially leading to damage of normal tissues.

Eosinophilia myalgia syndrome (EMS) A disease caused by marked promotion of eosinophil activity, resulting in a symptom complex of severe pain, inflammation of the tendons, fluid build-up in the muscles, and skin rash. The disorder has been linked to a contaminant of some commercial preparations of the amino acid L-tryptophan.

Epicoccum sp Epicoccum sp. - It is commonly found as a secondary invader in plants, soil, grains, textiles and paper products where Cladosporium and Aureobasidium are present. It is mostly saprophytic, or weakly parasitic. Epicoccum is frequently isolated from air and occasionally occurs in house dust. Reported to be an allergen but not in a high frequency. Due to the ability of this fungus to grow at 37°C, it can cause infection of skin in humans. Morphological characteristics are production of dark conidia, several-celled (15-celled), globose, verrucose, 15-25 microns in diameter, and in a fruiting body (sporodochium). Cultivation - Potato dextrose agar, 24°C.

Epicormic branching Branches arising from buds in bark along mainstem, most commonly occurring in trees under crown stress; also called watersprouts.

Epicotyl Portion of the stem of a plant embryo or seedling above the node where the cotyledons are attached.

Epidemic A widespread temporary increase in the incidence of an infectious disease (BMS). Although the term is commonly applied to the rise and fall of disease incidence within a single growth season, e.g. potato blight (Phytophthora infestans), yellow rust of wheat (Puccinia striiformis) or apple scab (Venturia inaequalis),neither the definition nor the concept excludes a similar sequence cover a period of several years or seasons. Furthermore, in some tropical regions 'seasons' are not readily definable.

A number of diseases caused by soil-borne pathogens provide examples of epidemics developing over periods of several or many years. A rapid increase in the levels of an infection. Typical of the microparasitic infections (with long lasting immunity and short generation times) an epidemic is usually heralded by an exponential rise in the number of cases in time and a subsequent decline as susceptible numbers are exhausted. Epidemics may arise from the introduction of a novel pathogen (or strain) to a previously unexposed (naive) population or as a result of the regrowth of susceptible numbers some time after a previous epidemic due to the same infectious agent. A change, usually a sudden increase, in a disease within a population.

A widespread and severe outbreak of a disease. A disease increase in a population.

Epidemic fadeout Parasite extinction occurring because numbers are so low immediately following anepidemic that it is possible for small stochastic fluctuations to remove all parasites.

Epidemic rate The amount of increase of disease per unit of time in a plant population.

Epidemiology The study of factors affecting the outbreak and spread of infectious diseases.

The study of factors influencing the initiation, development, and spread of infectious disease.

The study of the factors affecting the outbreak and spread of infectious diseases (BMS).

Epidermal cells Cells composing the epidermis which is the superficial layer of cells occuring on all plant parts.

Epidermis The outermost layer of living cells on the surface of a plant or animal.

Epigeous Fruiting above the surface of the ground.

Epigonium In mosses, a derivative of the archegonium which forms a hard, protective layer around the developing sporophyte. Later it breaks up, but the sporophyte retains a conical "dunce cap" or calyptra over its distal end. Strictly speaking, the term calyptra refers only to the "dunce cap," and the piece of the epigonium retained around the base (if any) is the vaginula.

Epigynous Of Oomycota, Having the antheridium above the oogonium and on the same hypha.

Epihypothallic Mode of fruit-body development in Myxomycetes subclass Stemonitomycetidae where the hypothallus forms under the plasmodium during fruit-body formation and sporangial stalks are hollow or fibrous.

Epinasty Downward bending of a petiole, so that the angle between its base and the stem becomes obtuse; the orientation of the lamina may be vertical, with the apex hanging downwards, or it may continue the curve of the petiole so that its upper surface faces inwards towards the stem. The petiole and lamina remain turgid, and the condition must be distinguished from wilting, in which the tissues become flaccid.

An abnormal downward-curving growth or movement of a leaf, leaf part, or stem.

Epiparasite An organism parasitic on another that parasitizes a third.

Epiphyllous On the upper surface of a leaf.

Epiphyte A plant that grows on another plant, or which uses a rock or host plant merely as a place of residence and obtains its moisture and nutrients directly from the air; an air plant. A type of growth form. No parasitism is involved. The epiphyte is rooted on the surface of the host. Vines rooted in the soil and climbing up on another plant are not epiphytes.

Epiphytotic Used for an epidemic (or epidemic disease) in plants (as opposed to 'epizoic', an epidemic disease in animals) by those who consider epidemic should be restricted to disease in man. The study of epiphytotics has been termed epiphytotiology or more frequently, but incorrectly, as epiphytology which is the study of epiphytes. While the terms epiphytotic and epizootic are occasionally useful to avoid ambiguity concisely, the use of epidemic in a broad sense is recommended.

An epidemic in a plant host population.

A widespread and destructive outbreak of a disease of plants. Epidemic.

The widespread and destructive development of a disease on many plants in a community or communities.

Epiptygma In nematodes: A vulval flap.

Episporium The spore wall layer external to the endosporium; the thick fundamental layer which determines the size and form of the spore.

Epithet One of the words which makes up the binomial of an organism.

Epitype A specimen selected as a standard for a species or lower taxon when all original material except for illustrations has been destroyed.

Epizoic Living on animals.

Epizootic An epidemic in an animal host population.

Epruinose Without pruina.

EPS (extracellular polysaccharides) Sugar polymers that contribute to the slimy appearance of bacteria probably promote colonization of plant tissues and disease development in bacteria.

Epstein-barr virus (EBV) One of the eight known types of human herpesviruses, also known as human herpesvirus 4 (HHV-4). It belongs to the gamma subfamily of herpesviruses. It commonly causes acute mononucleosis, and less commonly chronic mononucleosis. It some populations EBV is causally associated with life-threatening malignancies.

Equal Of a stipe, Having equal diameter from the apex to the base.

Equilibrium A state in which a system is not changing. A population size might be at a static equilibrium at which nothing is happening (there are no births or deaths) or a dynamic equilibrium at which different processes are balanced (there are the same numbers of births and deaths). More generally, the state to which a system eventually evolves, for example sustained periodic oscillations, might be called an equilibrium.

Eradicant Any chemical agent that eliminates particular pathogen(s) from diseased plants treated with that agent.

Eradicant fungicide A chemical substance that destroys a pathogen after it is established.

Eradication When a plant or area is infected with a pathogen, eradication can eliminate or reduce the disease threat. Rotation, sanitation, eliminating the alternate host, heat treatment, and use of certain chemicals can reduce or eliminate diseases.

Crop rotation is a common method in both commercial agriculture and home gardens. Knowledge of the pathogen and its host range are needed to plan effective rotations. Generally, rotation only reduces soil populations of fungi or nematodes if non-hosts are used. It does not eliminate pathogens altogether.

Removing plant debris (sanitation) is important in areas where pathogens may overwinter. Raking leaves, removing rotted fruit, picking up old vines, and pruning out dead wood or canes are all part of sanitation. Once collected, dispose of the debris by burning, burying, or composting. However, composting temperatures must be high enough to kill the pathogens.

Many species of rusts must infect two or more different hosts to complete their life cycle. Eliminating an alternate host may help reduce a rust disease. A famous example was the attempt to eradicate common barberry, the alternate host of wheat stem rust. This disease has caused major rust epidemics worldwide since ancient times.

Heat treatment can eliminate pathogens from greenhouse soil and from the surface of seeds and bulbs. Some fungicides have "kickback" activity that can stop certain fungal diseases if applied soon after infection has occurred. When caught in the very early stages, certain fungicides can eradicate Dutch Elm disease from American elms, for example.

Ergometrine An alkaloid of Claviceps purpurea sclerotia; used to induce labour.

Ergosterolergosterol A sterol lipid which occurs in fungal cell membranes, apparently in place of cholesterol. It is found in metazoans as a precursor to vitamin D2.

Ergot A disease of certain grasses and cereals, especially rye, caused by Claviceps spp.; a sclerotium, or resting structure, produced by Claviceps species and other closely related fungi in infected flowers of parasitized grain plants.

Ergotamine A 3-amino acid cyclopeptide derivative of lysergic acid from Claviceps purpurea sclerotia; ergotamine tartrate is used to manage migraine by causing vasoconstriction of cranial arterioles, thereby reducing the pulsation pressure and attendant headache.

Ergotism A disease caused by ingestion of grain contaminated with alkaloids of ergot fungi, often called the holy fire or St. Anthony's fire in the Middle Ages.

Eriophyid Any of a group of tiny mites characterized by having a slender body and two pairs of legs, rather than four.

Erose Of a margin, Irregularly incised.

Erosion When soil and other items are washed away due to weather, including rain and wind. Erosion is a problem when there are no plant roots holding soil in place. When soil erodes, it slowly crumbles away, changing the way the land looks. Erosion can become a big problem for people and their property. Erosion also pollutes water in streams, rivers, and lakes.

Erumpent Bursting or erupting through the substrate surface.

Erythrocytes Red blood cells.

Escape Plants in a given population that remain free of disease where it is prevalent, although they possess no natural inherent resistance to the disease.

Esophagus In nematodes: The portion of the alimentary canal between the buccal capsule, or stoma, and the anterior portion of the intestine.

Estivate When an animal "hides out" in the Summer. The heat is too much for certain animals and they don't want to dry out, so they hide and slow down their bodies until it rains or becomes cooler. Estivation is the equivalent of hibernation in the Winter.

Ethidium bromide (2,7-diamino-10-ethyl-9-phenylphenanthridinium bromide) A trypanocidal, bacteriostatic dye that binds to DNA and fluoresces under near-ultraviolet light; used for tracking nucleic acids.

Ethylene A plant hormone influencing various aspects of vegetative growth, fruit ripening, abscission of plant parts, and the senescence of flowers.

Etiolated (of plants) Excessively tall and lacking chlorophyll.

Etiolation Internode extension and the lack of green colour that result from growth of a plant in insufficient light or complete darkness. Yellowing and long, spindly growth as a result of insufficient light.

Etiology The study of cause; that phase of plant pathology dealing with the causal agent and its relations with the susceptible plant.

Eubacteria All bacteria which are not archaeobacteria; include all gram-positive bacteria with peptidoglycan walls, all gram-negative bacteria, all photosynthetic bacteria, all bacteria without a rigid cell wall.

Eucarpic With thallus divided into vegetative and reproductive structures.

Eukaryote An organism containing a membrane-bound nucleus and other organelles, i.e. all higher plants, animals, fungi, and protists.

Eukaryotic Having nuclei which are delimited by nuclear membrane, contain an even number of discrete chromosomes, and divide mitotically: the cytoplasm also contains organelles, such as mitochondria.

Eumycota The Kingdom to which the phyla Chytridiomycota, Zygomycota and Dikaryomycota belong.

Eusporangium Sporangia that arise from a group of superficial cells; this is the primitive character state in vascular plants.

Eutrophication The accumulation of excessive concentrations of plant nutrients in water courses leading to a bulkd up of algae.

Evanescent Present for a short time; breaking down.

Evaporate The process of a liquid turning into a gas or vapor.

Evapotranspiration Loss of water by evaporation and transpiration by the above ground parts of a palnt. Governed by meteoroligical conditions, especially when water supply is not limiting.

Evelate Without a veil.

Everbearing Term applied to strawberry cultivars that produce flowers and fruit all year as long as temperatures are favorable; often used synonymously with day-neutral.

Evergreen Opposite of deciduous, i.e. evergreen plants do not shed all their leaves annually, but shed and grow new leaves continually; most conifers are evergreen, but the evergreen habit is not restricted to conifers.

Evolution Darwin's definition: descent with modification. The term has been variously used and abused since Darwin to include everything from the origin of man to the origin of life.

Evolutionary tree A diagram which depicts the hypothetical phylogeny of the taxa under consideration. The points at which lineages split represent ancestor taxa to the descendant taxa appearing at the terminal points of the cladogram.

Excavate To dig. Woodchucks excavate a den in the ground; woodpeckers excavate a cavity in a tree.

Excentric Off-centre, to one side, any attachment intermediate between central and lateral.

Excessive mulch Mulches have many very beneficial uses which include water retention, winter protection, weed control, and lawnmower protection. However, damage from incorrect mulch applications around trees and shrubs is becoming more common as homeowners and professionals recognize these benefits. Problems may arise under certain circumstances: excessively deep applications of fine-textured organic mulches; mulches piled up against the trunk of young and/or thin-barked trees and shrubs; and plastic ground covers applied before the top-dressed mulch, especially when the plastic is in contact with heavy, clay soils.

Excessive mulching results in deterioration of fine roots, low oxygen levels to roots, formation of adventitious roots from stems covered by mulches, and mulch girdling. With mulch girdling, the trunk grows normally above the mulch line, but puts on little trunk growth below the mulch line. Symptoms of excessive mulching include: blackened roots, scorch, wilt, nutrient deficiency, early senescence, stunting, poorly anchored trees and shrubs, and excessive adventitious rooting off the trunk below the mulch line.

Exciple In an ascoma, Tissue or tissues enclosing the hymenium and subhymenium.

Exclusion As long as pathogens and host plants can be kept apart, disease will not develop. Many potentially-susceptible plants are grown in areas of the world still free of certain diseases. When plant pathogens are introduced into a new area, they can cause much more catastrophic epidemics than do the native pathogens. Host plants that evolve in the absence of a pathogen have no opportunity to develop resistance to the pathogen and are extremely vulnerable to attack.

The Dutch Elm fungus evolved in Europe where it causes little damage to European elms. However, this fungus devastated American elms when it was imported to the United States. Other devastating diseases that resulted from introduction of a pathogen from abroad include downy mildew of grapes in Europe and bacterial canker of citrus, chestnut blight, and soybean cyst nematode in the United States. Even today, many pathogens exist elsewhere in the world that would likely cause tremendous yield and quality losses if they were introduced into the United States.

Quarantines, inspections, and disease-free certifications help to prevent the spread of pathogens into a particular country, state or geographic area. The most visible use of this method is in California. At the state border, everyone must stop at an agricultural inspection station to be screened for fresh fruit, vegetables, and plants coming from areas known to have certain diseases or insects. To combat the devastating late blight fungus, the Montana Department of Agriculture began enforcing a quarantine on importation of seed potatoes and tomato transplants into Montana in 1996. However, this quarantine does allow importation of plants certified to be free of the late potato blight fungus.

Excretion The removal of a chemical from the body in urine, feces or expired air.

Excretory In nematodes: A tube or canal, lined with cuticle, that leads to the excretory pore.

Excretory pore In nematodes: The exterior opening of the excretory system, generally located on the ventral side of the body near the basal region of the esophagus; also known as the orifice of the cervical gland.

Exobasidial 1. Having the basidia uncovered;

2. separated by a wall from the basidium.

Exocrine glands Those which release their secretions through ducts, e.g. salivary glands, sweat glands, lachrymal (tear) glands.

Exodermis In plant roots, the layer of cells immediately inside the epidermis; cells may be dimorphic, and variably suberized, influencing entry of VAM fungi and pathogens.

Exogenous Arising on the outside of the generating structure.

Exogenous Formed externally to the parent cell, as are the meiospores of basidiomycetes.

Exonuclease A nuclease that sequentially removes nucleotides from one end of a strand of nucleic acid.

Exopathogen Nonparasitic organism whose extracellular toxic metabolites cause disease in plants.

Exoperidium Outer layer of a multilayered peridium.

Exoskeleton The supporting structure of an organism when it is on the outside of the body. Insects, crustaceans, and spiders all have exoskeletons. Birds, mammals, and reptiles have "endoskeletons," meaning they have bones on the inside.

Exosporium The spore wall layer external to the episporium and derived from it, but chemically distinct and frequently responsible for the ornamentation.

Exostome Of moss sporophytes, the outer wall of plates forming the peristome teeth.

Exotic Introduced from another country or area.

Exponential decay A decline in which the rate of decay is always proportional to the amount of material remaining; the constant of proportionality is the rate constant.

Exponential growth An increase in which the rate of growth is always proportional to the amount of material remaining; the constant of proportionality is the rate constant.

Expose To place a test plant where it can be expected to receive inoculum by natural means from a natural source.

Exposure Contact with a chemical. Some common routes of exposure are dermal (via skin), oral (by mouth) and inhalation (breathing in).

Extensive farming System of farming, using few inputs of bought in feed and fertilizer, usually carried out over wide area on large upland farms in the UK.

Extinction When all the members of a clade or taxon die, the group is said to be extinct.

Extracellular Outside the cells.

Extrachromosomal inheritance (also cytoplasmic or maternal inheritance) Inheritance of genes not located in the nucleus, i.e. those in mitochondria and chloroplasts

Extrafloral nectary A nectary located outside the flower.

Extramatrical VAM structures (mycelium, spores) developing outside the roots of the phytobiont.

Extranuclear inheritance Transmission of genetic information in the cytoplasm by DNA present in organelles, such as mitochondria.

Exudate Material that has passed from within a plant structure to the outer surface or into the surrounding medium, e.g. by diffusion and not usually through an aperture, as in root exudate, leaf exudate etc.

A substance (usually liquid) formed inside a plant and discharged from diseased or injured tissue. The presence of an exudate often aids in diagnosis (e.g., fire blight bacteria).

Exude To pass from within a plant structure to the outer surface or into the surrounding medium, e.g. by diffusion and note usually through an aperture. In ordinary English usage the verbs 'to ooze' and 'to exude' are virtually synonymous, but in plant pathology it is useful to distinguish the movement of substances through apertures of microscopic or macroscopic dimensions as 'oozing' and that through apertures of molecular dimensions over the whole surface of a structure as 'exudation'.

Eye A collection of several buds on the surface of a potato tuber, one of which will sprout and form a new stem when conditions are favorable.

F

F. SP. A taxonomic group within a pathogenic species defined in terms of host range, i.e. members of different formae speciales infect different groups of plants.

Facultative Able to live as a saprobe.

Facultative anaerobe Refers to an organism that normally grows aerobically but can grow anaerobically.

Facultative parasite An organism that is ordinarily saprophytic but under proper conditions may be parasitic.

Facultative saprophyte A mainly parasitic organism with the ability to survive for a part of its life cycle as a saprophyte and be cultured on artificial media.

Fairy ring A formation of certain mushrooms, such as blewits and Marasmius oreades, which grow in rings moving outward from a central point of origin.

Falcate (Of spores) Sickle-shaped.

Fall freeze damage Plants still growing actively in the fall may be damaged by unseasonably cold periods. This damage occurs before the plant has moved into its winter rest as indicated by development of fall color and normal leaf drop. Frozen leaves often hang on the tree until spring. Damage may range from no twig damage to death of the entire tree.

Planting trees and shrubs adapted to the climate is very important. Also, allowing trees to grow at a slower rate by not pushing them with fertilizer and excessive watering can minimize winter freeze damage.

Fallow Cultivated land that is allowed to lie dormant, with no crops growing on it, during a growing season.

False morel Gyromitra esculenta (Ascomycetes), which is poisonous.

False truffle A hypogeous, ectomycorrhizal gasteromycete, e.g., Rhizopogon.

Family Part of the system of biological classification: a group of closely-related genera. All whiteflies belong to the family Aleyrodidae. The name of a family of animals (e.g., Aleyrodidae) always has a capital initial letter and always ends in -idae. The family name is not placed in italics or underlined.

Farinaceous Mealy, resembling flour in form or smell.

Fasciation The proliferation of a shoot with incomplete separation of the elements, giving the appearance of a flattened coalescent bundle of shoots.

A distortion of a plant caused by an injury or infec-tion that results in thin, flattened, and sometimes curved shoots.

Fascicle Small group, bundle, or cluster.

Fasciculation Hyperplastic symptom characterized by a clustering of such plant organs as shoots into such structures as witches' brooms.

Fastidious In reference to prokaryotic organisms, having special growth and nutritional requirements.

Fasting Depriving a test vector of food for short periods. The term is inappropriate because fasting normally implies a voluntary abstinence from food by the subject, rather than an enforced deprivation, but its use in vector studies is now so common that a contra-recommendation is not justified.

Fasting, pre acquisition Fasting of a test vector prior to acquisition feeding. Synonymous with preliminary fasting.

Faveolate Of a surface, Honeycombed; alveolate.

Fecundity The capacity to produce offspring; the rate of offspring production.

Feeder roots The youngest roots with root hairs, important in absorption of water and minerals.

Feeding period The period that a test vector actually feeds on a virus source, or on a recipient plant.

Fenestrain nematodes A window or transparent spot; in some Heteroderidae, a thin-walled region of the vulval cone.

Fenestrate Having openings or translucent areas.

Fermentation Chemical changes in organic substrates caused by enzymes of living microorganisms.

Fern A plant that does not have flowers or seeds. Ferns have fronds (large type of leaf) and spores (a little bit like a seed, but works differently).

Fertigation Application of plant nutrients to a crop in irrigation water.

Fertilizer N Forms of fertilizer that include a source of N. The fertilizer is usually described in terms of its content of nitrogen, phosphorus and potassium, with one labelled 0:36:10 containing 0 nitrogen, 36 lbs P_2O_5 and 10 lbs K_2O per 100 lb fertilizer. Fertilizer N is usually in the form of urea, ammonium sulfate, potassium nitrate or anhydrous ammonia.

Fiber cell A long-walled plant cell which is often dead at maturity; fibers impart elasticity, flexibility and tensile strength to plant structure

Fiber Elongated plant cell with thick walls (in xylem and phloem), dead at maturity.

Fibril A fine hair or fiber.

Fibrillose Covered with silk-like fibres.

Fibromyalgia Also known as myofascial pain syndrome and fibromyositis. A group of common rheumatoid disorders (not involving the joints) characterized by achy pain, tenderness, and stiffness of muscles.

Fibrous Of a stipe, Usually rather thick and leaving a ragged edge when broken in two.

Fiddlehead A young frond on a fern plant. Fiddleheads are tender when they first grow and are eaten by animals, including deer.

Field A place of dense, shorter plants, including shrubs and grasses. Because hardly any trees grow in a field, plants that need a lot of sunlight are able to grow. A farmer's field has only the plants that the farmer wants to grow; a wild field, such as a meadow, has many different plants and animals.

Field capacity The state of saturated soil when all the soil moisture that is able to freely drain away has done so

Field guide A book that carefully describes different mushroom species so they can be tentatively identified.

Field immune (of plants) Not becoming infected by a pathogen in the field, although susceptible under experimental conditions.

Filament A long fine hair or fiber.

Filiform Long and slender; threadlike.

Fimbriate Of a margin or apex, Fringed with long, slender, hair-like processes (fimbriae) or hyphae.

Fin Body parts of fish used to swim and steer. Dorsal fins are on the back, caudal fins are on the tail, anal fins are underneath towards the back, and pectoral fins are underneath towards the front. Fin size, shape, and color can be used to identify different fish species.

Fish Cold-blooded, scaly animals from the Chordate Phylum. Fish live in water, breathe through gills, and move by swimming with fins. Most of them lay eggs in water. Some fish in our area include bass, bluegill, and catfish.

Fission Becoming two by division of the complete organism.

Fissitunicate A bitunicate "ascus has a distinctly bilayered wall, with the outer layer being rigid and the inner layer being expandable. As it matures, the thin outer layer splits and the thick inner layer absorbs water and expands upward. The ascus stretches up into the narrow neck of the ascoma, and the ascospores are expelled. These asci with a 'jack-in-the-box' design are also called fissitunicate."

Fistulose Hollow throughout its length.

Fitness The ability of an organism to survive and reproduce; the ability of an organism to pass its genes to the next generation.

Flabelliform Fan-shaped.

Flaccid Limp or flabby; nonturgid.

Flag Isolated, wilted or necrotic branch with dead leaves attached.

Flag leaf The leaf originating from the first culm node below the rachis; the uppermost leaf of the plant.

Flagellum A whip-like appendage which is the organ of locomotion of a motile cell; in fungi two types can be distinguished: whiplash, with a smooth continuous surface, and tinsel, with the surface covered with mastigonemes (hair-like processes); undulipodium.

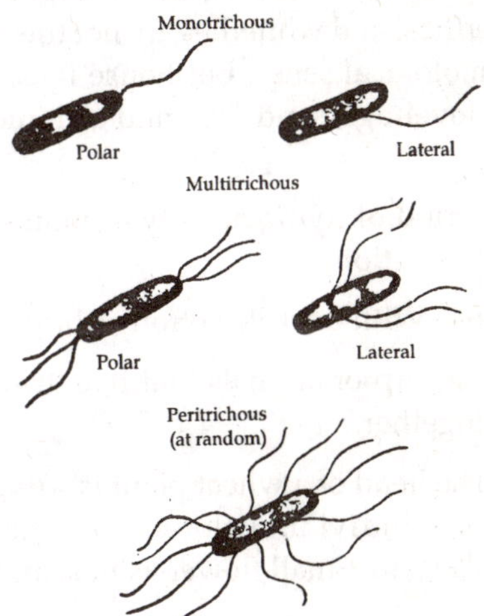

Fig. Various types of flagellation in bacteria.

Flagging The loss of turgor and the drooping of plant parts, usually following a water deficit.

Flail mower A mower that cuts turf by high-speed impact of inverted T-blades rotating in a vertical cutting plane relative to the turf surface.

Flavedo Outer part of the rind of citrus fruit, bearing oil glands and pigments.

Fleck A small, white to translucent lesion (spot) visible through a leaf.

White to tan necrotic lesions up to a few millimeters in length or diameter, usually confined to the upper surface of leaves.

Fledge When a baby bird first leaves the nest. A "fledgling" will often spend a day or two on the ground, practicing to run and fly. Fledglings are often attacked by predators.

Flesh In the pileus or stipe, The inner tissue, generally used at the macroscopic level; trama; context.

Flexuous Having turns or windings; capable of bending.

Flexure A turn or fold.

Flies Many families of 2-winged insects in the order Diptera (including tens of thousands of species worldwide). Dragonflies, butterflies, and whiteflies are not flies in the entomological sense, but house flies, horse flies, lovebugs, sand flies, and tachinid flies are.

Flight Period of flying activity of moths from one generation.

Flocculose With delicate cottony fibrils.

Flock A group of animals that live, travel, or feed together.

Floret The head of a wheat plant is composed of many individual florets in which the kernels form. Small flower as in a spikelet.

Flow diagram The diagrammatic representation of the structure of a system illustrating physical and information flows between compartments

Flowable A finely ground wettable-powder formulation sold as a thick suspension in a liquid. Flowables require only moderate agitation and seldom clog spray nozzles.

Flower The part of a plant that helps it grow new plants. The flowers have nectar and pollen in them. The nectar attracts bees and other insects, because it is a good liquid food. The pollen is a powder that must get to another flower to pollinate it. If a bee drinks nectar from a flower, it usually gets pollen on its body. Once this happens, the flower dies and a fruit grows there. Seeds inside the fruit fall out somewhere new and grow a new plant. Flowers often have bright colors to attract insects.

Flower break Break or stripe in flower color.

Flower bud A bud in which flower parts are contained.

Flower symptoms - Odd color changes (often in a mosaic pattern) and/or distortion (viruses).

Death of flower parts (fungi such as gray mold (Botrytis spp., bacteria)

Individual flowers or seeds converted into masses of black spores (corn smut).

Flowers that are green and smaller and more dense than normal (phytoplasma)

Fluorescent antibody One in which the gamma-globulin proteins are conjugated with a fluorescent dye, such as fluorescein isothiocyanate, to facilitate microscopic observation on the distribution of antigens in cells.

Focus A site of local concentration of diseased plants or disease lesions, either about a primary source of infection or coinciding with an area originally favourable to establishment, and tending to influence the pattern of further transmission of the disease (a MS).

Foliar application Application of a herbicide to the leaves or foliage of plants.

Foliar burn Injury to shoot tissue caused by dehydration due to contact with high concentrations of chemicals, such as certain fertilizers and pesticides.

Foliar Relating to leaves.

Foliose Describes a lichen thallus which has broad lobes free from the substrate.

Fontanelle A pore on the frontal region of the head of some termites through which secretions of the frontal gland are exuded.

Food chain Includes all businesses involved in the transformation of raw materials into food. It can be short e.g eggs purchased straight from the farm, or long and complex. eg transport/slaughterhouse/transport/processor/cannery/transport/retailer/consumer chain which beef destined for tinned casseroles may follow

Foot Basal area of a culm.

Forage Crops consumed by livestock

Forb A small, upright soft-stemmed or non-woody plant with broadleaves; the growth form of many common wildflowers.

Force of infection The per capita rate at which susceptibles are infected.

Forest A large place of dense trees, plants, animals, and other organisms. Older forests have taller trees than younger forests.

Forest decline Combination of biotic factors (e.g. insect pests, pathogens) and abiotic factors (e.g. drought, air pollution) that contributes to poor growth and death of populations of trees.

Forest edge The border of a forest and another area, such as a field, meadow, stream, marsh, or pond. Understory trees and shrubs grow taller here because they get more light. Many animals prefer to live at "edges."

Forewings Front pair of wings.

Form (latin, forma) 1. A subdivision of a species below the rank of variety. In this sense forms are distinguished on morphological grounds.

2. 'A subdivision of a species of a parasitic or symbiotic microorganism distinguished primarily by adaptation to a particular host'; forma specialis.

Form genus A non-phylogenetic category, equivalent to genus, distinguished on the basis of one or more morphological features. In the Deuteromycotina, form genera are used to classify anamorphs; such form genera are based primarily on the characteristics (including mode of development) of the conidia, conidiophores, and conidiomata.

Forma_specialis Literally "special form". An infraspecific taxonomic rank in which the taxa are distinguished on a physiological basis, particularly on the basis of adaptation to (or pathogenicity for) one or more specific hosts. In mycology, forma specialis is a taxonomic rank lower than form, subvariety, variety and subspecies, and higher than physiological race.

Formulation A pesticide as sold to personal and professional end users, usually a mixture of an active ingredient and several inert ingredients, and with a prominently-displayed brand name on its label.

Fornicate Arched; in Geastrum, having the fibrous and fleshy layers of the peridium becoming arched over the cup-like mycelial layer.

Fossil fuels Biological materials which have been subjected to long term geological effects. i.e coal, oil natural gas and peat

Foveate Pitted.

Foveolate Delicately pitted.

Foxfire The release of energy visible in the dark, caused by the presence of phosphorus chemical compounds, especially in Armillaria mellea (honey mushroom).

Fragmentation "Many fungi can reproduce by fragmentation. Any mycelium that is fragmented or disrupted, provided that the fragment contains the equivalent of the peripheral growth zone, can grow into a new colony. Many fungi are sub-cultured using this hyphal fragment technique. All of this weeks practical plates have been inoculated in this way with a cork bore taken from a colonized donor plate. Cut mycelial tips do not regenerate, but branches can form some distance from the damage point." Reproduction in the fungi.

Frass Solid larval insect excrement; mixed with wood fragments in woodboring or bark-boring insects.

Excrement of an insect, usually mixed with plant debris.

Free cell formation In an ascus, The process by which the eight nuclei, each with some adjacent cytoplasm, are cut off by walls in the immature ascus to form ascospores.

Free range System of (mainly) poultry keeping in which hens are allowed to range over a relatively large area

Free water Unbound water; often used to describe a film of water on a plant surface.

Freeze drying One of the methods used to preserve bacteria. Cells are suspended in a sucrose/peptone solution and dried under vacuum at -40°C. They are then sealed in vials while still under vacuum, and in this condition can survive for many years. Vials prepared in this way serve as a reserve supply of important strains.

Frond The leaf of a fern. Fronds often have many parts (leaflets) and are feathery.

Frost cracking Frost cracking is caused by extremely rapid temperature changes in bark and wood. As with sunscald, the bark and wood on the sunny side of the tree warms during the day. If a cold front moves in with a dramatic drop in temperature (i.e., from 30 F to -20F in a very short period of time), uneven contraction of the wood causes a crack to form suddenly. Sometimes the crack sounds like a gun shot.

Damage is most common in hardwood plants such as green ash. Frost cracking is not common and trees usually heal the cracks with few reprecussions.

Frost heaving Movement of the soil in the winter by freeze-thaw action.

Fructification 1. Synonymous with fruiting body.

2. The formation or development of a fruiting body.

Fruit The part of a plant that grows where a flower used to be, after the flower was pollinated and died. The fruit contains the seeds, which can grow new plants. Fruits are often fleshy with juices and nutrients for animals to eat. After animals eat the fruit, they poop the seeds out somewhere new, and the seeds can grow into new plants.

Fruit body A general term for the multicellular spore-bearing structures in both macrofungi and microfungi; alternative terms are carpophore, sporocarp and sporophore; hymenophore has a more restricted meaning.

Fruit symptoms Fruit decays, rots, and superficial spotting or russetting (fungi). Important diagnostic symptoms include specific color of rotted tissue, firmness of the tissue, and signs such as spores or fruiting structures.

Discolorations and malformations (viruses).

Discrete spots on fruit or soft rots in storage (bacteria).

Fruiting body A specialized structure, often macroscopic, on or in which spores are produced.

Any of various complex, spore-bearing fungal structures.

Fruticose Describes a lichen thallus which is shrub-like, erect or hanging, and branched.

Fry baby fish.

Fumigant A chemical toxicant (insecticide, fungicide, nematicide etc.) used in volatile form.

Vapor-active chemical used in the gaseous phase to kill or inhibit the growth of microorganisms or other pests.

Fumigation Application of vapor or gas, especially for the purpose of disinfecting or destroying pests.

Funariadiplolepidous Of moss sporophytes, having an inner (endostome) and outer (exostome) row of plates forming the peristome teeth.

Fungal Relating to fungi.

Fungi Fungi cause approximately 75% of the biotic diseases that occur in Montana. Fungi are organisms with a thread-like structure (hyphae). They enter plants through natural openings or wounds, or penetrate plants directly through intact tissues. Because they contain no chlorophyll, fungi do not produce their own food through photosynthesis.

More than 100,000 species of fungi now are known. Most of these fungi are strictly saprophytic, meaning that they consume dead organic matter such as fallen leaves and grass cuttings. Saprophytes do not attack living plants. Many mushrooms fall into this group as well as many fungi that live "invisibly" in the soil. Saprophytes are extremely important ecologically. Imagine what our world would look like if dead plants and animals and their waste did not decompose.

About 100 fungal species cause diseases in humans and animals, mostly affecting the skin and its appendages. However, more than 10,000 fungi cause diseases in plants. All plants are attacked by some kinds of fungi. In turn, each of the pathogenic fungi can attack one to many different kinds of plants.

Plant pathogenic fungi have developed specialized ways to take nutrients directly from living plants. An organism that obtains its food in this way is called a parasite. If a fungus or other living organism obtains nutrients only from living tissues, it is called an obligate parasite. Rusts, powdery mildews, and downy mildews are fungi that are obligate parasites. Non-obligate parasites live mostly as parasites, but under certain conditions they may feed on dead organic matter. Examples of non-obligate parasites are the fungi that cause potato late blight, Fusarium wilt, and damping-off.

The "body" of the typical fungus is made up of individual thread-like structures called hyphae. A mass of hyphae branches out in all directions and is known as a mycelium. Fungi reproduce through propagules called spores. Spores can survive adverse conditions and disperse to new areas through means which include wind, water, equipment, and insects. The shape, size, and color of spores and the structures in which they are formed are useful in fungal identification. A microscope and often cultural techniques are necessary for this level of identification.

Symptoms of fungal diseases include: leaf spots, blights, cankers, die back, root rot, damping off, soft rots, dry rots, anthracnose scab, and declines.

Fungi imperfecti An unfortunate and obsolete name for anamorphic fungi which are, or are suspected to be, the anamorphs of ascomycetes or basidiomycetes; better called 'conidial fungi.' They are no less perfect than the teleomorphs some of them possess.

Fungicidal Killing fungal spores or mycelium. Applicable to physical agents such as heat, ultraviolet light, X-rays, gamma-radiation etc., as well as to chemicals that are lethal at low concentrations.

Fungicide A pesticide that is used to kill fungi in plants, stored products or soil, or to inhibit their development.

(Chemicals that kill fungi for medical or veterinary applications are termed "antifungal agents"; they are classed as pharmaceuticals and so are not included in the Compendium.)

Fungicide, eradicant 1. A fungicide used to control disease after infection has occurred.

2. A fungicide applied to a substratum in which the fungus is already present (BMS).

Fungicide, protective A fungicide used to protect an organism against infection by a fungus.

Fungicide, systemic A fungicide which is absorbed through a plant surface and is translocated away from the site of application.

Fungistasis The prevention of fungal growth. The effect is reversible; if the inhibitor is removed or diluted, growth is resumed. In a broad sense the term can be applied to the non-germination of fungal spores due to the presence of auto-inhibitors or inhibitors from another organism or the substratum. It can also refer to the staling effect on hyphal growth.

Fungistat A chemical or physical agent that inhibits fungal growth, sporulation, or spore geminiation, but does not cause death.

Fungistatic Able to inhibit the growth and/or reproduction of at least some types of fungi.

Fungus An organism in the Fungi Kingdom. Fungi are similar to plants, but they cannot make their own food like plants do. The Fungi Kingdom includes mushrooms, mold, and yeast. Many fungi are mycorrhizal, meaning they have a relationship with trees. In this relationship the fungus grows on tree roots. It takes nutrients from the roots, but it also delivers nutrients to the roots from the soil.

Funiculus 1. A fine rope of hyphae;

2. the cord attaching the peridiole to the inner wall of a basidioma in some Nidulariales.

Furcate Forked.

Furfuraceous Powdered with bran-like particles; scurfy.

Furrow A groove in the bark of a tree, making the trunk look somewhat "wrinkled." Furrows can be shallow or deep, narrow or wide, depending on the species.

Fusarium oxysporum Fusarium oxysporum, also referred to as Agent Green, is a fungus that causes Fusarium wilt disease in more than a hundred species of plants. It does so by colonizing the water-conducting vessels (xylem) of the plant. As a result of this blockage and breakdown of xylem, symptoms appear in plants such as leaf wilting, yellowing and eventually plant death. Interest in Fusarium oxysporum as a pesticide was first raised after the discovery in the 1960s that it was the causative agent in the destruction of the Hawaiian coca population. The United States government was involved in a controversial program to use Fusarium oxysporum for the eradication of coca in Colombia and other Andean countries, but these plans were cancelled by president Bill Clinton who was concerned that the unilateral use of a biological agent would be perceived by the rest of the world as biological warfare. The Andean nations have since banned its use throughout the region. Use of biological agents to kill crops is potentially illegal under the Biological Weapons Convention.

Fusarium sp Fusarium sp. - (Aw 0.90). Commonly found in soil, plants, grains, and often times it is found in humidifiers. This fungus is the most common cause of mycotic keratitis. This mold has been isolated from skin lesions on burn patients, nail infections, ear infections, varicose ulcer, mycetoma , osteomyelitis following trauma, and disseminated infection. This fungus produces very harmful toxins, especially in storage of infected. crops.

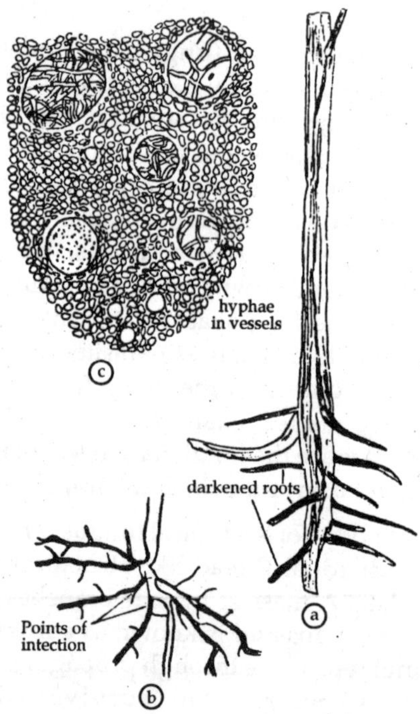

Fig. Pigeon pea wilt (Fusarium oxysporum f. sp. udum)

These toxins, known as trichothecene (scierpene) toxins target the circulatory, alimentary, skin, and nervous systems. Fusarium can also produce 1. Vomotoxin on grains which has been associated with outbreaks of acute gastrointestinal illness in humans.

2. T-2 Toxin and related trichothecenes are some of the deadliest known toxins. If ingested in sufficient quantity, T-2 toxin can severally damage the entire digestive tract and cause rapid death due to internal hemorrhage.

3. Fumosin, commonly found in corn and corn based products, with recently outbreaks of veterinary mycotoxicosis causing "crazy horse disease".

4. Zearalenone toxin which is similar in chemical structure to the female sex hormone estrogen and targets the reproductive organs. Morphological characteristics of this fungus include extensive cotton-like mycelium in culture, often with some tinge of pink, purple or yellow. Cultivation - Potato dextrose agar, 24°C.

Fusoid Spindle-shaped; tapering toward each end.

FWAG Farm and Wildlife Advisory Group.

G

Gall A localised proliferation of plant tissue producing a swelling or outgrowth, commonly having a characteristic shape, unlike any organ of the normal plant. Galls are usually formed in response to the action of a pathogen or pest.

A swelling or outgrowth of tissue induced by a pathogen or insect on a slant.

Outgrowth or swelling of unorganized plant cells produced as a result of attack by bacteria, fungi, or other organisms.

Gallery The horizontal tunnel dug just below the soil surface by a soil insect, i.e. a mole cricket nymph or adult. Termites also build galleries in wood by feeding on the cellulose. Carpenter ants often use termite galleries that have been abandoned.

Gametangium The sexual organ of the gametophyte generation – either an antheridium ("male")or archegonium ("female").

Gamete Cell, motile or non-motile, which can fuse with another compatible gamete to form a zygote (e.g., sperm,[motile], spermatium, [non-motile], egg, oosphere)

Gametogenesis The development of gametes.

Gametophore The structure on the gametophyte which actually produces the gamete.

Gametophyte The haploid phase (n=1) of a life cycle on which gametes (reproductive cells) are produced.

Gamma (g) glutamyl transferase One of a family of enzymes involved in transporting amino acids from the exterior to the cytoplasm of a cell. High serum levels of this enzyme serve as an indicator for liver disease.

Garden A piece of land where people grow flowers, vegetables, or other plants. Someone usually tries to keep certain plants in and others (weeds) out.

Gasteroid 1. Of basidiomycetes, Having angiocarpic development;

2. used by Oberwinkler & Bandoni for basidia from which basidiospores are not forcibly discharged (statismospores).

Gasterómycetes Holobasidiomycetes with basidioma closed at basidiospore maturity, hymenium present or absent, spore-shooting mechanism lost: basidiospores passively dispersed by a variety of interesting methods; a heterogeneous group.

Gel A matrix of polyacrylamide, agarose or similar material in which the electrophoresis of molecules is carried out.

Gel electrophoresis A type of electrophoresis in which the molecules in a sample moves through a gel composed of agarose or polyacrylamide.

Geminivirus Member of the only group of plant viruses with double particles, each particle containing circular or linear single-stranded DNA; this group is divided into two subgroups I and II (transmitted in a circulative manner by leafhoppers) and subgroup III (transmitted in a circulative manner by whiteflies or, in the case of some viruses in this group, mechanically transmitted).

Gemmae on a liverwortgemma A bud (or simply a group of cells) from which plants vegetatively reproduce.

Gene A unit of heredity in the chromosome; a sequence of nucleotides in a DNA molecule that codes for a polypeptide.

Gene expression The transcription of mRNA from the DNA sequence of a gene and the subsequent translation of that mRNA to give the protein gene product. Less strictly it can mean the transcription step alone.

Gene for gene concept The concept that corresponding genes for resistance and virulence exist in host and pathogen respectively.

Gene for gene hypothesis The hypothesis that corresponding genes for resistance and virulence exist in the host and pathogen, respectively.

General resistance Resistance that is effective against all biotypes of the pathogen.

Generalist natural enemy A natural enemy with a broad range of diet. Cattle egrets, for example, eat many kinds of insects, and are not specialized natural enemies of any of them. Generalist enemies are not used nowadays in classical biological control though some such uses did occur, especially in the 19th century, before biological control became a profession. Generalist natural enemies may, however, be purchased commercially and released to provide biological control of some pests.

Generation A group of offspring that are born together and live at the same time. Some insect species have more than one generation in a year. For example, a pair of ladybugs may give birth to one generation of thousands of young ladybugs. The first generation grows quickly and becomes adults. Those adults then mate with other ladybugs and give birth to a second generation in the same year.

Generative hyphae Hyphae that are thin-walled, branched, septate and with or without clamp connections.

Genet An individual or colony resulting from sexual reproduction.

Genetic Relating to heredity; referring to heritable characteristics.

Genetic code The nucleotide sequence of a DNA molecule (or, in certain viruses, of an RNA molecule) in which information for the synthesis of proteins is contained.

Genetic engineering Artificial manipulation of genes and chromosomes within plants, animals or other organisms to install desired traits or remove undesirable ones.

Genetic recombination The reassortment of genes produced as a result of crossing-over during meiosis.

Genetically modified organism Organism possessing a gene from another species; used to describe the organisms that have been the subject of genetic engineering.

Geniculate Bent abruptly at an angle, like a knee.

Genital papillae In nematodes: Tactile or sensory organs located on the male tail.

Genital primordium In nematodes: The initial cells of the reproductive system.

Genome The minimum complete set of genetic material in a gamete,zygote etc. In a virus, the nucleic acid component of the nucleocapsid.

Genotype 1. The hereditary potential of an organism forms the genotype;it is a theoretical unit because the physical unit that exists (the phenotype) is the result of the interaction of the genotype and the environment.

2. A single strain or species on which a genus is founded. This usage (by nomenclators) is one to be avoided.

Genus Part of the system of biological classification: a group of closely-related species. The word is singular (a genus), and its plural is genera (two genera). The genus name usually is placed in italics or is underlined, and it has a capital initial letter. Example: Periplaneta americana, the "Periplaneta" is the genus name.

Geocarp Fruit that ripens beneath the ground (e.g. peanut pod).

Geographic information system (GIS) Management system for data tied to precise locations.

Georeference Referring to the location of a specific point on the earth's surface by means of coordinates.

Geothermal energy Energy contained in the earths heat as steam water and hot rock

Geotrichum sp Geotrichum sp. This fungus is commonly found in dairy products and also in soils. This genus can sometimes be pathogenic to man. It is characterized by the formation of chains of colorless, slimy spores (conidia) through the fermentation of vegetative filaments. Some species of Geotrichum have strong odors.

Geotropic Growing vertically downward under the influence of gravity.

Germ pore A thinner and paler area or pore in the spore wall through which germination may occur; more than one germ pore may be present.

Germ slit A thin, often paler, line running along or across a spore and through which germination may occur.

Germ theory The theory that infectious or contagious diseases are caused by microorganisms (germs).

Germ tube The hypha produced by a germinated fungus spore, which with continued growth develops into the mycelium.

Germinate Begin growth (as of a seed, spore, sclerotium, or other reproductive body).

Germination In bacteria or fungi, the process by which a spore gives rise to a vegetative cell or hypha.

Germination by repetition The production and forcible release of a secondary ballistospore by a recently released basidiospore; characteristic of the Phragmobasidiomycetes.

Germination inhibitor A natural or synthetic substance capable of inhibiting the germination of a seed, spore or other plant propagule.

Germplasm Bearer of heredity material, often loosely applied to cultivars and breeding lines.

Ghost lineage A period of time during which a clade is inferred to exist, but is not known in the fossil record. As discussed in the text, Ascomycota and Basidiomycota are sister groups. Basidiomycota is known from the Early Devonian. This means that Ascomycota must have diverged from Basidiomycota at least this early. However, no definitive ascomycete remains are known older than the Missisippian. Thus, Ascomycota has a ghost lineage spanning the 60 My between these two periods.

Giant cells Large, usually multinucleate cells formed by abnormal cell fusions or failure of proper cell wall formation following growth and nuclear division. Associated with nematode feeding.

Gibberellin like substance A compound of unknown chemical structure, or lacking the gibbane ring skeleton of typical gibberellins, e.g. kaurene and steviol, but which are active in bioassay systems sensitive to gibberellins.

Gibberellins A series of substances resembling gibberellic acid both in biological properties and possessing a gibbane ring skeleton. Gibberellins have a broad spectrum of biological activity ranging from promotion of stem extension in intact plants, parthenocarpic fruit growth, flower initiation in long day plants and reversal of genetic dwarfism to the breaking of dormancy in seeds and plant organs.

Gill 1. The organs that underwater creatures use to breathe. Fish, tadpoles, and some insects get their oxygen from water through their gills.

2. The feathery parts of a mushroom under the cap. Gills release spores.

Girdle The act of removing a band of bark from around a tree trunk. Girdling kills the tree. Vines sometimes girdle a tree they grow on by wrapping around the tree trunk. Over time, the vine wears off the bark around the trunk.

Girdling Destruction of tissue in a ring around a twig, branch, or stem.

Girdling roots Girdling roots grow in a circular pattern around the trunk resulting in root death and loss of the anchorage system. Not all girdling roots are problematic. Only those that occur at the root collar area or above are considered chronically dangerous. Girdling roots that occur below the root collar area are not normally a threat to tree health or stability. It is nearly impossible to look at a developing girdling root and predict that it will cause problems for the tree. Problems are more likely in this situation though.

Genetically induced girdling roots: Certain tree species are prone to problematic girdling roots. The maples, especially Norway maple, are at the top of this list. Other species include poplars, the littleleaf linden cultivar "Greenspire", and the American beech. Culturally induced girdling roots: All trees develop root systems that could eventually result in girdling root problems, depending on how they were grown and/or planted. Pot-bound plants often develop girdling roots if the roots are not pruned at planting time. Girdling roots can develop from poor planting techniques, such as "twirling" a bare-root system into a planting hole that is too small for the roots, planting too deep, or planting in small soil volumes.

Symptoms include: leaning, early seasonal senescence, excessive twig dieback, thin appearance to crown, and little or no stem taper at ground level. Dig a few inches into the soil near the trunk and look for roots growing right next to the trunk in a circular pattern. . Generally, you can find girdling roots in this way without hurting the tree.

Gleba Fleshy spore-producing inner mass of a clestothecium (enclosed ascocarp), e.g. a puffball or stinkhorn. A gleba on the outside of an open ascocarp is referred to as a hymenium. The term is also used in a slightly different sense to mean the mass of spores left as the asci disintegrate.

Glebal chambers The cavities within the gleba, usually lined with hymenial elements; empty or gel-filled, e.g. in Melanogaster.

Gliocladium sp Gliocladium sp. This fungus occurs in soil or decaying plant matter as parasites of other fungi. Reported to be allergenic. Fungus that is structurally similar to Penicillium sp. but with conidia collecting in wet rather than dry masses. Cultivation - Potato dextrose agar, 24°C.

Global positioning system (GPS) A network of satellites generating signals which can be used by ground-based receivers to pinpoint current location coordinates (i.e., latitude and longitude).

Global warming Certain Gasses such as carbon dioxide, methane, nitrous oxides and chlorofluorocarbons (CFC's) absorb reflected infra-red radiation and trap it within the earths atmosphere. This heat which would normally be dissipated into space increases the 5temperature of the earth. An increase in the earths temperature would have profound effects on the worlds sea levels and climate. The control of theses gasses is therefore seen as a major international priority.

Globose Globe-shaped; spherical or nearly so.

Globulin A family of proteins found in abundance in plasma. They include the gamma globulins, which in turn include the various antibody molecules produced by the immune system.

Gloeocystidium A thin-walled, usually irregular cystidium with hyaline or yellowish and highly refractile contents.

Glomerulonephritis Kidney disease characterized by bilateral inflammatory changes in the glomeruli (tufts of capillary loops associated with the nephrons, the functional units of the kidney). The disorder is not caused by infection. It is considered an autoimmune disease.

Glucan A polysaccharide composed of repeating units of glucose.

Glucan, beta B-Glucans consist of linear unbranched polysaccharides of linked D-glucopyranose units. They are (together with chitin) an important component of the cell wall in Basidiomycota and Ascomycota.

Glucose A simple sugar which is actively transferred into the blood following the digestive breakdown of starch and other carbohydrates in the gut.

Glucoside A substance that, on decomposition, yields glucose and certain other compounds; some glucosides are defense compounds produced by plants, e.g. cyanogenic or phenolic glucosides.

Glue board A small cardboard sheet or boxlike apparatus having one or more surfaces coated with a sticky paste for capturing pests.

Glume Empty bract at the base of a spikelet (floret) .

Glycoalkaloid A bitter-tasting compound present in potato foliage and in the epidermis of potato tubers.

Glycogen An insoluble carbohydrate, similar to starch; a reserve food found in fungi, bacteria and animals.

Glycolysis A process in which sugar is changed anaerobically to pyruvic acid, with the liberation of a small amount of usable energy.

Glycoprotein A protein to which sugar molecules are bound, often in short chains ("oligosaccharides"). The free hydroxyl groups on the sugar molecules are often strongly hydrated, resulting in a matrix which contains considerable water and a surface with low friction. The hydroxyl groups also form hydrogen bonds with other sugars, creating a complex of flexible cross-links between individual protein molecules. Animal "gristle" is a good example of a glycoprotein matrix.

Gnat A term commonly applied to small flying insects, especially flies (order Diptera) or true bugs (order Hemiptera). Some fly and bug families include the term "gnat" in their approved common name.

Gnotobiotic Of the growth of an organism in the absence of other organisms, or in the presence only of known organisms. The term is applicable to the growth of 'germ-free' plants and animals as well as to microorganisms, for which the expression 'pure culture' is more commonly used.

Golgi body, golgi apparatus An intracellular organelle that consists of flat, disc-shaped vesicles often forming tubules at the edges; sites of accumulation and synthesis of cell products, e.g., wall material; in plants, sometimes called dictyosomes.

Gonad Sexual structure in animals that produces gametes.

Gonoplasm In Oomycota, the protoplasm in the middle of an antheridium which undergoes fusion with an oosphere.

Gossypol A substance poisonous to many animals, produced by numerous small glands in most cotton varieties.

Gout Formation of swellings at nodes or at the base of buds.

Gradual or direct metamorphosis Metamorphosis with three stages, namely egg, nymph and adult, as in the growth of grasshoppers.

Gradualism A model of evolution that assumes slow, steady rates of change. Charles Darwin's original concept of evolution by natural selection assumed gradualism.

Graft Transfer of aerial parts of one plant (e.g. buds or twigs - the scion) into close cambial contact with the root or trunk (the rootstock) of a different plant; a method of plant propagation; the joining of cut surfaces or growing roots of two plants to form a living union.

Graft transmission Transmission of a pathogen from one host plant to another through fusion of living tissue from the diseased host with living tissue of a healthy host.

Graft union Place where the rootstock joins the scion or top part of a grafted tree or vine.

Grain smut (Sphacelotheca sorghi) This disease also known as covered or kernel smut. The grains are transformed into white greyish sacs (smut sori), are slightly pointed to oval and filled with black powder (chlamydospores). Most of the grains are found infected. Perpetuate through contaminated seed. Control: 1. Use healthy seed,

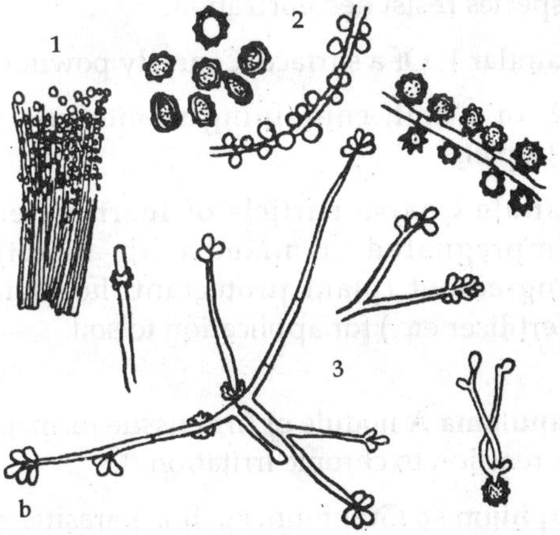

Fig. Grain smut of jowar (Sphacelotheca sorghi): (a) symptoms (b) a sorus magnified (c) columella (d) sorus not involving stamens (e) sorus with stamens involved (f) germinating spores.

2. Sow seed by broadcasting and water the field immediately thereafter,

3. Treat the seed with any one of the organic mercurial or copper seed dressing fungicides and

4. To collect and burn diseased heads.

Gram negative bacteria A common staining technique used with bacteria stains the organisms with crystal violet, then rinses them in alcohol. Some retain the dye and appear purple in color, while in others including the rhizobia, the stain washes out. To see such organisms they must be restained with some other dye. Bacteria that stain with crystal violet are termed Gram positive; those in which the dye washes out are called Gram negative.

Gram reaction The result of the Gram stain.

Gram stain An important bacteriological staining procedure discovered empirically in 1884 by the Danish scientist Christian Gram. When bacteria are stained with certain basic dyes, the cells of some species (Gram-negative species) can be easily decolorized with organic solvents such as ethanol or acetone. Cells of Gram-positive species resist decolorization.

Granular 1. Of a surface, Coarsely powdery;

2. of a cell, containing numerous oily droplets.

Granule Coarse particle of inert material impregnated or mixed with an active ingredient (plant protectant, herbicide, fertilizer etc.) for application to soil.

Granuloma A nodule of firm tissue formed as a reaction to chronic irritation.

Graphium sp Graphium sp. It is parasitic and commonly found as plant pathogen causing vascular diseases. Conidiophores are simple, colorless, produced in abundance, bearing oblong conidia that reproduce by budding. Mode of conidial development is variable in different species. Some species are imperfect stage of Ceratocystis. Cultivation - Potato dextrose agar, 24°C.

Grass A plant in the Grass Family (Poaceae). These are usually short plants with slender leaves. People grow different species of grasses for their lawns. Many animals eat the seeds and leaves. When allowed to grow wild, grasses get several feet tall.

Gravid Containing an egg or eggs; capable of depositing eggs.

Grazing A method of feeding by herbivores

Green manure A crop plowed under while still green and growing to improve the soil; sometimes used to enhance populations of antagonistic microorganisms for biological control.

Greenhouse effect Global warming due to build up of atmospheric CO_2

Gregarious Growing in groups, but not joined together.

Grid mapping Specifying locations of any site within a field or other management unit by using coordinates (i.e., latitude and longitude).

Grid system A means of assigning numbers to any location in the world using rows and columns(i.e., latitude and longitude); can be used to specify precise locations within a field.

Griseofulvin A chlorine-containing antifungal antibiotic from Penicillium griseofulvum, and P. nigricans; has been used as a systemic fungicide against plant pathogens and orally against dermtophyte infections in animals and humans.

Gross national product (GNP) The total value of goods and services produced by an economy over a particular period of time. GNP is made up of consumer and government purchases, private, domestic and foreign investments in the country and the total value of exports

Ground cover Any of various low, dense-growing plants, as ivy, pachysandra, etc., used for covering the ground, as in places where it is difficult to grow grass.

Group 1. Any taxon.

2. An assemblage of similar cultivars within a species or interspecific hybrid.

3. 'Congeries of organisms having common characteristics' , e.g. assemblages of related serotypes.

Growth form The overall morphology of a plant species, including its stature, leaf type, and habit. The general description of the type of growth exhibited by a plant, such as herbaceous, shrubby (bush-like) and arborescent (tree-like). The most basic grow forms are trees, shrubs, forbs, and graminoids; but there are many more specialized growth forms such as epiphytes, lianas, and stem succulents.

Growth inhibitor A substance inhibiting the growth of an organism. This all- embracing definition can include the activity of certain fungicides and also auto-inhibition by an organism. The term is difficult to define unless qualified and the dosage taken into account. With most substances the degree and kind of response depends on the level of inhibitors present. Substances present at high concentrations may be toxic, i.e. the growth is irreversibly inhibited, resulting in death of cells,while at low concentrations they may merely act as a growth retardant, the effect of which can be reversed by returning the organism to an inhibitor-free environment. The duration of exposure to the inhibitor also determines to some extent the type of response. At very low concentrations certain growth inhibitors function as growth stimulants, e.g.coumarin.

Growth phases of bacteria The growth of bacteria in culture follows a definite course which may be divided into four phases: 1. lag phase, the period starting from inoculation into a fresh medium, during which metabolic activity and cell size increase but there is little or no cell division.

2. log phase, or exponential phase, a period when cell division occurs at a constant rate such that log number of cells plotted against time is in a linear relationship.

3. stationary phase, a period when the viable count remains constant as the number of new cells counterbalances the number of dying cells.

4. decline phase, the period during which the viable count falls.

Growth promoters Substances given to livestock to increase growth rates

Growth regulator A pesticide constructed to mimic insect hormones that control molting and the development of some insect systems affecting the change from immature to adult. In most cases this pesticide prevents the insect from becoming a sexually mature adult, and perhaps even causes its death.

Growth retardant A substance that decreases the rate of growth (of a fungus, higher plant etc.)

Growth stages 1. Tillering stage when a plant produces additional shoots from a single crown, as in wheat.

2. Jointing stage when the internodes of the stem are elongating.

3. Boot stage when the seedhead of a plant begins to merge from the sheath, usually applied to grain crops.

Grub An insect larva; A term loosely applied, usually to larvae of Coleoptera; larva is thick-bodied with well-developed thoracic legs but no abdominal prolegs.

Guard cells Paired, specialized, epidermal cells that contain chloroplasts and surround a stoma.

Gubernaculum In nematodes: Spicule guide; sclerotized accessory piece.

Guiding ring In nematodes: Sleeve-like structure that surrounds and guides the stylet in genera of the Dorylaimoidea. Position varies among the genera from near apex to posterior portion of stylet.

Gum A complex polysaccharide of variable composition, swelling in water to form a gel or a viscous solution.

Gum, wound Gum produced in the wood of hardwoods and some conifers in response to injury, infection or other 'irritation'.

Gummosis Emission of gum from a tissue or organ; production of gum may be external, constituting a characteristic symptom, as in cherry bacterial canker (Pseudomonas mors-prunorum) and cucumber gummosis (Cladosporium cucumerinum), or it may be internal, resulting in the filling of lysigenous cavities in xylem or phloem, or occlusion of vessels as a response to wounding or infection by vascular pathogens.

Guttation Exudation of watery, sticky liquid from hydathodes, especially along leaf margins.

Guttulate Of a spore, Containing one or more oil-like globules.

Gymnocarpous Of Basidiomycota, Where the hymenium of the developing fruit-body is exposed from initiation to maturity.

Gyromitrin A heat-labile, carcinogenic, cellular toxin produced by Gyromitra esculenta (the False Morel); breaks down to monomethylhydrazine (MMH), which is also extremely toxic.

H

Habit The manner of growth.

Habitat Food, water, shelter, and space together make up an organism's habitat. Without any one of these four things, the organism would not be able to survive.

Hadromycosis A term coined by G. H. Pethybridge for a fungal disease in which infection is more or less confined to the xylem. The term vascular wilt disease is applicable in most of such diseases, and is preferred. half-leaf test.

Half life Used of plant virus inoculum for the time taken to lose half its infectivity on ageing, or in heat inactivation studies. Also used to express the progressive decline in transmission of some viruses by their vectors as the period between acquisition and inoculation feeding increases.

Hallucinogen A psychoactive substance which causes disturbances of perception, e.g. psilocybin.

Halonate Of a spore, Having a transparent coat; of a leaf spot, having concentric rings.

Halophilic Capable of living in a salt-rich environment such as sea water, saline soil, salt-preserved food, etc. and often preferring such an environment.

Hamanatto An Oriental food obtained by fermenting whole soybeans with Aspergillus oryzae; tao-cho (Malaysia); tao-si (Philippines); tu su (China).

Hamate Hooked.

Hamathecium A neutral term for all types of hyphae or other tissues between asci, or projecting into the locule or ostiole of an ascoma; subdivided into interascal pseudoparenchyma, paraphyses, paraphysoids, pseudoparaphyses, peri-physoids and periphyses.

Haploid A nucleus is haploid if it contains only one copy of each non-redundant gene. The great majority of animals, and higher plants, are diploid. That is, the form we usually observe has two copies of each gene, one from each parent. In lower plants there are numerous exceptions to the rule. Furthermore, the gametophyte generation is always haploid.

Harborage site Small cracks or areas where insects or other pests remain safely hidden during their periods of rest. Harborage sites are usually located near food and water sources.

Hardiness Ability to withstand stress.

Hardpan A hardened soil layer, caused by cementation of soil particles.

Hardwood A tree made of tough material. Many hardwood trees are used by people to make houses, furniture, tools, or other objects.

Hartig net The intercellular hyphal network formed by an ectomycorrhizal fungus in the surface layers of a root; the effective interface between the symboints.

Haulm The stems and leaves of a crop left after harvest.

Haustorium A simple or complex structure formed by an interaction between a branch of a fungal hypha and a host cell into which it has penetrated without causing lethal injury, generally assumed to provide a means by which a fungus absorbs nutrients from the host cells.

A modified mycelial branch that grows into a plant cell, makes intimate contact with the protoplast, and absorbs food.

HAV/HBV/HCV Hepatitis A, B and C virus. These very different viruses all cause the liver disease hepatitis. Hepatitis B and C are blood borne, while Hepatitis A is an enterovirus which is faeco-orally transmitted. Other known hepatitis viruses include Hepatitis D and Hepatitis E.

Hazard The likelihood that an injury will result from the use of a pesticide. A hazard constitutes both toxicity and exposure.

Head The place where the brain and sense organs of most animals is located. With insects, the head is the first of three sections (head, thorax, abdomen). It contains eyes, antennae, and mouthparts.

Head blight The destructive damage to an inflorescence or seed head, e.g. head blight of wheat caused by Fusarium graminearum Schwabe.

Headage payment Payment made to farmers based on the number of ewes or suckler cows kept on the farm

Heading Term used to describe the emergence and formation of a grass flowering structure (head).

Heart rot A decay of the heartwood, sometimes extended to a decay in the centre of a plant organ.

Heartwood Central cylinder of nonfunctional xylem in a woody stem.

Heat unit Synonym for degree-day.

Hedge A group of shrubs grown in a line by people to make a barrier or fence. Some birds and other small animals use the hedge for nesting or shelter.

Hefted sheep Hill or mountain sheep that are grazed in the same area that they were born, which is unfenced, they do not stray. A heft is a flock of such sheep sold with the farm

Helicoid Spiral; coiled like a helix.

Helicospore A septate or non-septate asexual spore; with a single axis curved through at least 180 degrees, but may describe one or more complete rotations, in two or three dimensions; any protuberances, other than setulae, not more than 1/4 the length of the spore body.

Helminthosporium sp Helminthosporium sp. Found as saprophytic or parasitic. Mycelium dark, often in substrate; conidiophores single or clustered, tall, erect, brown, simple; conidia develop laterally through pores beneath septa while conidiophore still growing, single, sub-hyaline to brown, obclavate. Cultivation - Potato dextrose agar, 24°C.

Fig. Leaf spot of rice caused by Helminthosporium oryzae: symptoms.

Helminths Worms, in particular the five groups of parasitic worms: Monogeneans (flukes), Digeneans (flukes, schistosomes), Cestodes (tapeworms), Nematodes (roundworms) and Acanthocephalans (spiny-headed worms).

Helminthosporium oryzae Helminthosporium oryzae causes disease in rice and led to the Bengal famine of 1942.T he Helminthosporium disease of rice occurs in Java, Sumatra, Japan, China, India, Straits and Federated Malay States, Italy, the Philippine Islands, and in Louisiana in the southern United States. The severity of the disease varies in different places, probably on account of climatic conditions. In the Philippines, the losses among seedlings may be 10 to 58 percent. In Louisiana no actual count was made. From the writer's observations, the disease occurs on all varieties of rice in culture there, although it is practically unrecognized. The disease occurs as leaf-spots on seedlings and older plants, as seedling- and leaf-blight, as infections of culms and sheaths, and to some extent as infection of the glume. The disease causes total blasting of kernels in cases of severe infection of the panicles. In all cases, many of the apparently healthy grains carry the mycelium. When the invaded parts of the rice die, sporulation takes place and conidia are spread by the wind, causing secondary infections. The Helminthosporium disease of rice in the Philippines and in Louisiana is caused by Helminthosporium oryzae Breda de Haan. The different strains have certain morphological and physiological differences.

Hematologic Having to do with the blood.

Hemiparasite Parasitic flowering plant that contains chlorophyll when mature (photosynthetic) and obtains water with dissolved nutrients by connecting to the host xylem via the haustorium, e.g. true mistletoe.

Hemizonid In nematodes: Lens-like structure situated between the cuticle and hypodermal layer on the ventral side of the body just anterior to the excretory pore; generally believed to be associated with the nervous system.

Hemizonion In nematodes: A companion structure to the hemizonid; it is smaller and located posterior to the hemizonid.

Hemoglobin A red pigment present in nitrogen-fixing root nodules that maintains oxygen flow to the bacteria. It is similar in overall composition and function to the hemoglobin in our blood.

Hepatic Concerning the liver, as in hepatoma (liver cancer).

Hepatitis B virus A small DNA virus capable of causing both acute and chronic liver disease, possibly by eliciting tissue damage by the immune system. The virus may also be a risk factor for hepatic carcinoma. It is often transmitted through sexual activity or through exposure to contaminated blood.

Hepatitis C virus An RNA virus related to the pestiviruses and flaviviruses. It is capable of causing both acute and chronic liver disease. As with hepatitis B, the liver damage resulting from this infection may be the result of immune reactivity against virus-infected liver cells.

Herb Any non-woody vascular plant; a category of plants including both forbs and graminoids (grasses). Hence the term "herbaceous" layer. Herbaceous: having this sort of growth form.

Herbaceous Describing primary, soft, nonwoody tissue, as a plant or plant part; having the characteristics of an herb.

Herbicide A material used specifically to manage or prevent damage caused by weeds. Sometimes generalized to be synonymous with pesticides.

Herbicide damage Tree damage from herbicide sprays or granules applied to turf or surrounding cropland is very common in Montana. Affected leaves become distorted, twisted, curled, or cupped, and margins may turn brown. Trees may be killed by this treatment, but more often are stunted for a year.

Herbicide safener A chemical that protects crops from injury by herbicides, but does not prevent the herbicide from killing weeds.

Herbivore An animal that eats mostly plants.

Herd immunity Specifically, the mechanism by which an infection may be eradicated from a population although some susceptibles still remain, because the remainder of the population is immune and thus transmission is reduced.

More generally, the immunological status of a population of hosts and its effect on transmission rates.

Heritability A measure of the ability of a characteristic to be passed from one generation to the next.

Hermaphrodite An animal that has both male and female body parts. When hermaphroditic animals, such as earthworms and slugs, mate, each animal is then able to lay eggs.

Herpesvirus A family of large DNA viruses that infect a wide range of animal species. Eight distinct types have been associated with a variety of human diseases.

Heteroauxin Synonymous with auxin, in the sense of IAA and no longer used. The first-claimed naturally-occurring auxins a and b, the cyclopentene acids, which led to IAA being named the ' different ' auxin, have never since been isolated and their existence is in doubt.

Heterobasidium A basidium showing septation, usually a phragmobasidium.

Heteroecious Completing the life cycle on two unlike hosts, e.g. a rust fungus such as Puccinia graminis on **wheat and barberry**; an aphid, Phorodon humuli, on damson and hop.

Requiring two taxonomically different hosts to be able to complete the entire life cycle, as in the rust fungi.

Requiring two or more unrelated hosts for completing the life cycle of a rust.

Heterogametes Male and female gametes that are morphologically distinguishable.

Heterogamy Plasmogamy between morphologically different gametes.

Heterokaryon A cell having two or more genetically different nuclei, sometimes as a result of anastomosis .

Heterokaryosis The result of forming a heterokaryon of a fungus; the condition of a hypha or cell having two or more genetically distinct haploid nuclei.

Heterokaryotic The condition of being a heterokaryon.

Heterokont Having flagella of more than one kind, as in the motile cells of the Oomycota (and of many other Chromista, including the brown algae).

Heterologous titre The dilution end point of an antiserum when tested against an antigen other than that used in its preparation.

Heteroploid Having a complement of chromosomes differing from that characteristic of the species.

Heterosis Increased vigor of hybrid offspring when compared to either parent.

Heterosporous Having two types of spores: megaspores and microspores.

Heterothallic Describes fungi in which two genetically distinct but compatible mycelia must meet before sexual reproduction can take place.

Heterothallism The type of sexual reproduction where conjugation is possible only through the interaction of different thalli.

Heterotopy Hyperplastic symptom in which an organ develops in a position other than its normal one.

Heterotrichous (filament) A growth pattern of algae with prostrate filaments for attachment and erect, spreading filaments for photosynthesis.

Heterotrophic Unable to derive energy from photosynthesis or from inorganic chemical reactions, and so dependent on energy-containing organic compounds derived from the current or prior existence of other organisms.

Heterozygous Having alternate forms (allele) of a gene on homologous chromosomes.

Hi test Haemagglutination inhibition test, a serological test used to detect antibodies specific to a particular family of viruses which possess the ability to agglutinate red blood cells e.g. measles, rubella and influenza.

Hibernaculum A shelter occupied during the winter by an insect, notably peach twig borer.

Hibernate To be inactive during Winter. Many insects; most reptiles and amphibians, and some mammals hibernate.

Hierarchy A series in which each element is categorized into successive ranks or grades with each level subordinate to the one above.

Higher fungi Members of the Ascomycota, the Basidiomycota and the Fungi Anamorphici.

Hilar appendix In basidiospores, The projection which connects the spore with the sterigma; apiculus.

Hilum In a basidiospore, The mark or scar on the apiculus or hilar appendix at the point of attachment to the sterigma; in a conidium, the scar on the spore left after detachment from the conidiogenous cell.

Hirsute Bearing coarse, rough, longish hairs.

Histopathology The study of pathology of cells and tissues; the microscopic changes characteristic of disease.

Histoplasmosis Deep-seated mycosis of humans caused by Histoplasma capsulatum (Hyphomycetes); common in the Mississippi Valley.

HIV (human immunodeficiency virus) The cause of Acquired Immune Deficiency Syndrome (AIDS).

Hoary Covered with a very short, closely interwoven hairs.

Holdfast Algal organ for attachment.

Holobasidiomycetes Basidiomycetes in which the basidia are not subdivided by septa.

Holobasidium A basidium not divided by primary septa but which may contain in any of its parts one or more adventitious secondary septa associated with localised withdrawal of cytoplasm.

Holoblastic Of conidiogenesis, Mode of blastic production of cell walls in which, following completion of any developmental stage, the fungus in a new stage lays down wall layers which are continuous with all of the wall layers used in the previous stage.

Holocarpic Where the entire thallus functions as the fruit-body.

Holocarpic reproduction In fungi, reproduction in which the entire fungal body is segmented into spores.

Holoendemic An infection whose prevalence is fairly uniform throughout a region, country or continent. Often used in the malaria literature.

Holomorph Any fungus considered in its entirety, i.e., including all latent or expressed (anamorphic or teleomorphic) forms.

Holoparasite Parasitic flowering plant that lacks chlorophyll (nonphotosynthetic) and must rely totally on the contents of the xylem and the phloem of the host plant, e.g. broomrape (Orobanche spp.).

Holothallic Of conidiogenesis, Thallic conidiogenesis with both outer and inner walls of the conidiogenous cell contributing to the formation of the conidium wall.

Holotype The single specimen designated or indicated as "the type" by the original author at the time of publication of the original description.

Homing endonuclease An enzyme which catalyzes the insertion of an intron into chromosomal DNA. Homing enzymes may be coded by the intron itself, or may come from elsewhere. Homing enzymes may work in a variety of ways. For example they may act as, or work with, a reverse transcriptase to create a DNA copy of an intron RNA. Alternatively, they may directly ligate the intron transcript (RNA) into the DNA od the "host," relying on the host DNA repair machinery to convert the inserted sequence into double-stranded DNA.

Homokaryon A hyphal cell, mycelium, organism, or spore in which all the nuclei are genetically identical.

Homokaryotic Refers to a homokaryon.

Homologous 1. with the same evolutionary origin though now often different in form and/or function;

2. (of chromosomes) bearing genes for the same characters.

Homologous titre The dilution end point of an antiserum when tested against the antigen used in its preparation.

Homology Two structures are considered homologous when they are inherited from a common ancestor which possessed the structure. This may be difficult to determine when the structure has been modified through descent.

Homosporous Having one type of spore.

Homospory A condition in which an organism produces only one type and size of spores, (microspores).

Homothallism The type of sexual reproduction where conjugation can occur without the interaction of two differing thalli.

Homozygous Having identical alleles at the same locus on homologous chromosomes.

Honey fungus The agaric, Armillaria mellea, which is parasitic on trees and causes serious root rots.

Honeydew A mixture of sugars and other plant-derived chemicals excreted by some species of aphids and by some species of insects in families related to aphids. When these insects feed on plants, honeydew drips from them onto plant leaves or onto the ground. Fresh honeydew may be fed upon by other insects as an energy source. Ants of many species are avid feeders on honeydew, and they may even guard the aphids against predators and parasitoids to protect this energy source. Unconsumed honeydew on plant leaves promotes growth of a black fungus called sooty mold, which may become so dense that it interferes with the metabolism of the plant.

Horizontal resistance Resistance which is effective against all biotypes of the pathogen.

Horizontal survey A study of a community, perhaps stratified by age, sex, ethnicity etc., but at one point in time or over a short time interval.

Although a snapshot, horizontal surveys of prevalence and intensity within different age classes of a community can nevertheless provide valuable information on the rate at which hosts acquire infection through time, provided that the host and parasite populations have remained approximately stable for a period of time (ie stable endemicity). Same as cross-sectional study, converse of longitudinal study.

Horizontal transmission Transmission occurring generally within a population, but not including vertical transmission.

Hormone A chemical substance produced in one part of a an organism and transported in minute quantities to induce a growth response in another part, e.g. in plants, auxins, cytokinins, and gibberellins.

Horticultural oils Highly refined petroleum (or seed derived) oils that are manufactured specifically to control pests on plants.

Horticulture The cultivation of fruit, flowers, vegatables and shrubs, also used to describe the commercial production of such crops on general farms

Host An organism harbouring a parasite (BMS). A living organism harbouring another organism or virus dependent on it for existence (APS). The individuals of a species which are capable of harbouring a parasite, even though a particular individual or population of that species may not at the time be parasitized, can also be described as hosts; however, in these circumstances it is preferable to refer to the organism as a potential host. The term is used so commonly in preference to suscept to denote an organism harbouring a pathogen, that we are not inclined to make a contra-recommendation, and indeed have used it in this way throughout the Guide. Nevertheless, it must be stated that the conceptual parallel of 'host and parasite' is 'suscept and pathogen'.

The plant on or in which a parasite lives and from which it obtains its food.

A living organism serving as a food source for a parasite.

A plant that is invaded by a parasite and from which the parasite obtains its nutrients.

Host diversity Plants from a particular crop, pasture or prairie species are not all the same. This may be because of cross pollination between plants, or because of genetic mutations. Such differences can mean that some plants in a species will be resistant to particular diseases or are better in their ability to fix nitrogen. Because of such differences and their potential importance, the US maintains a strong germplasm collection program.

Host feeding Adult parasitoids that obtain nutrition from hosts by consuming host body fluids.

Host plant Living plant attacked by or harboring a parasite or pathogen and from which the invader obtains part or all of its nourishment.

Host plant resistance Host plant resistance is one of the preferred methods for minimizing the damage caused by pests and and associated viruses, because it does not require the complete elimination of the pest to be effective. HPR protects the crop by making it less suitable for the pest or by making the crop tolerant to the pest. The result is less crop damage. An example of this is the use of hairy-leaf cotton rather than smooth-leaf (glabrous) cotton to reduce the impact of whiteflies on yield.

Host range The range of plant species known to serve as hosts for a given parasite also, the range of action of a phage in terms of the species, strains etc. of bacteria in which it can grow.

Host specific A term used to describe pathogens or insects that attack only certain species of hosts.

Host specific toxin A metabolic product of a pathogenic microorganism which is toxic to the host of that pathogen. Relatively few host specific toxins are known; one example is victorin, the polypeptide toxin produced by Cochliobolus victoriae in Victoria blight of oats.

HPV (human papillomavirus) Causing genital warts and genital cancers.

HSV (herpes simplex virus) Types one and two, HSV-I and HSV-II, are among the causes of cold sores and genital ulcers.

Hulle cellshülle cell In some ascomycetes, "thick-walled globose cells (Hülle cells) develop by budding at the tips of specialized hyphae. Hülle cells envelop the developing cleistothecium and may serve as nurse cells."

Human herpesvirus 6 A virus of the herpesvirus beta-subfamily, discovered in 1985, that infects more than 95% of people by the age of 2 years. It has been causally associated with roseola, mononucleosis-like illness, inflammation of lymph glands. There is also suggestive evidence for a role in multiple sclerosis.

Humicolous Living in or on soil.

Humus The organic fractions of soil in which decomposition is so far advanced that its original form can't be distinguished.

Huperzia, with microphylls and eusporangiamicrophyll A leaf with only one vascular bundle.

Hyaline Transparent or nearly so; translucent; often used in the sense of colorless.

Hybrid The first generation offspring of a cross between two individuals differing in one or more genes

Hybridization 1. In molecular biology: The formation of stable duplexes between complementary sequences by way of Watson-Crick base-pairing.

2. Cross-breeding.

Hybridoma A cell produced by the fusion of an antibody-producing cell and a lymphoma (cancer) cell for production of monoclonal antibodies.

Hydathode A specialized leaf structure with one or more openings through which water is discharged from the interior of the leaf to its surface.

Hydrogen fluoride (HF) A common pollutant from volcanoes and several industries such as brick factories, aluminum smelters, and phosphate fertilizer factories.

Hydroids The water storage or water vascular cells of Bryophyta. Analogous to xylem in tracheophytes.

Hydrological cycle The cycle of water evaporated from water surfaces and plants and precipitated as rain

Hydrolysis Any chemical reaction in which water is one of the reactants. A chemical decomposition in which molecules of a substance are split into simpler compounds through reaction with the elements in water.

Hydroponic culture Growth of plants in an aerated liquid medium that supplies all nutrients needed for growth.

Hydroseed To apply seed in a water mixture by pumping through a nozzle that sprays the mixture onto a seedbed. The water mixture may also contain amendments such as fertilizer and mulch.

Hydrosis Necrotic symptom of disease characterized by water-soaking of tissues.

Hygrophanous Having a water-soaked appearance when wet, and noticeably changing colour on drying.

Hygroscopic Tending to absorb moisture; of a fruit-body, opening and discharging spores in dry air.

Hymenium "The hymenium is the layer of cells containing the spore-bearing cells (usually basidia or asci) of the fungus. The hymenophore is a collective term for the fleshy structures that bear the hymenium. Thus, in a gilled mushroom, all the gills constitute the hymenophore, and the hymenium is the layer of cells on the surface of those gills. Although the hymenophore may be convoluted and enclosed within the fruiting body, the hymenium still has to be, in some sense, on the 'outside' of the hymenophore in order for either of these structure to qualify for their names. Otherwise, you have a gleba, inside a peridium." hymenium.

Hymenomycete Common name for a member of the Euholobasidiomycetes where the hymenium is exposed at maturity and basidiospores are ballistosporic; also commonly used for a member of the Basidiomycota.

Hymenophore A spore-bearing structure (a fruit-body), especially a basidioma, or that portion of it bearing the hymenium.

Hyper Prefix, beyond, over, above; usually implying excess or exaggeration.

Hyperauxiny A condition in plants in which the level of auxin in the tissues is higher than normal. Hyperauxiny may result from infection by fungi, viruses, bacteria and nematodes and by chemical and environmental stimuli. The auxin involved is commonly IAA and appears to be of host rather than of pathogen origin, e.g. in bacterial wilt of tobacco caused by Pseudomonas solanacearum. It may also result from the inactivation or destruction, by the pathogen, of auxin-destroying enzymes of the host.

Hyperparasite A parasite upon another parasite.

Hyperparasitoid A parasite that lives on or in another parasite. Insect The class of small animals with three pairs of jointed legs and one pair of antennae, at least in the adult phase. Mole crickets, tachinid flies, and sphecid wasps all have this arrangement in the adult phase, and mole cricket nymphs also have it. However, some insect larvae (grubs) are legless. Spiders and ticks have four pairs of jointed legs, centipedes and millipedes have many more pairs, and these are not insects.

Hyperplasia A condition in which a tissue is enlarged as a result of excessive production of cells, as in crown gall caused by Agrobacterium tumefaciens.

The abnormal increase in the number of cells without their enlargement.

Hypersaprotroph Asaprotroph found only on substrata previously invaded by other saprotrophs.

Hypersensitivity The violent reaction of an organism to attack by a pathogenic organism or a virus resulting in prompt death of invaded tissue, thus preventing further spread of infection (APS, BMS). A means whereby plants genetically susceptible to infection by a particular parasite avoid serious injury by that parasite. Hypersensitive is a qualification of sensitive and implies a response to a low level of inoculum. It is the extreme case of susceptibility and as such is the antithesis of immunity, although the two terms are often mistakenly used as synonyms, e.g., 'the immune reaction'. Care must be taken to avoid confusion with the term used in the animal sense of 'an extreme sensitivity due to an antigen-antibody reaction excessive growth due to an enlargement of individual cells,as in root galls of Cruciferae (together with hyperplasia,) caused by infection with Plasmodiophora brassicae.

Hypersensitive reaction and pathogenicity (hrp) gene Gene required for elicitation of the hypersensitive (HR) response in resistant plants and causation of disease in susceptible plants; a type of avirulence gene.

Hypersensitive response (HR) Rapid and localized cell death at the site of infection in resistant interactions between plants and pathogens.

Hyperthyroidism A condition marked by excessive secretory activity of the thyroid gland.

Hypertrophy The excessive enlargement or development of an organ or tissue, with increase in cell size but without increased cell division.

Hypha Is a long, branching filament that, with other hyphae, forms the feeding thallus of a fungus called the mycelium. Hyphae are also found enveloping the gonidia in lichens, making up a large part of their structure. A typical hypha consists of a tubular wall, usually made of chitin, which surrounds, supports, and protects the cellss that compose a hypha. For most fungi, a cell within a hyphal filament is separated from other cells by internal cross-walls called septa (singular septum). Some forms of parasitic fungi have a portion of their hyphae modified to form haustoria that are able to penetrate the tissues of a host organism. Similar, yet mutualistic forms of penetrating hyphae are called mycorrhizae and are important in assisting nutrient and water absorption by plants." Hypha

Hyphal body Portion of mycelium that becomes separated from remainder of thallus.

Hyphal peg 1. A compound, hyphal, fasciculate projection extending beyond the general level of the hymenium, consisting of two or more parallel or interwoven hyphae, encrusted or gelatinized.

2. A projection from a hypha.

Hyphal sheath (syn. mantle) Dense hyphal mass of ectomycorrhizal fungus enclosing the short feeder roots of plants.

Hyphidium A little- or strongly-modified terminal hypha in the hymenium of hymenomycetes.

Hyphochytriomycota Phylum of Chromistan fungi with anteriorly uniflagellate zoospores; hence, hyphochytrid.

Hyphomycetes Conidial anamorphs (mostly ascomycetous, some basidiomycetous) producing exposed conidiophores, not enclosed in any protective structure, and usually liberating conidia in large numbers.

Hyphopodium Stalked, thick-walled, lobed cells that stick to plant surfaces; sometimes used to describe the infection structures produced by ectotrophic hyphae of certain root-infecting fungi such as Gaeumannomyces.

Hypo Pefix, below, under, beneath, lower; used either in place or degree.

Hypocotyl That part of the axis of a developing embryo just below the attachment of the cotyledons.

Hypodermal Pertaining to the hypodermis.

Hypodermis In nematodes: A thin tissue layer beneath the cuticle that thickens to form the dorsal, lateral, and ventral chords, which extend the length of the body.

Hypoendemic A term from the malaria literature used to mean (roughly) an area with little transmission.

Hypogynous Of Oomycota, having the antheridium below the oogonium and on the same hypha.

Hyponasty More rapid growth of the lower side of an organ than of the upper side.

Hypoparathyroidism A condition caused by the reduction or absence of secretions of the parathyroid gland.

Hypophyllous On the under surface of a leaf.

Hypoplasia The under-development of a tissue or organ resulting from reduced cell division, and caused by a pathogenic agent or by various growth regulating substances and inhibitors.

The underdevelopment of cells, tissues, or organs.

Hypothecium Medullary exciple; the hyphal layer below the subhymenium in an apothecium.

Hypothesis A concept or idea that can be falsified by various scientific methods.

Hypothyroidism Diminished production of thyroid hormone, leading to thyroid insufficiency.

Hypovirulence A reduced level of virulence in a strain of pathogen resulting from genetic changes in the pathogen or to the effects of an infectious agent on the pathogen.

Hysterothecioid Long, thin and cleft like a hysterothecium (ascoma in the Hysteriaceae); lirelliform.

Hysterothecium A sexual fruiting structure of the Ascomycete fungi, usually football-shaped or elongate in appearance and occurring on infected needles.

I

Iatrogenic Related to an abnormal state or condition produced in a patient through inadvertant or erroneous treatment.

Ibotenic acid A metabolite of Amanita muscaria; changes to muscimol, a hallucinogen, when the basidiomata are dried.

Icosahedral Having 20 faces, as a polyhedral-shaped virus particle.

Idiomorphs Nucleotide sequences that occupy the same locus in different strains but are not related in sequence or common descent.

Idiopathic Denoting a disease of unknown cause.

Imaging tests Any of a variety of methods for observing the internal anatomy of the body, ranging from simple x-rays to complex three-dimensional scanning techniques using nuclear magnetic resonance, positron emission, and other techniques.

Imbricate Overlapping, like the shingles of a roof.

Immarginate Without a well-defined edge.

Immersed Embedded in the substratum.

Immobilization This occurs when inorganic ions are assimilated by soil organisms and are bound organically again

Immune Cannot be infected by a given pathogen.

Immune globulin A crude preparation of antibody molecules collected from pooled multiple blood donations, used as a means for passively transferring antimicrobial resistance to susceptible individuals.

Immune suppressants Agents that block or restrict the activity of one or more components of the immune system, usually leading to increased susceptibility to infectious disease.

Immunity Freedom from disease, due to lack of qualities permitting or to possession or acquirement of qualities preventing the operation of the pathogenic factor. Freedom from attack by a pathogenic organism or virus due to lack of qualities corresponding to its requirements or to possession or acquirement of additional qualities unfavourable to it (APS). Exemption from infection. This is not the same as freedom from disease, although the term has frequently been used in this sense; a plant which is infected but is not showing symptoms may be resistant or tolerant but it is not immune. Immunity implies total exclusion of the potential pathogen. The term immune reaction is commonly used as a synonym for hypersensitive reaction, but because this reaction requires at least a transient infection we recommend that the use of 'immune reaction' be discontinued.

A relationship between a plant and a causal agent in which the plant does not become diseased. 1. A state in which a host is not susceptible to infection or disease, or

2. the mechanisms by which this is achieved.

Immunity is achieved by an individual through one of three routes: natural or innate immunity genetically inherited or acquired through maternal antibody, acquired immunity conferred after contact with a disease, and artificial immunity after a successful vaccination.

Also termed specific immunity, resistance or specific resistance.

Specific immunity is divided into cellular immunity acting via the direct involvement of T cells and humoral immunity involving antibodies and B cells.

Immunoassay A detection method based on antibodies specifically selected to react with the substance to be detected (the antigen).

Immunogenicity The ability of a vaccine to stimulate the immune system, as measured by the proportion of individuals who produce specific antibody or T cells, or the amount of antibody produced, say. Not the same as efficacy.

Immunopathology Damage to the host caused by its own immune response against a pathogen.

Immunosuppressant A substance, such as the fungal metabolite, cyclosporine, (produced by the hyphomycete, Tolypocladium niveum) which partially or completely suppresses the immune system; used to prevent rejection of transplanted organs.

Immunosuppression A reduction in the capacity of the immune system. Caused by infection (eg HIV), drug treatment, pregnancy and malnutrition among others. Immunosuppressed individuals are commonly referred to as immunocompromised.

Immunosupressant Chemical or activity that suppresses the natural immune responses in animals, including humans.

Immununity, acquired A non-inherited immunity acquired by some predisposing treatment, modifying the chemistry of the plant. The protection afforded to some species by pre-inoculation with non-virulent microorganisms or viruses prior to inoculation with virulent ones is not immunity if any degree of infection by either strain is involved.

Immununity, natural Immunity due to innate genetic characters.

Imperfect fungi Group of fungi without a sexual stage; also, the asexual stage of ascomycetes and basidiomycetes.

Imperfect state The state characterised by the production of asexual spores (conidia) or by the absence of spores the anamorph.

In planta In a plant.

In situ In its original place or environment.

In vitro Literally 'in glass'; frequently in the sense 'not under natural conditions', e.g. in the laboratory; in experimental culture; (of chemical reactions), not in living cells.

In vitro studies Studies of chemical effects conducted in tissues, cells or subcellular extracts from an organism.

In vivo Literally, in that which is alive. Properly applicable to physiological and biochemical activities in living cells and tissues and to the growth of parasites, pathogens etc. on host plants, but commonly used more loosely in the sense of growth in natural conditions, in contrast to growth in culture, i.e. in vitro.

In vivo studies Studies of chemical effects conducted in intact living organisms.

Inamyloid Not reacting in iodine-containing solutions such as Melzer's Reagent and therefore appearing yellow or hyaline.

Incidence The rate at which new cases of infection arise in a population. Classically measured as an attack rate.

Incipient Early in development (of a disease or condition).

Incisure In nematodes: A longitudinal cuticular cleft that divides the lateral fields; sometimes called involution or line.

Incitant A term used commonly, but incorrectly, to describe the causal agent of a disease. Properly, the term should refer to some other factor which promotes the pathogenic action of the causal agent.

Incite To cause a disease.

Inclusion body Structure developed within a plant cell as a result of infection by a virus, often useful in identifying the virus.

Incompatible Of mating types, strains, etc., not cross-fertile.

Incomplete dominance The interaction of alleles of a gene that produces an intermediate phenotype, as in the production of pink flowers when red and white alleles are present in a heterozygous individual.

Incorporate To mix a material such as an herbicide into the soil by mechanical action.

Incubate When a bird sits on its eggs to keep them warm as baby birds grow inside. Sometimes the male or female, or both, incubate the eggs. Incubation usually lasts for a week or more. Turltles and snakes will often bury their eggs to incubate underground by themselves.

Incubation period The period of time between penetration of a host by a pathogen and the first appearance of symptoms on the host.

The time that elapses between infection and the appearance of symptoms of a disease. Not the same as the latent period.

Time between infection by a pathogen and appearance of symptoms.

Indehiscent Pertaining to fruit that does not split open at maturity.

Indeterminate Having a growth pattern in which stems continue growing indefinitely; with flower clusters, the opening of the lower (lateral) flowers first, and the terminal one opening later.

Indexing Any procedure for demonstrating the presence of known viruses in suspect plants.

Determining presence of disease in a plant by removing buds or other parts for inoculation of a susceptible indicator plant that exhibits specific symptoms of a transmissible disease.

Indicator host A plant species that gives characteristic symptoms to a specific virus. Used in virus diagnosis.

Indicator plant One which reacts to a certain pathogen or environmental factor with obvious symptoms and is used to assist in the detection of the pathogen or the environmental factor. Synonymous with indicator host when applied to pathogens.

Indigenous Organisms that are native to a specific environment.

Indirect life cycle A life cycle which requires one or more intermediate hosts before the definitive host species is reinfected.

Indirect penetration Penetration of plant tissues by a pathogen through natural openings (e.g. stomata) or wounds.

Indirect transmission Transmission of a parasite through an indirect life cycle.

Induced Produced in response to a stimulus.

Induced systemic resistance (ISR) Reduced disease symptoms on a portion of a plant distant from the area where the inducing agent is active, caused by the triggering of active plant defenses against a variety of pathogens; used to describe increased resistance in plants induced by certain rhizobacteria.

Indurated Hardened.

Indusium A cover; in Phallales, a veil which hangs from the base of the receptacle.

Inert ingredient A substance that is not an active ingredient and that is included in a formulation for reasons other than pesticidal activity. Functions of inert ingredients include diluting the pesticide, making it safer, making it more effective, making it easier to measure and mix, making it easier to apply, and making it more convenient to handle.

Despite being called "inert" and not being toxic to pests, these substances can be biologically or chemically active, and can cause environmental and health problems. In some countries, they are listed on the label.

Infect To enter and establish a permanent or temporary parasitic relationship with an organism. It is undesirable to use the term in the sense of inoculate, which may not lead to infection; it should not be used in the sense of contaminate.

Infection The entry of an organism or virus into a host and the establishment of a permanent or temporary parasitic relationship.

The establishment of a parasite within a host plant.

The presence of a parasite within a host where it may or may not cause disease.

Process in which a pathogen enters, invades, or penetrates and establishes a parasitic relationship with a host plant.

Infection court The site of invasion of a host by a parasitic organism or virus. Stomatal cavities, wounds and hydathodes may be infection courts for bacterial infection of leaves.

Infection cushion Organized mass of hyphae formed on the surface of a plant from which numerous infective hyphae develop.

Infection focus Initial site of infection, generally with reference to a population of plants.

Infection peg A structure formed by deposition of lignin, callose, cellulose; suberin etc. around a hypha which is penetrating the wall of a living host cell. A lignituber is an infection peg apparently composed of lignin.

Infection period The time required for infection to occur under conducive environmental conditions, usually hours of leaf wetness and temperature.

Infection thread The specialized hypha of a pathogenic fungus that invades tissue of the susceptible plant.

Infection, inapparent Any infection that gives no overt signs of its presence.

Infection, latent An inapparent infection that is chronic and in which a certain host-parasite relationship is established.

Infectious Capable of producing propagules that disperse from one host to another and infect it.

Infectious disease A disease caused by a pathogen that multiplies and can be transmitted from plant to plant.

Infectious period The time period during which infecteds are able to transmit an infection to any susceptible host or vector they contact.

Infective 1. Of an organism or virus able to attack a host and cause infection. The term is frequently qualified, e.g. weakly or strongly infective.

2. of a vector carrying or containing a pathogen and able to transfer it to a host plant causing infection. For virus diseases, not synonymous with viruliferous.

Infest To over-run the surface of a plant, or to be dispersed through soil or other substrate. Used especially of insect and other animal pests. When used with reference to microorganisms or virus particles on plant surfaces there is no implication that infection has occurred. The infesting organism may or may not be a contaminant: infest and contaminate are not synonymous.

Infestation Presence in numbers (e.g., of insects, mites, or nematodes). Do not confuse with "infection," a term that applies only to living, diseased plants or animals.

Infiltration rate The rate at which ponded water on a soil surface enters the soil profile.

Inflorescence The flowering portion of a shoot including the spikelets, and any supporting axis or branch system.

Infrared photography A picture of an area using film which is sensitive to wavelengths higher than those visible to the human eye; used to detect plant stress.

Infrastructure The availability of roads, power supplies, education and health facilities, for example which all industries share and for which they do not directly pay for

Infundibuliform Funnel-shaped.

Ingest to obtain food by engulfing it (as opposed to absorbing, which is what fungi do).

Ingress The act, by a plant pathogen, of gaining entrance into the tissues of a susceptible plant.

Ingroup In a cladistic analysis, the set of taxa which are hypothesized to be more closely related to each other than any are to the outgroup.

Initial inoculum Inoculum, usually from an overwintering source, that initiates disease in the field, as opposed to inoculum that spreads disease during the season.

Injection The introduction of liquid (or inoculum) by means of a hypodermic needle etc. The term should not be used in the same sense as infiltration.

Injury Damage caused by transitory interaction with an agent such as an insect, chemical, or unfavorable environmental condition.

Innate Being an integral part of the tissue; not readily removable from the adjacent tissue.

Inner bark In older trees, the living part of the bark, the phloem.

Inoclum Pathogen or part of a pathogen that can cause infection.

Inoculant A commercial preparation, often but not always based on peat, that is used to introduce rhizobia into soils. Inoculants may be seed applied or introduced directly to the soil.

Inoculate To introduce a microorganism into an environment suitable for its growth; to bring a parasite into contact with a host.

Inoculation The act of inoculating; the placement of microorganisms or viruses at a site where infection is possible (the infection court).

Inoculation access time The length of time a vector is allowed to spend on a test host in transmission experiments. It is not implied that the vector feeds throughout this time. Synonymous with test access time but not with inoculation feeding time.

Inoculation threshold period The minimum time necessary for a vector to feed on a test plant in order that transmission may occur.

Inoculative Of a vector carrying or containing a pathogen and able to introduce it into a host plant.

Inoculum Material containing microorganisms or virus particles to be introduced into or transferred to a host or medium. The term can also refer to potentially infective material available in soil, air or water and which by chance results in the natural inoculation of a host.

The pathogen or its parts that can cause infection. That portion of individual pathogens that are brought into contact with the host.

The amount of parasite to which an individual host is exposed at transmission.

Pathogen or pathogen part (e.g., spores, mycelium) that infects plants.

Inoculum density A measure of the number of propagules of a pathogenic organism per unit area or volume.

Inoculum potential A term used by plant pathologists with many different interpretations. As defined by J. G. Horsfall it referred to the amount of a parasite available for infection in the soil. It was measured by planting susceptible hosts in the soil and counting the number that became infected. G. A. Zentmeyer, P. P. Wallace & J. G. Horsfall expanded the original concept to include a combination of environment and pathogen in which the pathogen was considered to be part of the environment, and inoculum potential was then defined as the ' disease-producing power of the host environment '. S. D. Garrett considering soil infesting fungi, attempted to equate the infection potentials of different inocula under differing conditions as 'the energy of growth ofa fungal parasite available for infection of the host at the surface of the host organ to be affected '. This definition is perhaps the most widely known, especially amongst plant pathologists in the UK. A. E. Dimond & J. G. Horsfall defined inoculum potential as the number of independent infections that are likely to occur in a given situation in a population of susceptible healthy tissues. This is considered to be the resultant action of the environment, the virulence of the pathogen, host susceptibility and the amount of inoculum used. Infection pressure may be considered synonymous with Dimond and Horsfalls' definition of inoculum potential.

The underlying concept of inoculum potential with all these authors has been to relate the infective activity of different pathogens under various conditions. The use of the term to describe the number of infective units (propagules) at the plant surface is clearly of little value since all viable inoculum is potentially infective and results obtained for one situation involving only a single variable would be quite different from those from another. Similarly it cannot be measured by counts of diseased plants unless all the conditions which could affect infection could be specified and eproduced; clearly a very difficult if not impossible task. The minimum quantity of inoculum required to produce infection in a given situation is better described in terms of numerical threshold of infection. Inoculum potential sensu Garrett, while recognising that different inocula have different potentials to infect, suffers from the disadvantage that the ' infective energy ' of a parasite cannot easily be measured. In some ways inoculum potential used in this sense is similar to virulence, but the latter can be considered as a genetic attribute of the pathogen independent of the number of infective propagules used or of modifying effects of the environment. The term as used by Garrett is valuable as a philosophical concept in plant infection even though it cannot be directly quantified.

Inoperculate Describes sporangia, esp. some unitunicate asci, which lack a specialized operculum or cap that opens to permit spore discharge.

Inorganic Containing no carbon; generally used to indicate materials (for example, fertilizers) that are of mineral origin.

Insect Six-legged animals in the Arthropod Phylum. Insects also have an exoskeleton (hard part is on the outside) and three main body sections (head, thorax, and abodmen). Most insects also have wings, though not all of them fly. The Insect Class includes bees, ants, beetles, butterflies, moths, flies, crickets, fleas, and many others.

Insect attractant A chemical that lures insects to a trap, thereby removing them from crops, animals or stored products.

Insect growth regulators (IGRs) Pesticides that kill or control insect pests by mimicking or interfering with the natural hormones that regulate the insect's developmental stages (e.g., egg, larva, pupa, adult). Examples: methoprene, which arrests the development of fleas at or before the pupal stage

Insect life cycle There are two basic kinds of life cycles in insects. Some (for example grasshoppers) have three phases: egg-nymph-adult. Others (for example wasps and flies) have four phases: egg-larva-pupa-adult.

Insect repellent A chemical that deters an insect from landing on a human or an animal.

Insecta One of the classes of animals in the Phylum Arthropoda. They have three body divisions and six legs, and may or may not have wings. The true insects.

Insecticide A material used specifically to manage or prevent damage caused by insects. Sometimes generalized to be synonymous with pesticides.

Fig. Insecticide

Insomnia Inability to sleep even in the absence of external impediments, during the period when sleep should normally occur.

Instantaneous rate In a short time interval (e.g a week), the number of events (e.g. births in the UK)taking place during the interval is approximately proportional to the length of the interval. The constant of proportionality is the rate at which these events occur. This argument becomes more and more exact as the length of the time interval becomes shorter and shorter. When the time interval has shrunk to an instant, the rate has become an 'instantaneous rate'. A bad term; just call it a rate.

Instar The insect itself during the time between molts in the larva or nymph, numbered to designate the various periods; i.e., the first instar is the insect between the egg and first molt.

Integrated control The complementary use of biological and chemical methods to control pests and pathogens.

Integrated pest management (IPM) An ecosystem approach to plant or insect problem management that determines if pest control is needed, when it should be initiated, where it should be applied, and what strategy and mix of tactics to use. It focuses on prevention of pest problems; utilization of the ecosystem's ability to control pests; establishment of levels of pests that can be tolerated; reliance, whenever possible, on nonhazardous biological and cultural controls; and minimized reliance on pesticides, avoidance of highly toxic pesticides, and utilization of pesticide applications that minimize damage to the natural controls and integrity of the ecosystem.

Integument The outermost layer(s) of an ovule which will develop into the seed coat; most seed plant ovules have one integument, angiosperm ovules have two integuments.

Intensity 1. (traditional) The mean parasite burden within all the infected members of the host population. Also called mean abundance.

2. (newer) The mean parasite burden within both infected and uninfected hosts.

It is important to indicate which usage is adopted, since they give different statistics, unless the prevalence is 100%. Macroparasites, and infections like malaria, are usually measured in terms of intensity.

Intensive farming A system of farming with the aim to produce the maximum number of crops in a year with a high yield from the land available and to maintain a high stocking rate of livestock.

Interaction Action (of viruses) mutual action between two or more agencies. This term has been employed by virus disease workers and the following classes of phenomena have been grouped under it: partial or complete suppression of one virus; synergistic association between two viruses; abnormal change in concentration of one virus.

Intercalary Formed or situated somewhere between apex and base of a given structure.

Intercalary Of growth, between the apex and the base; of cells, spores, etc., between two cells.

Intercellular Lying or growing between the cells.

Intercrop To grow two or more crops simultaneously on the same area of land.

Interference (of viruses) The action of one virus in modifying the usual effect of another virus. Such viruses are often related.

Internal transcribed spacers Regions of a DNA sequence that separate genes for certain ribosomal RNAs (rRNAs). These genes appear to be nearly identical among a wide variety of species, but the DNA spacers between them are quite variable and may be species-specific. This makes them good potential targets for "primers" for polymerase chain reaction (PCR) amplification that is used to identify a wide variety of organisms.

Internode The portion of a stem between two successive nodes.

Intersex An individual more or less intermediate in phenotype between male and female, displaying secondary male or female characters.

Interveinal Between (leaf) veins.

Intervention Under the EC price support mechanism for agricultural produce was set up in 1972. Its functions included licensing trade in a range of agricultural products outside the EC, intervention buying and the operation of other market support arrangements. Intervention buying or support buying was designed to ensure that farm incomes were not unduly suppressed by over production within the EC. When the free market price of specified commodities falls someway below the target price, the board intervenes in the market, by purchasing and storing the produce. The commodity is then subsequently resold within or outside the EC, disposed of as food aid or destroyed.

Intolerant Killed rapidly when invaded by a given pathogen. The use of this term is not recommended. It should not be used as a synonym of sensitive (BMS).

Intracellular Lying or growing within the cells.

Intramatrical spores Another name for the vesicles produced in host roots by most endomycorrhizal fungi.

Intraparietal Of position of pigment, on the inner portion of the wall, usually in the form of spirals, rings or irregular clumps.

Intraveinal Associated along or within veins.

Introduce When a new organism is brought to an environment where it did not live before. Introduced species can cause great problems for native species and for people.

Intron An intron, generically, is a piece of DNA which is transcribed into RNA but does not form part of the final gene product. In one fashion or another, it must be spliced out of the RNA before the RNA can be used for translation, ribosome formation, etc. Group I introns are large introns whose RNA product has an extremely complex secondary structure. The RNA itself (with or without cofactors such as nucleotides or metal ions) catalyzes its own removal from the initial RNA transcript. That is, it functions as a ribozyme, an enzyme made up of RNA. Most Group I introns then catalyze their own circularization. Mobile Group I elements may also encode a homing endonuclease, or use an endonuclease native to the host, to insert a DNA copy of the transcribed intron into the homologous location on a sister chromosome, or into another copy of the "home" gene in the case of genes present in multiple copies. The point of insertion is never random, and tends to be quite specific to a particular gene. A Spliceosomal Intron, at the opposite extreme, is a short (c. 40 bp) intron which serves as a substrate for splicing enzymes which remove the intron and probably also perform various regulatory roles, including adding the 3'-poly(A) "tail" which acts as a shipping label, specifying that the nuclear RNA is a messenger RNA which can be exported to the cytoplasm for translation. Various regulatory roles and one possible template for the evolution of the system. Group I and spliceosomal introns are likely to be related, with Group I introns gradually becoming simplified and more dependent on the host cell until they are eventually reduced to spliceosomal introns or disappear altogether.

Intumescence Blisters produced on leaves and other plant parts under conditions of high moisture and restricted transpiration.

Invagination Retraction, under force of pressure, of an outer surface toward the inside.

Invasion In plant pathology: Spread of a pathogen through tissues of a diseased plant.

Invertebrate An animal without a backbone. Invertebrates include insects, arachnids (spiders & ticks), gastropods (snails & slugs), crustaceans (crayfish & isopods), centipedes, worms, and others.

Involute with the edge rolled downwards or inwards.

Ion An atom or molecule that carries a positive or negative electrical charge due to losing or gaining electrons through a chemical reaction.

IPM A combination of strategies to reduce losses due to pests and pathogens based on environmental and economic considerations.

Iron deficiency damage Iron deficiency occurs commonly in certain Montana trees and shrubs such as maple, rose, and raspberry. Symptoms include yellowed leaves surrounding leaf veins that remain green. Although iron levels are sufficient in many soils, this iron may not be released to the plant in high pH soils.

Spraying a solution iron chelate on the leaves in the early season often provides a rapid improvement in the plant color. However, this foliar iron often must be reapplied again even within the same growing season. Iron can also be injected into the tree trunk for a longer-lasting effect. Soil applications of iron generally are not effective since the higher pH quickly renders this iron unavailable to the plant.

Irradiation Exposure to some form of radiant energy.

Isidia Simple or branched protuberances of the cortex in lichens, which may break off and act as a form of vegetative propagule.

Isogamete One of two sex cells (gametes) that are alike in form.

Isogamy The condition in which gametes are morphologically similar, as in the members of the Zygomycotina.

Isogenic Two or more organisms or cells having identical genotypes.

Isolate 1. To separate a microorganism from host or substratum and establish it in pure culture;

2. a single pure culture made by direct isolation from fresh material and any subcultures made from it. The first culture of the series may be distinguished as the primary isolate (3) and line.

Isolate, virus A virus that has been obtained from an infected plant by inoculation of a host or in procedures for separation of viruses or virus strains. Use of this term carries no commitment as to the purity or identity of the virus that is isolated, in contrast to its use in connection with fungi and bacteria.

Isolation 1. The process of separating a microorganism or virus from a substratum or host.

2. maintenance of healthy plant material at sites away from sources of inoculum; or maintenance of stocks of infected material at sites remote from potential hosts.

Isoline One line in a series of genetically similar plant lines that carry different specific genes for resistance to a particular pathogen.

Isometric Usually used for virus particles to describe those that are icosahedral in structure and appear approximately round.

Isotype Any duplicate of the holotype of a taxon.

Isozyme One of a number of enzymes that catalyse the same reaction(s) but differ from each other in primary structure and/or electrophoretic mobility. (Also isoenzyme.)

Isthmoid With a constriction.

Isthmus In nematodes: Relatively narrow portion of esophagus just anterior to the basal region.

ITS Regions of a DNA sequence that separate genes for certain ribosomal RNAs (rRNAs). These genes appear to be nearly identical among a wide variety of species, but the DNA spacers between them are quite variable and may be species-specific. This makes them good potential targets for "primers" for polymerase chain reaction (PCR) amplification that is used to identify a wide variety of organisms.

J

Jelly fungi Wood-inhabiting basidiomycetes with gelatinous basidiomata; Orders: Tremellales, Auriculariales (Phragmobasidiomycetes); Dacrymycetales (Holobasidiomycetes).

Jointing Elongation of rice internodes before flowering.

June bearing A term applied to short-day strawberry cultivars.

June census Annual farm returns on the 4th June each year provide the basis of MAFF published statistics regarding holding size enterprises, labour etc.

Juvenile An immature form that appears similar to but usually smaller than the adult and is not sexually mature (e.g. insects with gradual metamorphosis, nematodes)

Juvenile hormone A biochemical that occurs in insects and regulates their development. Can be used to control some insects by preventing larvae from developing into adults.

K

Karnal bunt Karnal bunt is a disease of the seed and symptoms become evident only when the grain fully develops. The pathogen converts the infected ovary into a sorus where a mass of dark brown colored teliospores are produced.

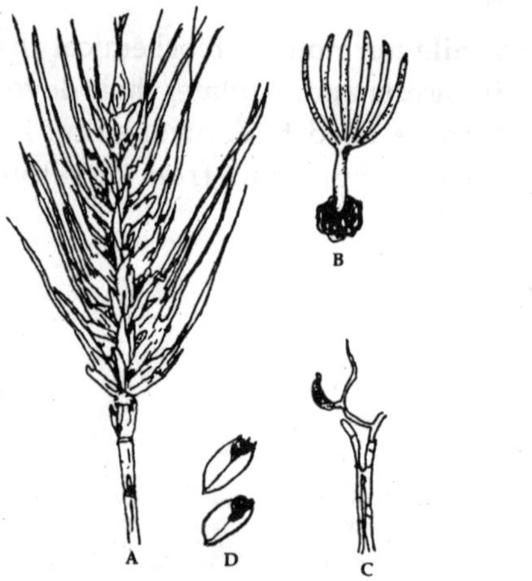

Fig. Karnal bunt of wheat (*Neovossia indica):* (a) infected ear, (b) germinating spore with sporidia (c) paired sporidia germinating to form secondary sporidium (d) partially infected grains.

If the host is vulnerable and the environment favorable, the entire grain turns into a sorus. During threshing the teliospores get released, and the severely infected grain gets blown off. In partially infected grain the embryonic tissue is not invaded and such KB infected seed do germinate and produce a weak plant that has poor survival chances. The infected grain emits a fishy odor due to trimethylamine and the wheat products from severely KB infected grain are unpalatable.

Karyogamy The fusion of two sexually compatible haploid nuclei after cell fusion, i.e. after plasmogamy.

Karyology The study of the behaviour of nuclei in eukaryotes.

Karyotype The chromosomal constitution of a eukaryotic cell in terms of the number, size amd morphology of the chromosomes at metaphase.

Katsuobushi A Japanese fermented food; cooked bonito fish is fermented by Aspergillus glaucus until dry; the product is shaved into ribbons and used for flavouring other foods.

KB (kilobases) An abbreviation for 1000 base pairs of DNA.

Keratin The protein that is the main component of skin, hair, feathers and horns.

Keratinophilic Capable of decomposing keratin, as are many of the fungi that cause superficial mycoses in man.

Ketjap An Oriental fermented food; Indonesian soy sauce; black soybeans are fermented for 2-3 days by Aspergillus oryzae; the root of the words Ketchup and Catsup.

Ketoconazole (nizoral) An antifungal antibiotic.

Kinetin 6 Furfuryl amino purine isolated from animal and plant DNA, capable of promoting cell division in plant tissues at 1 x 10-3 ppm.

Kinetosome Basal body; self-replicating organelle at the base of flagellum, similar to a centriole.

Kingdom The largest group for classification of living things (taxonomy). There are five kingdoms that scientists put all living things (organims) into: Animal Kingdom, Plant Kingdom, Fungi Kingdom, Protist Kingdom, Moneran Kingdom. Each kingdom is then divided into smaller groups called a phylum.

Kinins The original class name for substances which promote division in cells of various plant species, usually grown as callus in aseptic culture. The prefix cyto- (cytokinins) has been added to avoid confusion with a group of peptides, also called kinins, which cause muscular contractions in animal systems.

Klendusity A special kind of disease escape in which a susceptible plant avoids disease because of an intrinsic property of the plant itself that greatly reduces the chances of its being inoculated, even though there may be an abundance of inoculum in the area.

Klendusity The failure of a susceptible host to become infected, in the presence of the pathogen, because of qualities preventing or hindering the operation of a vector or other inoculating agent, e.g. a variety that is resistant to an aphid will not become infected with viruses for which the aphid is a vector.

Klendusity may be considered a form of disease escape but the term should not be applied to other forms of disease escape.

Ability of an otherwise susceptible variety of plant to escape infection because of the way it grows (e.g., early-maturing plants escape late-season diseases).

Knot A localized abnormal swelling; a gall.

Koch's postulates Three criteria were proposed by Koch for proving the pathogenicity of an organism; 1. the suspected causal organism must be constantly associated with the disease;

2. it must be isolated and grown in pure culture;

3. when inoculated into a healthy plant it must reproduce the original disease. A fourth criterion was appended by E. F. Smith.

4. the same organism must be reisolated from the experimentally infected plant.

Koji A 'starter' consisting of Aspergillus oryzae cultured on roasted wheat or barley (for hamanatto) or A. orzyae or A. soyae cultured on rice (for miso), which is then inoculated on the appropriate substrate.

Kombucha The following have been isolated from 'Kombucha tea' Bacterium xylinum, Bacterium xylinoides, Bacterium gluconicum, Saccharomyces ludwigii, Saccharomyces apiculatus varieties, Schizosaccaromyces pombe, Acetobacter ketogenum, Torula varieties, Pichia fermantans and other yeasts. In other words, this is a really mixed bag which has occasionally had very deleterious effects.

Krebs cycle The series of reactions that results in the oxidation of pyruvic acid to hydrogen atoms, electrons, and carbon dioxide; the electrons, passed along electron-carrier molecules, then go through the phosphorylation and terminal oxidation processes; also called the tricarboxylic acid cycle or TCA.

L

Label A printed statement affixed to the pesticide container by the manufacturer, listing the contents, directions for use, and precautions. A pesticide label must be approved and registered by the U.S. Environmental Protection Agency and the Utah Department of Agriculture.

Labial In nematodes: Pertaining to the lips.

Labiate Having lips; lobed.

Laccate Polished, varnished, shiny.

Lace bugs A large group of small insects (less then 5 mm in length) that have an elaborately sculptured dorsal surface as adults. The immatures are usually spiny. These bugs are plant feeders and their feeding causes a yellow spotting of the leaf. Some do considerable damage to trees.

Lacerate Appearing as if roughly cut or torn.

Lacewings There are several familes of lacewings. The adults have large wings with many veins, giving this group its name. The immatures are predaceous on many pest species and as such are important generalist natural enemies. The larvae of some species are called antlions and trashbugs.

Laciniate Of a margin, appearing as if cut into bands.

Lacrymoid Tear-drop shaped.

Lactate An ester of lactic acid produced during non-respiratory glucose combustion. It may accumulate during some disease conditions, leading to lactate acidosis.

Lactate dehydrogenase An enzyme important to the process of glucose combustion in the body, and an important mechanism for cellular energy production.

Lactiferous hypha A hypha that contains latex or is homologous to hyphae that contain latex.

Lacuna A gap or cavity.

Lageniform Swollen at the base and narrow at the apex; gourd-shaped; flask-shaped.

Lake A large body of water. Usually fed by several streams. Larger and deeper than a pond.

Lamella In agarics, A vertical, radiating plate covered with hymenium, on the ventral side of a pileus.

Lamellula In agarics, a short gill that does not extend all the way from the pileal margin to the stipe.

Lamina Expanded part of a leaf.

Lanceolate Shaped like the head of a lance, narrow and tapering towards the apex, or, sometimes, each end.

Lancet Any piercing mouth structure.

Land race Plant stocks selected by farmers on a local basis over many years.

Larva One of the phases in the life cycle of some insects, such as butterflies, wasps, and flies. Larva is the singular form of the word, and larvae is the plural form. A young insect which quits the egg in an early stage of development and differs fundamentally in form from the adult; e.g., a maggot is the larval form of a fly. Many insect larvae are very different in appearance from the adults. Insect larvae do not have wings. Insect larvae feed, grow, and molt several times.

Larvicide An insecticide that kills the larvae of insects.

Latent Producing no visible symptoms (generally refers to an infection or a pathogen).

Latent infection 1. Infection unaccompanied by visible symptoms.

Latent period 1. Elapsed time between infection and the appearance of disease symptoms.

2. the elapsed time between phage infection or induction and lysis of bacterial cells.

3. the period after acquisition of virus by a vector before it becomes infective

The time from infection to when the individual is infectious to others. In helminths it is termed the pre-patent period. Not the same as the incubation period.

Latent virus A virus that infects a plant without causing macroscopic symptoms.

Lateral At the side; of stipe attachment, at the side of the basidioma.

Lateral bud Bud that develops in the axil of a leaf.

Lateral field In nematodes: An interruption of the transverse striae by longitudinal cuticular thickenings situated on top of the lateral chords. The field is divided by longitudinal striae (incisures) and at times by transverse markings (areolation).

Lateral meristem Same as vascular cambium.

Lateral shoot A shoot originating from a vegetative bud in the axil of a leaf or from the node of a stem, rhizome or stolon.

Latex A hyaline or coloured juice exuding from an injured or cut portion of a fruit-body.

Latitude Angular distance, in degrees, minutes and seconds, from the equator.

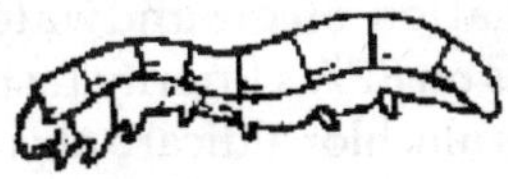

Fig. Larva

Layby application An application, usually of fertilizer or herbicide, after the crop is well established; especially, an application at the latest time in the season when it is still possible to pass through the field with a tractor.

LC50 Median lethal concentration; the concentration of a pesticide necessary to kill 50 percent of the test organisms. Defined in terms of a particular species of test animal. Usually expressed in terms of hours of exposure to parts per million (ppm) or milligrams per liter (mg/l) concentration of the pesticide. Used to quantify the toxicity of substances in water to fish or aquatic organisms, or the inhalation toxicity of airborne substances to animals.

LD50 Median lethal dose; the amount of a pesticide that is lethal to 50 percent of exposed test animals. This number is defined in terms of a particular species of test animal (e.g., rat, rabbit) and a particular route of exposure (e.g., dermal, oral). Usually expressed in milligrams of pesticide per kilogram of body weight (mg/kg.)

Pesticide active ingredient LD50 (oral, rat) (in mg/kg)

B.t. >10,000

Chlorpyrifos 82 - 270

Cyfluthrin 16 - 647

2,4-D 350 - 500

Diazinon 66 - 697

Glyphosate 4,320

Insecticidal soap >10,000.

Leaching The movement of a pesticide chemical downward through soil as a result of water movement, potentially causing contamination of groundwater. Selected Chemicals on EPA's Priority List of Leaching Pesticides alachlor aldicarb atrazine baygon chlorothalonil dacthal/DCPA methomyl picloram propoxur.

Leaching fraction The proportion of applied irrigation water that is added to meet the crop's leaching requirement.

Leaching requirement The amount of water in excess of a crop's evapotranspiration requirement that is needed to maintain maximum yield by leaching harmful salts from the root zone.

Leaf Small, flat, green part of a plant. Leaves are important because they use sunlight to turn carbon dioxide (a waste gas) into oxygen (the gas all animals need to breathe).

Leaf area index The ratio between the total leaf surface area of a plant and the surface area of ground that is covered by the plant.

Leaf base An interconnected system of parenchymous strands with many air spaces that extend throughout the vegetative organs of some arborescent lycopsids. Essentially, parichnos tissue is parenchyma with lots of air space between cells. This tissue probably aided photosynthetic gas exchange in much the same way as do the lenticels of modern trees.

Leaf curl Distortion resulting from unequal growth or expansion of leaf tissues, e.g. along the two sides of the midrib, or of the palisade and mesophyll layers.

Leaf dip Inspection of a sample of sap from a suspected virus-infected leaf using an electron microscope.

Leaf litter Dead leaves, bits of bark, and other dead plant matter. Leaf litter is the main shelter for many small animals, including insects, centipedes, and isopods.

Leaf margin The outer edge of the leaf; leaf margins may be smooth, lobed, indented, etc.

Leaf roll Curving of the leaf lamina, generally towards, and parallel with, the midrib.

Leaf scorch When adequate temperature and moisture suddenly turns hot and dry, leaf scorch may occur. Affected trees generally have yellow or brown leaf edges. The affected leaves may remain on the tree or the leaves may drop early. Watering during hot, dry weather may prevent or alleviate this problem.

Leaf sheath Lower, tubular part of a grass leaf that clasps the culm.

Leaf spot Dead or discolored areas on the leaf tissue that are fairly small and are usually circular. Picture: Impatiens Necrotic Spot Virus on gloxinia.

Leaf spot of turmeric In Turmeric, leaf spot and leaf blotch are the most important diseases. They effect the crop in its fast growing period and causes yield losses upto 60%.

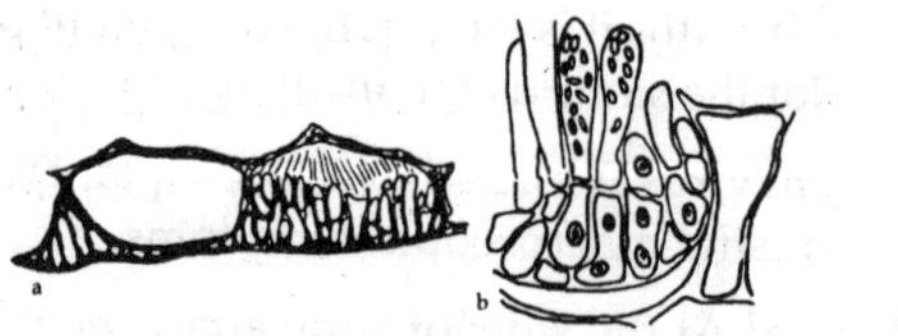

Fig. Leaf spot of turmeric caused by Taphrina maculans: (a) mycelium in the walls of epidermal cells (b) asci and ascospores.

Leaf symptoms Leaf discoloration or yellowing in localized or distinct patterns (viral).

Necrotic (dead) areas on leaves, often containing fruiting bodies (fungi)

Necrotic areas on leaves, often with water-soaked margins (bacterial)

Small rusty-red, brown or black spots and stripes (fruiting bodies of rusts and smut fungi)

Leaf distortion (elongated, dwarfed, thickened, etc.) (viral).

Leaf galls (fungi such as peach leaf curl and oak leaf blister, insects).

White, powdery substance on leaves (powdery or downy mildew).

Wilting (vascular wilt fungi, root rot pathogens, bacteria, drought).

Leafhopper Mobile insects with sucking mouthparts (order Homoptera).

Leaflet One of the separate blades or divisions of a compound leaf.

Leafminers Several very different insects can be leafminers: flies, beetles, or moths. In all cases it is the immature that bores through the leaf usually leaving a noticable trail. One beetle is a very serious pest of black locust, while one small moth is a serious pest of citrus.

Leafroll Disease symptom in which the edges of the leaf roll or turn up; often a symptom of virus infection.

Least toxic Some use the term to mean those chemicals which have the lowest acute toxicity as measured by an oral or dermal LD50. Others use it in a broader sense to denote those chemicals with the lowest acute and chronic toxicities and the least potential for disrupting the ecosystem in the short or long term.

Leathery wing Somewhat thickened but flexible front wings, found on grasshoppers and earwigs.

Lecanorine Lichen apothecia in which algal cells are found in the thickened margin of the fruit body. The margin is frequently the same colour as the thallus.

Lecideine Lichen apothecial ascomata with no algal cells in the margin.

Lectin A plant protein that binds to certain sugar residues in glycoproteins.

Lectotype One of a series of syntypes that, after publication of the original description, is selected and designated through publication to serve as the type.

Legitimate Of taxonomic names and epithets: Published in accordance with the Code of Nomenclature.

Legume The collective common name for a large family of dicotyledonous plants (peas, beans, clovers, soybean, etc.) that have irregularly shaped flowers, produce pods and fruit of a particular shape, and form nitrogen-fixing root or stem nodules in symbiosis with rhizobia.

Lenticels Natural openings in the surface of a tuber or stem, similar to leaf stomata, that can open and close and allow gas exchange.

Lenticular Shaped like a biconvex lens.

Lepidopterous Belonging to the insect order Lepidoptera. This order contains the butterflies and moths.

Leprose Describes a kind of lichen thallus in which there is no organized cortical layer, and the thallus tends to break up into a mass of powdery soredia (mixed algal-fungal propagules).

Leptoderan In nematodes: Caudal alae that do not meet posterior to the tail tip.

Leptoid Cells involved in solute storage and transport in Bryophyta. Analogous, possibly homologous, to the phloem of tracheophytes.

Leptosphaerulina sp Leptosphaerulina sp. Ascomycete found on leaves and herbaceous stems. Its asexual state is Pithomyces. Ascoma an ostiolate, immersed in host tissue, apez erupent at maturity. Asci few, bitunicate, thick-walled, 8-spored. Ascospores are colorless, variable in shape from oblong to ellipsoid or short cylindric, transverse and longitudinal septa, sometimes longitudinal septa lacking in some spores, with a thin gelatinous sheath, sometimes becoming brownish with age. Cultivation - Corn meal agar, 24°C.

Leptosporangium Sporangium developed from a single superficial cell. Leptosporangiate: having leptosporangia. This is a derived character state in the fern clade; compare to eusporangium.

Lesion A localised area of diseased or disordered tissue.

A local injury or delimited diseased area.

A wound; a well-marked, but limited, diseased area; a break or rupture through a tissue, especially a surface tissue.

A localized area of discolored, diseased tissue.

Lesion test A test to determine the ability of an organism to cause a necrotic lesion on plant material, e.g. bean pods, usually from a high inoculum dose. This test is not indicative of ability to infect in the field.

Lesion, primary Used in plant virology for lesions developing in inoculated leaves at the initial points of infection.

Lesion, starch A localised area of starch accumulation or depletion around the initial points of infection in virus-inoculated leaves. Such areas are revealed by staining killed decolourised leaves with iodine.

Leucocytes White (non-haemoglobin-containing) blood cells of many types, including phagocytic macrophages and antibody-producing lymphocytes.

LFA Less Favored Area

Liability A broad word for legal obligation, responsibility, or debt.

Liable Responsible for something (such as harm done to another person); having a duty or obligation enforceable in court against you by another person.

Library In molecular biology: A set of cloned fragments together representing the entire genome.

Lichen A dual organism in which a fungus (usually an ascomycete) maintains a green alga or a cyanobacterium captive within its thallus in a symbiosis that approaches balanced parasitism. Some lichens consist of three or more symbionts.

Life cycle The period of time that it takes for an organism to live its entire life. Some animals, such as insects, change their shape several times during their life cycle. These changes are called "metamorphosis."

Life expectancy Longevity, the average length of life of individuals in a population.

Lignicolous Growing on or in wood.

Lignification Hardening of tissue through the deposition of lignin in the cell wall.

Lignin A polymer of phenylpropanoid units that is an important constituent of wood. It is resistant to biodegradation by most organisms, but can be degraded by many basidiomycetes.

Lignin structures Lignin is formed by the removal of hydroxyl groups from sugars, creating phenolic compounds and short-chain alcohol ligands. Lignin polymers are heavily cross-linked. There is great variation in lignin, even within the same plant. Undoubtedly, it has evolved to have a somewhat random structure to foil enzymatic attack. Lignin is, in fact, extremely difficult to dissociate – so much so that even as late as 1980 almost nothing was known of the detailed chemistry of lignin, the second most common biopolymer on the planet. Fortunately, considerable progress has been made since then. The basic monomer of lignin can be thought of as 4-alkylcatechol. The alkyl ligand is a 1 to 3 carbon chain which may be substituted with hydroxy or keto ligands at any or all positions. Both phenol groups and the alkyl substitutions may cross-link to other monomers or their side chains, yielding a fairly light, but strongly cross-linked and randomly ordered mass, i.e., wood. What is more, the chemical structure virtually guarantees that any intermediate degradation product will be a highly toxic phenolic compound. Lignin was a rather early product of plant evolution. The necessary implication is that, contrary to most current thought on the subject, land plants were subject to biological attack (although not necessarily attack by herbivorous animals) by at least the Middle Devonian, and probably even earlier.

Lignituber A peglike proliferation of the host cell wall induced by, and ensheathing, a penetrating fungal hypha and apparently composed of lignin.

Ligulate Strap-like in shape.

Ligule In many grasses, a short membranous projection on the inner side of the leaf blade at the junction where the leaf blade and leaf sheath meet.

Likelihood Given some data (for example maternal antibody levels), and a model for how a process supposedly generating it occurs, (for example, exponential decay with rate m), and for how variance is likely to occur in observations of the process (for example sampling error), the likelihood that a particular model (ie a particular m plus the error in observation can generate the observed data.Picking the parameter value which is most likely to have generated the observed data is the method of maximum likelihood.

Limbate Edged with another colour; of a volva, adnate to the base of the stipe and having a narrow, free, membranous margin.

Lime Granular deposits of calcium carbonate found in the capillitium, peridium and stalk of some Myxomycetes, sometimes as large, irregular masses (lime knots).

Limestone A rock, mostly of carbonate, usually crushed and spread on soil to reduce acidity.

Limitation A time limit. For example, a 'statute of limitations' is a law that sets a maximum amount of time after something happens for it to be taken to court.

Limoniform Lemon-shaped.

Line 1. an inbred homozygous strain.

2. 'An assemblage of sexually reproducing individuals of uniform appearance propagated by seeds or spores, its stability maintained by selection to a standard '.

3. isolate.

4. a subdivision of a physiologic race.

In plant pathology the term should be used only in the sense of (2), as applying to host plants.

Lineage Any continuous line of descent; any series of organisms connected by reproduction by parent of offspring.

Linear A process is linear if doubling the starting conditions doubles the outcome. For example, the number of new schistosome infections establishing in mice is directly proportional to the number of cercaria the mice are exposed to, over a certain range. Outside this range, the response is nonlinear: there is no longer a direct proportionality.

Linear differential equations are ones whose solutions are linear. This property allows us to solve linear differential equations completely: most nonlinear differential equations can't be solved analytically.

Lipids Molecules containing a fatty acid with a long hydrocarbon chain, e.g., triglycerides (ester of a fatty acid plus glycerol), phospholipids (triglyceride containing 2 fatty acid chains and phosphate as a polar group).

Lips In nematodes: Cuticular structures (usually six: two subdorsal, two lateral, two subventral) surrounding the mouth opening; lips may be fused in pairs.

Lirellate Long and thin, as in the ascomata of some lichens.

Litter A group of young; usually mammals.

Lobe A rounded part that sticks out. Many leaves have lobes. Sometimes they look like fingers on a hand. You can often tell what kind of tree it is by counting the lobes on a leaf and looking at their shape.

Lobed margin (leaves) Margin indented one quarter or more of the distance from the margin to the mid-vein or (where this is lacking) to the long axis of the leaf; compare entire margin and toothed margin leaves

Lobulate With small lobes.

Local infection An infection affecting a limited part of a plant.

Local invasion That involving only a localized area of the plant.

Local lesion Small, restricted lesion, often the characteristic reaction of differential cultivars to specific pathogens, especially in response to mechanical inoculation with a virus.

Local lesion assay A quantitative estimation of the infectivity of a preparation of a pathogen, e.g. a suspension of bacteria or viruses, from the number of lesions produced in inoculated leaves. Half-leaves are the common experimental unit, and the leaves of an opposite pair differ less than leaves from different nodes: hence the terms half-leaf test and opposite leaf test.

Local lesion host A host which develops local lesions on inoculation with a virus.

Local necrosis The death or disintegration of cells and tissues in a localized area of an organ.

Localized dry spot A dry spot of sod amid normal, moist turf that resists rewetting by normal irrigation and rainfall. It is associated with many factors, including thatch, fungal activity, shallow soil over buried material, or elevated sites in the terrain.

Locational reference Tying data collected from a yield monitor, sensor or other sampling methods to a specific geographic position.

Locular Containing chambers or hollows.

Locule Generally, in botany, a locule is a small chamber in which a reproductive structure develops. In the ascomycetes, it refers to a small chamber in the hymenium (generalized reproductive tissue) in which a perithecium develops.

Locule, alternatively loculus A chamber, compartment or other limited volume which is used to accumulate or store either gametes or spores. Typically, the locule contains only gametes or spores. Thus a sporangium is a spore-producing organ, but a locule within the sporangium is where the spores are accumulated before release.

Loculoascomycetes A group of Ascomycotina with bitunicately discharging asci, producing ascospores that are generally septate and borne in unwalled locules (pseudothecia) in ascostromatic ascomata with an ascolocular ontogeny. Not accepted by some authors.

Locus 1. A specific location;

2. The position on a chromosome occupied by a particular gene, hence often used synonymously with gene.

Lodge To fall over, such as can happen with corn plants infested with rootworms.

Lodging Applies particularly to cereals where the crop has been flattened by the wind or rain, or if the stems are unable to support the weight of the grain.

Loose kernel smut Loose kernel smut, caused by the fungus Sporisorium cruentum (synonym Sphacelotheca cruenta) is less widespread than covered kernel smut. Loose kernel smut attacks all groups of sorghums, including johnsongrass, although certain varieties in some groups are immune or highly resistant. Sudangrass is usually not infected.

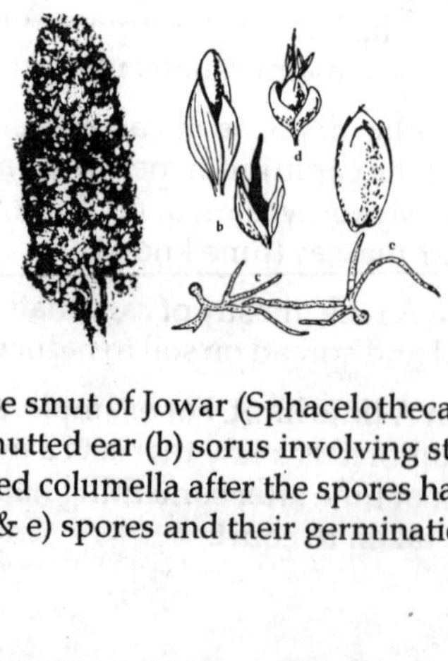

Fig. Loose smut of Jowar (Sphacelotheca cruenta) : (a) smutted ear (b) sorus involving stamens (c) exposed columella after the spores have fallen (d & e) spores and their germination.

Loose smut (Ustilago tritici) Loose smut is probably the most obvious of all wheat diseases. At flowering, all parts of the head and grain except the rachis (backbone) are replaced by black spore masses. This disease occurs throughout the world and reduces yields proportionate to the incidence of smutted heads. The disease does not cause complete yield loss, but losses as high as 27% have been reported, and more commonly, losses of 15% are recorded.

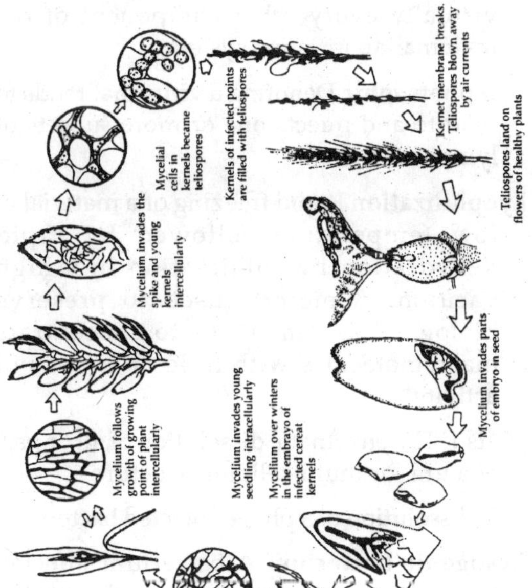

Fig. Life cycle of loose smut of wheat (Ustilago tritici).

This disease is more prevalent in Western Australia than the rest of Australia and especially in barley. In contrast to seed borne bunts, loose smut has little effect on seed quality. The loose smut fungus survives as a dormant fungal thread inside the embryo of wheat seed. The pathogen is activated when the infected seed germinates, and it extends toward the growing point of the plant. Evident from flowering onwards when the plant begins to form the head, the fungus invades all of the young head tissue except for that of the rachis (backbone). Production of plant growth hormones by the fungus results in infected plant heads reaching flowering earlier than healthy heads. The head produced by the infected plant contains black spore masses in place of the grain. The spores are loosely held and are easily spread by wind onto neighboring healthy plants. Because flowering of infected heads occurs earlier than healthy heads, production and release of spores occurs when the rest of the crop is flowering. Spores are blown by the wind into the flowers of the healthy plants. The spores enter the ovaries and become part of the developing grain. In this way, seed for the following year becomes infected.

Longicorn An insect having antennae as long as or longer than the body; specifically belonging to the family Cerambycidae.

Longitude Angular distance, in degrees, minutes and seconds, from the prime meridian which extends from the North to the South Pole and passes through Greenwich, England.

Longitudinal study A study taking place over time. If individuals are followed, this is a longitudinal cohort study. If individuals are not followed, but classes (usually age classes) are restudied, this is a longitudinal cross-sectional study.

The converse of a horizontal study.

Look alikes Mushrooms that closely resemble and may be mistaken for other mushrooms.

Looper A caterpillar in which some or all of the middle abdominal prolegs are wanting and which moves by placing the posterior part of the abdomen next to the thorax, forming a loop of the intervening segments, then extending the anterior part of the body forward.

Lower fungi Members of the Myxomycota, Oomycota, Hyphochytriomycota, Chytridiomycota and Zygomycota.

Lowland farming Farming on low lying land as distinct from uplands, usually more intensive

LSD (lysergic acid diethylamide) A powerful hallucinogen derived from Claviceps sclerotia.

Lubricous Slippery, of an oily smoothness.

LUCA (last universal common ancestor) The last common ancestor of all extant species: Eubacteria, Archaea, and Eukarya.

Lumen In nematodes: Triradiate canal or duct of the esophagus. In fungi: The space bounded by tissue or wall, as the central cavity of a cell.

Lunate Crescent-shaped; half-moon shaped.

Lungs Body organs used by animals to breathe oxygen from the air. Animals with lungs must be able to breathe air, although some can hold their breaths under water for a long time.

Luteovirus Member of a group of plant viruses with isometric particles containing one molecule of linear RNA, mainly confined to the phloem, and usually not mechanically transmitted but transmitted in nature by aphids in a circulative manner.

Lygodium annulus Specialized structure of some fern sporangia that is involved in the opening of the sporangium. In some mosses, on drying out the cells of the annulus contract and the sporangium ruptures, releasing the spores. In others, the annulus bears the peristome teeth, which perform a similar function.

Lyme disease A tick-transmitted inflammatory disorder that begins with a characteristic skin rash, and may be followed weeks to months later by neurologic, cardiac, or joint abnormalities.

Lymph node Secondary immune organs distributed at discrete locations throughout the body. These organs play a central role in the activation and trafficking of immune lymphocytes in the body.

Lymphocyte Small white blood cells that are uniform in appearance, but very diverse in function. Collectively, they are responsible for antibody production, direct cell-mediated killing of virus-infected cells and tumor cells, and for the regulation of virtually every other component of the mammalian immune system.

Lymphotrophic Denoting a virus that tends to bind to and infects one or more subsets of lymphocytes.

Lyophilization Rapid freezing of a material at low temperature followed by rapid dehydration by sublimation in a high vacuum. A method used to preserve biological specimens or to concentrate macromolecules with little or no loss of activity.

Lysis 1. The enzymatic dissolution of all or part of a uni- or multi-cellular structure.

2. dissolution of a phage-infected bacterium.

Lysogenic bacterium A bacterium infected with phage, which is integrated with the bacterial genetic material. All progeny are likewise infected and have the capacity to produce phage (by occasional cells undergoing lysis spontaneously or after application of inducing agents, e.g. ultraviolet light). Lysogenic bacteria are not sensitive to the phage they carry.

Lysogenization The infection of a bacterium by phage to produce a lysogenic clone or subclone of descendants. Lysogenization occurs some time after, or some bacterial generations after, the initial infection.

M

Macerate To soften by soaking; to cause disintegration of tissues, etc., by separation of cells, e.g. by microbiological action, solutions of pectolytic enzymes, chelating solutions such as ethylenediaminetetraacetic acid (EDTA), etc. The term is frequently used incorrectly instead of comminute.

Macroconidia The larger of two types of conidia produced by certain fungi, such as Fusarium spp.

Macroconidium The larger, and generally diagnostic conidium of a fungus which also has microconidia.

Macrocyclic Describes rust fungi which produce all 5 developmental stages — basidiospores, spermatia, aeciospores, urediniospores, and teliospores.

Macrocyst In Myxomycota, An encysted aggregate of myamoebae; the resting form of a young plasmodium; the alternative to the sorocarp in some cellular slime moulds (Dictyosteliomycetes).

Macronutrients Inorganic chemical elements required in large amounts for plant or fungal growth, e.g., nitrogen, carbon, potassium, calcium, phosphorus, magnesium, and sulphur.

Macroparasites Typically, the parasitic helminths and arthropods. In general, parasites which do not multiply within their definitive hosts but instead produce transmission stages (eggs and larvae) which pass into the external environment. Immune responses elicited against macroparasites generally depend on the number of parasites present in a given host and tends to be of a relatively transient nature. The key epidemiological measurement is generally the number of parasites per host.

Such parasites are often found in a highly aggregated distribution.

Macroscopic Visible to the naked eye, without the aid of a microscope.

MAFF Ministry of Agriculture Fisheries and Food.

Maggot Larvae of the higher diptera, especially of some families related to houseflies.

Magic mushrooms Typically, hallucinogen-containing species of Psilocybe, but also spp. of Gymnopilus, Panaeolus, Conocybe, and Amanita muscaria.

Magnetic resonance imaging The use of nuclear magnetic resonance of protons to produce cross-sectional proton density images of internal structures of the human body.

Malabsorption syndrome Syndromes resulting from impaired absorption of nutrients from the bowel.

Malaise A feeling of general discomfort or uneasiness, an out-of-sorts feeling, often the first indication of an infection or other disease.

Mammal repellent A chemical that deters mammals from approaching or feeding on crops or stored products.

Mammal Warm-blooded, usually hairy animals from the Chordate Phylum. This class of animals breathes air, gives live birth, and feeds milk to their young. Human beings are members of the Mammal Class, as well as dogs, cats, deer, mice, squirrels, raccoons, bats, opossums, and others.

Mammillate Breast-shaped; apex conical with a papillate tip.

Mandible The first pair of jaws in insects. These are stout and jawlike in chewing insects, or needle- or sword-shaped in piercing-sucking insects

Mannitol A polyhydric alcohol, often found as a storage compound in ectotrophic mycorrhizal mantles.

Manoxylic Wood type that contains abundant parenchyma; typical of cycads; compare with pycnoxylic.

Mantle A compact layer of hyphae enclosing short feeder roots of ectomycorrhizal plants; connected to the Hartig net on the inside, and to the extramatrical hyphae on the outside; acts as a sink for nutrients.

Map datum Standardized system for map production which allows one to reliably use any map produced under that system. More than 100 different map datums are currently in use.

Marginal veil In agarics, An incurving proliferation of the pileal margin which protects the developing hymenium.

Marginate 1. With a well-defined edge;

2. of a lamella or stipe base, where the lamella-edge is differently coloured to the face; where a bulbous stipe base is well-marked by a distinct rim.

Marker-assisted selection A procedure used in plant breeding in which genetic markers that are (a) easy to identify and (b) linked to desirable genetic traits that are difficult to identify (such as disease resistance) are used to aid in selection from a population.

Marsh An area with soft, wet land (also called "wetland"). Marshes are very important because they help clean polluted water, and because many animals and plants live there.

Masked symptoms Virus-induced plant symptoms that are absent under some environmental conditions but appear when the host is exposed to certain conditions of light and temperature.

Masked virus One carried by a plant which does not show symptoms of its presence.

Mass action transmission Transmission of infection which occurs at a rate directly proportional to the number or density of both susceptibles and infecteds present.

Some authors reserve the name mass action for transmission processes of the form b X Y/N, which we associate with STD-type transmission, and describe transmission rates of the form b X Y, as pseudo-mass action; the two are equivalent if the population size is unchanging.

Mass median diameter (MMD) The figure dividing a total volume of spray into two equal parts: one half of the mass of spray is contained in droplets of smaller diameter than MMD, and the other half is contained in droplets of larger diameter.

Mastigoneme In a tinsel flagellum, Flimmer; tinsel; a fine hair-like projection that extends laterally from undulipodia.

Mastigote Of motile cells, with one or more flagella; mastigote stage a stage in the life cycle which has flagella.

Material safety data sheet (MSDS) An information sheet provided by a pesticide manufacturer describing chemical qualities, hazards, safety precautions and emergency procedures to be followed in case of a spill, fire or other emergency.

Maternal immunity Immunity for a neonate provided by IgG antibody generated by a mother and passed across the placenta to the unborn offspring. This provides short lived protection (with a typical half life of 3-6 months) to the neonate. Also known as immunity.

Maternal inheritance (also cytoplasmic or extrachromosomal inheritance) Inheritance of non-nuclear genes, i.e. those in mitochondria and chloroplasts.

Mathematical model A formal framework to convey ideas about the components of a host-parasite interaction.

Construction requires three major types of information: (a) a clear understanding of the interaction within the individual host between the infectious agent and the host, (b) the mode and rate of transmission between individuals, and (c) host population characteristics such as demography and behaviour.

Mathematical models can aid exploration of the behaviour of the system under various conditions from which to determine the dominant factors generating observed patterns and phenomena. They also aid data collection and interpretation and parameter estimation, and provide tools for identifying possible approaches to control and for assessing the potential impact of different intervention measures.

Mating disrupter A chemical that interferes with the way that male and female insects locate each other using airborne chemicals (pheromones), thereby preventing them from reproducing.

Mating types Compatible strains, usually designated + and - or A and B, necessary for sexual reproduction in heterothallic fungi.

Matrix The material in which an organism or organ is embedded.

mature plant resistance.

Matsu take Tricholoma matsutake; An important edible fungus in Japan; grows in ectomycorrhizal association with pines. Tricholoma magnivelare, the North American pine mushroom, is also highly prized.

Matter A type of substance. For example, "plant matter" is material made of plants or plant parts.

Mazaedium A dry, powdery mass of mature ascospores, sometimes mixed with sterile threads, formed on ascomata in some Caliciales and Onygenales (Ascomycota).

Meadow Another name for field, especially one growing wild.

Mealybugs These insects of the family Pseudococcidae are plant feeders. The name is derived from the mealy or waxy secretions that covers the bodies of these insects. Some are important plant pests.

Mechanical control Control of pests by physical means such as the use of screens or row covers.

Mechanical injury Injury of a plant part by abrasion, mutilation, or wounding.

Mechanical inoculation Of plant viruses, a method of experimentally transmitting the pathogen from plant to plant; juice from diseased plants is rubbed on test-plant leaves that usually have been dusted with carborundum or some other abrasive material.

Mechanical transmission Transmission of a virus from an infected host to a healthy one without the intervention of a vector, either by physical contact and friction of one plant against the other or experimentally, e.g. by rubbing sap from an infected plant on the leaves of a healthy one.

Meconium Fecal pellet excreted by a larva before pupation.

Median effective concentration (EC 50) The concentration of a material required to bring about a prescribed effect, e.g. retardation of growth, in 50% of the individuals in a population of a given species.

Median effective dose (ED 50) The dose of a material required to bring about a prescribed effect in 50% of the individuals of a population of a given species.

Median latent period (LP 50) The time taken for 50% of a population of viruliferous vectors to complete their latent period.

Median lethal concentration (LC 50) The concentration of toxicant required to kill 50% of a large group of individuals of one species.

Median lethal dose (LD 50) The dose of a toxicant killing 50% of a large group of individuals of one species.

Medium A mixture of organic and/or inorganic chemical compounds and water that provides the nutrients needed for the growth of a microorganism in vitro; for higher plants, a mixture of fertilizers and other components in which a plant is growing.

Medium culture A substance or solution for the culture of microorganisms. Defined medium – of a prescribed composition, used for determining the biochemical capabilities of the organism, e.g., auxotrophs; complete medium – containing all nutrients required for growth; minimal medium – the simplest chemically defined medium on which the wild type (prototroph) of a species will grow and which must be supplemented by one or more specific substances for the growth of auxotrophic mutants derived from the wild type; selective medium medium containing certain chemical components which restrict the growth of some microorganisms but encourages the growth of others.

Medulla Central part of an organ.

Megagamete The "female" gamete, analogous to an egg and sometimes referred to as an ova.

Megagametophyte In heterosporous plants and in seed plants, the female gametophyte produced by a megaspore.

Megaphyll leafmegaphyll A leaf with branched veins and a flat area for receiving sunlight.

Megasporangium A sporangium that produces megaspores.

Megaspore A large, haploid (N=1) spore of a heterosporous plant that produces a megagmetophtye (female gametophyte)

Meiosismeiosis A process common to almost sexual reproduction in eukaryotic cells. The homologous chromosomes of a diploid nucleus first exchange homologous genes (alleles) on a roughly random basis, so that the resulting chromosomes carry a mixture of genes from each parent. The nucleus then divides normally (by mitosis) to yield two diploid daughter nuclei. Finally, the nuclei divide again, but now without DNA replication, to yield four haploid cells.

Meiosporangium A thick-walled diploid sporangium of certain Blastocladiales producing uninucleate, haploid zoospores (meiospores); more recently, used much more widely for any sporangium (e.g. ascus, basidium) in which meiosis results in spore production.

Meiospore A uninucleate, haploid spore arising directly by meiosis.

Meixner test For amatoxins: 1. express fluid from agaric onto newsprint,

2. dry,

3. add a drop of concentrated hydrochloric acid,

4. blue colour developing in 1-20 mins. indicates presence of amatoxin; higher levels produce colour sooner. Not always reliable, in my experience.

Melanin Brown-black pigment; common in sclerotia and other survival structures.

Melanism Brown-black pigment.

Melting out A disease (primarily of Kentucky bluegrass) caused by the fungus Drechslera poae. It is characterized by the appearance of dark leaf spots, often followed by the killing of irregular areas of turf.

Melzer's reagent Used to elicit amyloid or dextrinoid reactions in spores, asci, hymenial tissues, etc.; chloral hydrate 100 g, potassium iodide - 5g, iodine - 1.5 g, distilled water - 100 mL.

Membranous Parchment-like or membrane-like.

Memnoniella sp Memnoniella sp. Cellulolytic fungus very closely related to Stachybotrys. Both fungi have a worldwide distribution and often found together, and commonly found in soil. Recent studies on mycotoxins revealed that Memnoniella echinata can have toxicity similar to that of some isolates of S. chartarum. In terms of their chemical products, both S. chartarum and M echinata produce phenylspirodrimanes, but these two organisms differ in that the former produces macrocyclic and trichoverroid trichothecenes and the latter produces griseofulvins. Both produce varying amounts of simple trichothecenes. Thus, it is suggested that Memnoniella should also be considered potentially dangerous in indoor air. The conidiophores are dark, simple, bearing at apex a cluster of thick, short phialides; conidia of Memnoniella echinata are very similar to those of Stachybotrys, dark, 1-celled, globose. The major difference between the two fungi is that the conidia are in long persistent chains (aggregated in slimy heads in Stachybotrys). Also the aerodynamic diameter of Memnoniella is smaller and it would be expected to have an even greater potential to penetrate deep into lungs than the conidia of Stachybotrys. Cultivation - Corn meal agar, 24°C.

Merge To combine two or more data sets into one usable set.

Meristem A plant tissue containing cells that are actively dividing (or have the potential to do so). Meristem found in the tip of roots and stems, the apical meristem, is responsible for the growth in length (primary growth) of these organs. The cambium is a lateral meristem that is responsible for increase in girth (secondary growth) in perennial plants. Some plants also have intercalary meristem, as in the stems of grasses, for example. These are responsible for their continued growth after cutting or grazing has removed the apical meristem of the shoots. Tiscali.

Meristem culture Aseptic in vitro culture of a plant or plant part from a portion of the meristem; a method used to produce pathogen-free plants.

Meristem tip The meristem dome and first pair of leaf primordia which is the usual unit (0.1-0.5 mm long) excised for aseptic culture, e.g. in attempts to produce virus-free plants. Sometimes referred to erroneously as the meristem, or ambiguously as the shoot tip.

Merosporangium In Zygomycota, a cylindrical sporangium from the swollen end of a sporangiophore in which a single row of sporangiospores is produced; occasionally with only one sporangiospore.

Mesoamerica The region of Mexico and Central America that is one of several possible centers of domestication for Phaseolus vulgaris. Bean varieties from this area tend to be small seeded compared to those from the Andean region of South America.

Mesoendemic A term from the malaria literature used to mean (roughly) an area with some transmission.

Mesoperidium The middle layer of a multilayered peridium.

Mesophyll Parenchymal tissue between the upper and lower epidermis of a leaf.

Mesothorax The second or middle segment of the thorax; it bears the middle legs and the front wings (if present).

Mesozoic The geological era comprising the Triassic, Jurassic and Cretaceous period.

Messenger RNA A chain of specific ribonucleotides that codes for a specific protein; template for the assembly of amino acids into protein; in cells mRNA is transcribed from DNA, but some RNA viruses function directly as mRNA.

Metabasidium The developmental stage of the basidium in which meiosis (or nuclear division) occurs. If a probasidial remnant is left, it may be separated from the metabasidium by a septum.

Metabolism The biochemical processes by which a particular substance is broken down by an organism.

Metabolite Any chemical participating in metabolism; a nutrient.

Metachromatic A reaction where certain spore walls and hyphae turn reddish to violet when placed in cresyl blue.

Metacorpus The swollen posterior portion of the corpus; sometimes called the median bulb.

Metamorphosis The "cycle of change." Some animals go through metamorphosis, or stages of a life cycle. With insects, most usually go through a four stage cycle (1 = egg, 2 = larva, 3 = pupa, 4 = adult). This is called complete metamorphosis. For other insects, incomplete metamorphosis is when the cycle only has three stages (1 = egg, 2 = nymph, 3 = adult). Aquatic insects go through incomplete metamorphosis because life is too difficult in the water to have a pupa (resting) stage. Besides insects, other animals that go through metamorhposis include frogs (egg, tadpole, adult) and some salamanders. Ticks have their own metamorphosis (egg, larva, nymph, adult).

Metaplasia Changed condition of a structure or organ; hyperplastic class of symptoms characterized by overdevelopment other than that due to hypertrophy or hyperplasia. (e.g., abnormal starch accumulation, virescence, etc.).

Metathorax The third segment of the thorax; it bears the hind legs and the second pair of wings (if present); sometimes hard to see as it is either closely united with the mesothorax or appears to be part of the abdomen.

Metazoans Multicellular animals.

Metula A branch of a conidiophore bearing phialides.

Metulae Special cells at the apex of a complex conidiophore which support conidiogenous cells, as in many species of Penicillium, Leptographium.

Mica A mineral consisting of shiny, transparent, platelike chemical crystals.

Micaceous Covered with glistening particles.

Microbial Pertaining to or relating to microbes, or microorganisms.

Microbial pesticides Pesticides whose active ingredients are living microorganisms (e.g., bacteria, fungi, algae, viruses, parasitic nematodes) or the toxins that they produce. Usually highly selective in their action. Most must be consumed by the pest organisms in order to be effective. With a few exceptions (e.g., milky spore disease), microbial pesticides must be applied repeatedly to have continued effectiveness - they usually do not become permanently established in the environment as natural predators or parasitoids can. Examples: B.t., milky spore disease, steinernema feltiae (nematode), nuclear polyhedrosis virus.

Microbody Cellular organelle bounded by a single membrane, and containing enzymes; derived from endoplasmic reticulum.

Microclimate Weather conditions on a small scale, e.g. at the surface of the plant or within a crop.

Microconidia The smaller of the two types of conidia produced by certain fungi.

Microconidium The smaller of two kinds of conidia formed by certain fungi.

Microcyclic Of rust life cycle, Producing only telia and sometimes pycnia; not producing all five spore states.

Microcyst In Myxomycota, An encysted myxamoeba or swarm spore.

Microflora The combination of all microorganisms in a particular environment.

Microgamete The "male" gamete, analogous to sperm.

Microgametophyte In heterosporous plants and in seed plants, the male gametophyte produced by a microspore.

Microgram One-millionth of a gram: written as 'g'.

Micron A millionth of a meter (or, a thousandth of a millimeter).

Micronutrients (microelements) Elements such as zinc, boron, iron and molybdenum that are essential for plant and microbial growth, but are only needed in very small amount. In the case of molybdenum, this may only be ounces per ha.

Microorganism An organism of such small size that it can only be seen as an individual organism with the aid of microscope.

Microparasites Typically, viruses, bacteria, fungi and protozoa. More generally, parasites which multiply within their definitive hosts. Microparasites are characterized by small size, short generation times, and a tendency to induce immunity to reinfection in those hosts that survive. The duration of infection is usually short in relation to the lifespan of the host, but there are important exceptions, such as the slow viruses. The key epidemiological variable, by contrast with macroparasites, is whether or not the individual host is infected.

Micropores Extremely narrow connectives penetrating the septa of some mycelial yeasts, e.g. Geotrichum.

Micropropagation Generation of new, disease-free potato plants from tiny pieces of meristem tissue.

Micropyle In nematodes: A minute opening in the membrane of an egg through which the spermatozoa enter.

Microsclerotia Very small sclerotia, such as those produced by the Verticillium wilt fungus.

Microsclerotium A small clump of dark-colored, more or less thick-walled cells, each of which is viable; produced in culture and rarely in the xylem of host plants.

Microscopic Very small; seen only with the aid of a microscope.

Microsporangium A sporangium that produces microspores.

Microspore A small, haploid (N=1) spore of a heterosporous plant that produces a microgametophyte (male gametophyte).

Microsymbiont The bacteria associated with legumes in the formation of nitrogen fixing nodules.

Microtubule Microtubules are protein structures found within cells. They are generally long and form a structural network (the cytoskeleton) within the cell's cytoplasm, but in addition to structural support microtubules are used in many other processes as well. They form a substrate on which other cellular chemicals can interact, they are used in intracellular transport, and are involved in cell motility. The assembly and disassembly of microtubules into their subcomponent tubulin is one way in which cells can change their shape. A notable structure involving microtubules is the mitotic spindle used by eukaryotic cells to segregate their chromosomes correctly during cell division. Microtubules are also responsible for the flagella of eukaryotic cells (prokaryote flagella are entirely different). Microtubules are straight, hollow cylinders have a diameter of about 25 nm are variable in length but can grow 1000 times as long as they are thick. Microtubules are built by the assembly of dimers of alpha tubulin and beta tubulin. Microtubules grow at each end by the polymerization of tubulin dimers (powered by the hydrolysis of GTP), and shrink at each end by the release of tubulin dimers (depolymerization). However, both processes always occur more rapidly at one end, called the plus end. The other, less active, end is the minus end. Microtubules participate in a wide variety of cell activities. Most involve motion. The motion is provided by protein "motors" that use the energy of ATP to move along the microtubule. From The Cytoskeleton.

Middle lamella Pectin-rich layer between the walls of adjacent plant cells; this layer is attacked and dissolved by damping-off fungi.

Middorsal In the middle of the upper side or dorsum.

Midrib The central vein of a grass leaf, extending from the stem to the leaf tip.

Migrate To move from one area to another. Sometimes animals, especially birds, move to a new area when the seasons change. Usually birds migrate south for the Winter, then back north for the Summer. This is so they can survive at a comfortable temperature and have plenty of food.

Migratory Migrating from place to place on a plant or from plant to plant when feeding.

Mildew 1. in popular use, a visible growth of a mould on a plant surface or other substrate, though in the case of food products the term mould is commonly used.

2. in professional colloquial use by plant pathologists, mildew is usually synonymous with powdery mildew, a growth on a plant surface by a member of the Erysiphales; a disease so signalized, or a powdery mildew fungus.

Downy mildew, refers to infection by a member of the Peronosporaceae; a disease so signalized,or a downy mildew fungus. Dark mildews, is caused by one of the Meliolaceae.

Milk stage The early stage of grain development when the grain is filled with a milky liquid.

Milligram One-thousandth of a gram: written as 'mg'.

Mills period The period of apple leaf wetness satisfying requirements for leaf infection by Venturia inaequalis ascospores at a stated temperature.

Mimic 1. A species that looks very similar to another species. For example, the Viceroy butterfly looks a lot like the Monarch butterfly. Monarchs are poisonous to most predators, but Viceroys are not. Since the Viceroy is a mimic of the Monarch, it gets protection from predators.

2. To copy or behave like another species. Some birds, such as mockingbirds and starlings, mimic the songs of other birds.

Mimicking symptoms Symptoms similar to those caused by pollutants but induced by other abiotic or biotic causal agents.

Miner The larval stage of an insect which makes galleries or burrows between the upper and lower surfaces of leaf tissue.

Mineral oils Synonymous with horticultural oils.

Mineral soil A soil consisting predominantly of, and having its properties determined by, mineral matter.

Mineralisation The conversion of bound organic nitrogen into the mineral (ionic or inorganic) form required for plant uptake. The inorganic nitrogen is then converted in a process known as nitrification from ammonium salts to nitrites, and from nitrites to nitrates by nitrifying bacteria which obtain their energy from this oxidation process. Once free nitrate is formed, the rapid recycling process offers many options, the nitrate may be assimilated by plants, denitrified or leached.

Minimal medium A type of culture medium lacking specific growth factors; it does not support the growth of some or all auxotrophic strains of a given organism but permits the growth of prototrophic strains.

Minituber A small tuber produced under greenhouse conditions on a small potato plant generated by micropropagation.

Minnow Any of many species of small fish. Minnows are an important food source for larger fish and other animals. They are often used as bait for fishing

Miridae A family of abundant true bugs commonly called plant bugs or leaf bugs. These are mostly plant feeders, but some are predaceous on other insects.

Miso An Oriental food product, used for soups and as flavouring agent, composed of rice and cereals + soybeans fermented by Aspergillus oryzae and Saccharomyces rouxii.

Mites These arthropods are not insects, but are in the same order as ticks. They are very small and require a strong hand lens to see. Most are plant feeders and can do considerable damage. Some are important predators of other spider mites.

Miticide A pesticide intended to kill mites.

Mitochondria Intracellular organelles concerned with the Krebs cycle and electron transport: the chief source of ATP in non-photosynthesizing cells.

Mitochondrion An organelle responsible for most of the oxidative metabolism in the cell.

Mitosis A sequence of cellular events that culminates in the division of a eukaryotic nucleus into two genetically similar or identical nuclei whose ploidy is the same as that of the parent nucleus. Mitosis occurs during asexual cell division.

Mitosporangium A thin-walled diploid sporangium of certain Blastocladiales producing uninucleate diploid zoospores (mitospores); more recently used more widely for sporangia containing spores produced by mitosis.

Mitospore A spore receiving a nucleus or nuclei resulting from mitotic division; characteristic of asexual reproduction.

Mitosporemitosporangium A fungal organ which develops spores by mitosis, i.e. asexual spores.

Mixed grazing More than one type of animal grazing the same area at the same time

MLO Archaic term for phytoplasma; plant-parasitic pleomorphic mollicute (prokaryote with no cell wall) found in phloem tissue; cannot yet be grown on artificial nutrient media.

Model A simplified representation of a system.

Modify environmental factors Factors such as moisture and heat, and, in the case of certain organic materials that decay, to gradually improve soil quality. Plant derived in (organic) or synthetic materials may be used.

Mold Any microfungus with conspicuous, profuse, or woolly superficial growth (mycelium and/or spore masses) on various substrates; especially an economically important saprobe; molds commonly grow on damp or decaying matter and on the surface of plant tissues.

Mole cricket (gryllotalpidae) A family of insects related to crickets and grasshoppers. In this family the nymphs and adults are adapted for digging. There are three major developmental phases: egg-nymph-adult. There is no pupal phases. The nymphs molt six to eight times during their growth. Some of the genera included are Gryllotalpa, Nemocurtilla, and Scapteriscus.

Mollicute A proposed trivial name for any member of the class Mollicutes. The use of this name could permit "mycoplasmas" to be used specifically for members of the genus Mycoplasma.

Mollicute One of a group of prokaryotic organisms bounded by flexuous membranes and lacking cell walls.

Molluscicides Chemical substances which kill snails or other molluscs.

Mollusk An animal in the Mollusk Phylum. Mollusks have soft bodies, which are usually inside a shell. Most of them live in water. This phylum includes clams, mussels, snails, and slugs.

Molt To shed a coat (fur), feathers or skin. In the Summer, mammals molt their winter coats. Birds often molt old feathers for new ones. Insect larvae and crayfish molt their exoskeletons (outer skin or shell) for newer, larger ones.

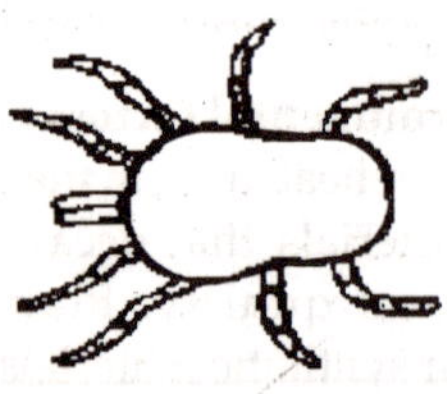

Fig. Mite

Monilia sp Monilia sp. Its mycelium is white or gray, and abundant in culture. Conidiophore branched; conidia pink, gray, or tan in mass, 1-celled, short cylindric to rounded, in acropetalous chains. Cultivation - Potato dextrose agar, 24°C.

Moniliform Contracted at short, regular intervals like a string of beads.

Monilioid Describes closely-septate hyphae with swollen cells, constricted at the septa.

Monitoring Carefully watching and recording information on the activities, growth, development, and abundance of organisms or other factors on a regular basis over a period of time, often utilizing very specific procedures.

Monocarpic A plant that produces reproductive propagules only once in its lifetime.

Monocentric Having derived from, or pertaining to, a single centre; with one centre of growth and development.

Monoclinous With the antheridium on the oogonial stalk.

Monoclonal antibody Antibody produced from clones of a single antibody-producing cell.

Monocot A member of a group of plants whose seedlings have a single cotyledon; often known as grasses.

Monoculture Condition in which one species is grown in an extensive pure stand; describes most agricultural situations.

Monocyclic Of a disease or pathogen: Producing one generation of inoculum and one cycle of infection during a single growing season.

Monodelphic Of nematodes: Possessing one genital tube or ovary.

Monoecious Male and female sexual apparatus present in the same individual. Same as hermaphrodite; opposite of dioecious.

Monogastric Non ruminant animal having only one stomach, e.g. pig, human

Monogenic Of nematodes: Producing offspring of only one sex.

Monogenic resistance Resistance conferred by a single gene.

Monogyne Having only one queen in the colony. For example, some ant species are monogyne, while others are polygyne.

Monokaryon A haploid mycelium (or cell) which contains only one type of nucleus.

Monokaryotic Having one nucleus per cell.

Monomer A simple molecule that can form polymers by combining with identical or similar molecules.

Monomethylhydrazine (MMH) A chemical used for rocket fuel. It is poisonous if inhaled or ingested. It is released from some Helvella and Gyromitra mushrooms by drying or boiling in water, and is a break-down product in the human body from a chemical named gyromitrin which is found in these mushrooms.

Monomitic Describes basidiomata constructed of only one kind of hypha, the generative type.

Monomorphic Having only one form, or one size.

Mononematous With conidiophores borne singly.

Monophyletic A group descended from a single common ancestor within the group being studied. The term is also used in a slightly different sense. A "monophyletic group" usually refers to a clade, i.e. an organism and all of its descendants. Thus Sinraptor and Allosaurus are (very likely) monophyletic within Allosauroidea. However they are not, by themselves, a clade, because they do not include all descendants of their last common ancestor.

Monopodial Of branching patterns, Monopodial branching is a pattern of branching in which there is one main shoot with lateral branches emerging from it. Many firs and spruces exhibit monopodial branching patterns, therefore the Christmas tree image is a good example of monopodial branching. Branching in which there is a clear central branch with lateral branches that develop some distance behind the end of the central branch.

Monosaccate A pollen grain with one saccus or buoyant bladder.

Monotrichous Having only one flagellum.

Monotype The sole species of a newly proposed genus.

Monoxenic culture A culture containing one species of organism growing in the presence of one other species of organism.

Morbidity State of ill-health produced by a disease.

Morel A large, edible ascoma of the genus Morchella (Pezizales, Ascomycetes); fruits in Spring.

Morph Form.

Morphology The study of the form of organisms; form and structure of organisms.

Mortality rate The per capita death rate in a population. The mortality rate is the reciprocal of the population life expectancy.

Mosaic A leaf symptom in which numerous small areas of discolouration stand out against a background of a different tint, tending to have a clearly defined boundary delineated by veins. A pattern of green angular areas against a predominantly yellow background may result; alternatively, the areas bounded by the veins may be chlorotic giving a yellow-on-green mosaic. If discrete areas of colour later coalesce a mottle symptom may result.

A diseased condition where different portions of a leaf vary in amounts of chlorophyll, thus giving the leaf a mottled appearance; usually caused by viruses.

Disease symptom characterized by nonuniform foliage coloration, with a more or less distinct intermingling of normal and light green or yellowish patches. Usually caused by a virus.

Moss Very short, small green plants from the Bryophyta phylum.

Mottle A leaf symptom in which small but numerous areas of discolouration, commonly chlorotic irregularly shaped and without sharply defined boundaries, stand out against a background of different tint, the pattern being unrelated to the vein network.

Irregular, diffuse patterns of chlorotic areas interspersed with normal green leaf tissue.

An irregular pattern of light and dark areas.

Mould A mycelial microfungus or a visible growth of such a fungus. Blue mould or green mould is caused by Penicillium spp.; grey mould, by Botrytis cinerea; sooty mould is caused by one of the Capnodiaceae but the term has been used incorrectly for growth of Cladosporium spp. and other fungi on foliage, etc.

Mouthparts The name given to all parts of the mouth when taken together (collectively).

Movement protein For viruses, a protein encoded by the pathogen that is required for spread within an infected plant.

Mowing frequency The number of times a turf-grass area is mowed per week, month, or growing season. (The reciprocal of mowing frequency is mowing interval the number of days, weeks, etc., between successive mowings.)

Mowing height The distance above the ground surface at which the turf grass is cut during mowing.

Mowing pattern The pattern of back-and-forth travel while mowing turf. Patterns may be changed regularly to distribute wear and compaction, to avoid creating "grain," and to create visually aesthetic effects, especially for spectator sports.

MRNA A form of RNA that carries information to direct the synthesis of protein.

MtDNA Each mitochondrion has its own small DNA genome which does not undergo recombination. In higher plants and in animals, both of which have well-differentiated microgametes (sperm) and macrogametes (eggs), the mitochondria of offspring are all decended from the mitochondria in the macrogamete. It is not obvious that this should be the case in Fungi, given their more egalitarian mating practices; and we are unsure if the same rule applies.

Mucilaginous Composed of mucilage.

Mucor sp Mucor sp. Often found in soils, dead plant material (hay), horse dung, fruits and fruit juice. It is also found in leather, meat, dairy products, animal hair, and jute. It is almost always in house dust, frequently in air samples and old dirty carpets. Wood chips and sawdust are often attacked by M. plimbeus causing "wood chips disease" and "furrier's lung". Accumulated dust in ventilation ducts may contain high concentrations of viable Mucor spores. Asthmatic reactions to Mucor have been described. It is a Zygomecete fungus that may be allergenic (skin and bronchial tests). It is an opportunistic pathogenic organism and it may cause mucorosis in immune compromised individuals. The sites of infections are the lung, nasal sinus, brain, eye, and skin. Infection may have multiple sites. This organism and other Zygomycetes will grow rapidly on most fungal media. Conidia (aplanospores) are globose to ellipsoidal,7-8 microns in diameter, yellowish brown and slightly rough-walled, and are produced in sporangia that are developed around a piriform columella with typical projections. Identification is based on the way sporangia are formed. Cultivation - Potato dextrose agar, 24°C.

Mucormycosis A disease of man or animals caused by one of the Mucorales, e.g., Absidia corymbifera, but sometimes also applied to infections caused by members of the Entomophthorales.

Mucronate In nematodes: Ending in a sharp point.

Mulch A protective covering that is spread on the ground around plants to inhibit evaporation and weed growth, control soil temperature, enrich the soil, or prevent the dispersal of pathogens. It may be organic material such as leaves, peat, or wood chips, or inorganic material such as plastic sheeting.

Multicellular Composed of many cells.

Multicomponent virus A virus in which the genome needed for full infection is divided between two or more particles (e.g., cowpea mosaic virus, brome mosaic virus, cucumber mosaic virus).

Multigenic resistance Resistance conferred by several genes.

Multiline A mixture of similar lines grown in fixed proportions with the intention of making it possible to substitute improved lines for less effective ones as the opportunity and need arises.

Multilocular With several to many internal spore-producing cavities or chambers.

Multinucleate Having more than one nucleus per cell.

Multiparasitism When two or more species of parasitoids attack the same host.

Multiparticulate virus Plant virus in which the genome is divided into more than one piece of nucleic acid, which are encapsidated in more than one particle.

Multipartite virus Plant virus in which the genome is divided into more than one piece of nucleic acid.

Multiple chemical sensitivity (MCS) A controversial and poorly understood syndrome that causes affected individuals to react adversely to low levels of a wide range of chemicals and environmental agents, including pesticides, perfumes, household cleaners, and more. Symptoms may include severe headaches, joint pains, fatigue, heart arrhythmias, shortness of breath, and more.

Multiple chemical sensitivity disorder A controversial diagnosis of an allergy-like sensitivity to an unusually broad range and number of substances. This condition has not been subjected to rigorous scientific scrutiny, and there is considerable doubt as to whether or not it actually exists.

Multiple infection An infection in which an individual is infected by parasites of more than one species.

Multiple sclerosis A slowly progressive central nervous system disease characterized by disseminated patches of demyelination in the brain and spinal cord.

Multipolar Describes yeasts in which each successive blastic conidium arises from a different point on the mother cell.

Multiseptate Having many septa or crosswalls.

Mummification The drying up and shriveling of fruits and other plant parts.

Mummy Unharvested nut remaining on the tree (also called sticktight); the crusty skin of an aphid whose inside has been consumed by a parasite.

Muriform Patterned like a brick wall; of a spore, with intersecting septa in more than one plane.

Muscardine diseases Diseases of certain insects, esp. silkworms, caused by hyphomycetes; green muscardine – Metarhizium anisopliae; white muscardine – Beauveria bassiana; yellow muscardine – Paecilomyces farinosus.

Muscarine Toxic quaternary ammonium compound found in species of Clitocybe and Inocybe; causes perspiration-salivation-lacrymation syndrome.

Muscicolous Growing on mosses or liverworts.

Muscimol A hallucinogenic derivative of ibotenic acid; formed in Amanita muscaria when basidiomata are dried.

Mushroom The "flowering" part of a fungus. The main body of a fungus is a network of tiny thread-like parts, called mycelium. Mushrooms grow from the mycelium during a particular season and release tiny dust-like spores. Spores travel by wind, water, or animal and hopefully start a new fungus in a new location.

Mutagen Any agent that is capable of causing genetic mutations (changes in DNA) of either somatic or germinal cells.

Mutation An alteration in genetic structure, such as a change in the base sequence of DNA, or structural or numerical chromosome abnormalities. Mutations carried in germ cells are inherited by future generations and may contribute to genetic disease. Mutations in somatic cells may be implicated in the development of cancer or other diseases.

Mutualism (mutualistic symbiosis) A kind of symbiosis in which both or all partners gain from the association, e.g. mycorrhizae.

Myalgic encephalomyelitis A synonym for chronic fatigue syndrome in common usage in the United Kingdom and Canada.

Mycangia Special organs of some wood-inhabiting beetles and wood-wasps in which they carry their symbiotic ambrosia fungus.

Mycelia sterilia A group in the Fungi Imperfecti that does not produce spores, but may produce chlamydospores or sclerotia, e.g. Rhizoctonia and Sclerotium; subsequent research has defined sexual reproduction for some species.

Mycelial cord A discrete filamentous aggregation of hyphae which, in contrast to a rhizomorph, has no apical meristem.

Mycelium A network of numerous hyphae which develop within or along the substrate. "The spore and initial hyphae are haploid; that is, they contain only one copy of each chromosome. When a haploid mycelium meets another haploid mycelium of the same species, and they are sexually compatible, the two mycelia join together and each cell receives a nucleus from the other mycelium. This process is called diploidization. Some authors seem to want to reserve the term mycelium for hyphae that have undergone diploidization (the earlier, haploid stuff is just hyphae), but most people seem to use it for both phases." hypha.

Mycetomes Pouches at the beginning of the midgut in anobiid beetles in which their endosymbiotic yeasts are stored.

Mycobiont The fungal component of a symbiotic relationship, as in a lichen.

Mycoherbicide A herbicide in which the active ingredient is a living fungus that infects a weed and then kills or inhibits it. A type of microbial pesticide. Mycoherbicides are not included in the Compendium.

Mycoinsecticide A preparation of entomopathogenic fungi used to kill insects.

Mycologist A person who has been professionally trained to study fungi.

Mycology The study of fungi.

Mycoparasite A fungus which attacks other fungi (sometimes called hyperparasite).

Mycoparasitism One fungus living on another.

Mycophagist One who has a special appreciation for eating mushrooms.

Mycophagous Feeding on fungi.

Mycoplasma 1. The name of a genus of organisms which have a unit membrane but, unlike bacteria, lack a rigid cell wall and are therefore highly pleomorphic. Mycoplasmas are the smallest free-living micro-organisms, having ribosomes and both RNA and DNA. Many plant diseases originally attributed to viruses are now known to be caused by mycoplasmas.

2. a member of the genus Mycoplasma.

Mycoplasmalike organism Archaic term for phytoplasma; plant-parasitic pleomorphic mollicute (prokaryote with no cell wall) found in phloem tissue; cannot yet be grown on artificial nutrient media.

Mycorrhiza A mutualistic or weakly pathogenic association of a fungus and the roots of a plant. There are a number of types, some of which are: ectomycorrhiza, where the fungus is on the surface of the roots and forms a Hartig net; endomycorrhiza, where the fungus invades the cells of roots where it is frequently confined to well-defined layers; ectendomycorrhiza, where there is a Hartig net but the fungus also penetrates the root cells; and vesicular arbuscular, where the hyphae in the cortical cells of the root may be coiled or branched.

Mycorrhizal relationship An example of mutualism where a species of fungus (mushroom) wraps itself around a trees' roots. The mushrom pulls nutrients from the tree to grow, and at the same time delivers nutrients from the soil to the tree.

Mycoses Diseases of humans or animals caused by fungi (e.g., ringworm, histoplasmosis, coccidioidomycosis, athlete's foot).

Mycosis An infection by a parasitic fungus, or a disease so caused.

Mycotoxicosis Any disease of man or animals resulting from the ingestion of mycotoxins.

Mycotoxin A toxin produced by a fungus. The term is usually reserved for fungal metabolites that are toxic to man and/or animals and are produced by molds growing on foodstuffs (e.g., aflatoxins, ergot alkaloids).

Mycotrophic Refers to green plants having mycorrhizae.

Mycovirus A virus that replicates in cells of fungi.

Myoglobin The oxygen-transporting protein of muscle, resembling blood hemoglobin in function.

Myxamoeba A naked cell capable of amoeboid movement; characteristic of the vegetative phase of myxomycetes and such Plasmodiophoromycetes as Plasmodiophora brassicae.

Myxomycetes Ubiquitous, type I allergen. Often found on decaying plant material, however occasionally found indoors. Dispersed by wind in the dry phase, while the wet amoebic phase is motile. Myxomycetes exhibit characteristics of protozoans and fungi. Indistinguishable from smuts under 600x microscopy.

Myxomycota The slime molds, a class of fungi characterized by amoeboid vegetative protoplasts, plasmodia, and by brightly colored spore bearing capillitia.

N

Nabam A dithiocarbamate herbicide and soil fungicide, available commercially in liquid form for mixing with water for irrigation (the compound is related to dichlorvos).

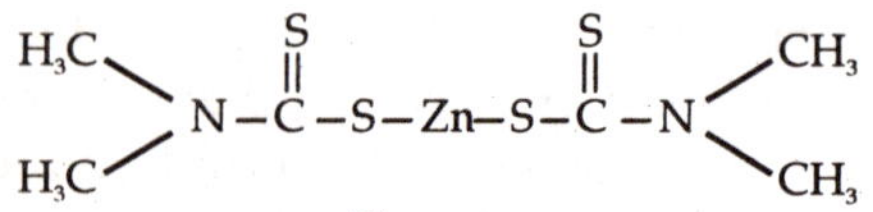

Fig. Nabam

Naiad Another name for nymph. Naiad is used to refer to the nymph stage of aquatic insects, such as dragonflies.

Naked Smooth; not ornamented in any way by hairs or other structures.

Nanometre One-billionth of a metre: written as 'nm'.

Narcolepsy A sudden, uncontrollable disposition to sleep occurring at irregular intervals, with or without obvious predisposing or exciting cause.

Narrow range oil A highly refined petroleum or seed-derived oil that is manufactured specifically to control pests on plants, also called horticultural oil.

Native Originally from a place. Native plants and animals are organisms that were here before people came. Sometimes new species are introduced to a place. These new species are called "invaders" or "non-natives."

Natural control The suppression of pest populations by naturally occurring biological and environmental agents.

Natural enemies Predators, parasites, or pathogens that are considered beneficial because they attack and kill organisms that we normally consider to be pests.

Natural killer cell (NK) A lymphocyte which, unlike other lymphocytes, does not require specific activation by foreign antigen. They are considered to play a "front line" role in controlling infection, curbing infection until a specific, coordinated immune response can be mounted.

Natural pesticide A pesticide in which the active ingredient is a biochemical or some other naturally-occurring substance, as opposed to a synthetic pesticide in which the active ingredient has been manufactured. It must not be assumed that natural pesticides are safe or environmentally friendly.

Natural regulation This is the action of nature without intervention by man. Populations of pests may be held in check by weather and/ or by natural enemies to a greater or lesser extent. If this natural regulation did not exist, then damage to crops would be very much worse than it currently is. The worst-case scenario of total destruction of crops does not often occur, so natural regulation, a gift of nature, is worth a lot of money. The aim of biological control (an action by man) is to use whichever of the natural enemies that can be manipulated to improve on the suppressive effects provided by nature.

Navicular Boat-shaped.

Necrosis Death of tissue accompanied by dark brown discoloration, usually occurring in a well-defined part of a plant, such as the portion of a leaf between leaf veins or the xylem or phloem in a stem or tuber.

Necrotic Dead.

Necrotroph 1. An organism that kills part or all of another organism before deriving nutrients from it (usually applied to plant pathogens).

2. An organism that derives nutrients from dead plant or animal tissues, whether or not it is responsible for the death of those tissues.

Necrotrophic Growing by first killing the host organism or mycelium.

Nectar A mixture of sugars and trace amounts of other chemicals produced by plants in nectaries, usually located in the flowers. Nectar is fed upon by insects as an energy source, and is used by honey bees to make honey.

Nectary A nectar-secreting gland in a flower.

Needle The leaf of a pine tree. Pine leaves are long and skinny, not flat and rounded like most leaves.

Needle cast (of conifers) Disease symptom caused by fungi resulting in premature drop of needles.

Negative sense RNA Ribonucleic acid complimentary to the positive or plus sense; not translated into protein.

Negligence The failure to exercise a reasonable amount of care in a situation that causes harm to someone or something.

Nematicide A pesticide that is used to kill nematodes in plants or soil.

(Chemicals that kill nematodes for medical or veterinary applications are termed "anthelmintics"; they are classed as pharmaceuticals and so are not included in the Compendium.)non-target organism

Any organism that a pesticide is not intended to control.

Nematodes Nematodes are nonsegmented roundworms. Plant parasitic nematodes are small, measuring less than 0.5 mm in length. Therefore, a microscopic is needed to view them. A tremendous number of non-parasitic nematodes exist in the soil and environment. In comparison to this, the number of plant parasitic nematodes is very small. The presence of a spear-like mouthpart called a stylet differentiates parasitic nematodes from non-parasitic ones. All plant parasitic nematodes are obligate parasites and use the stylet to extract nutrients from the host plant.

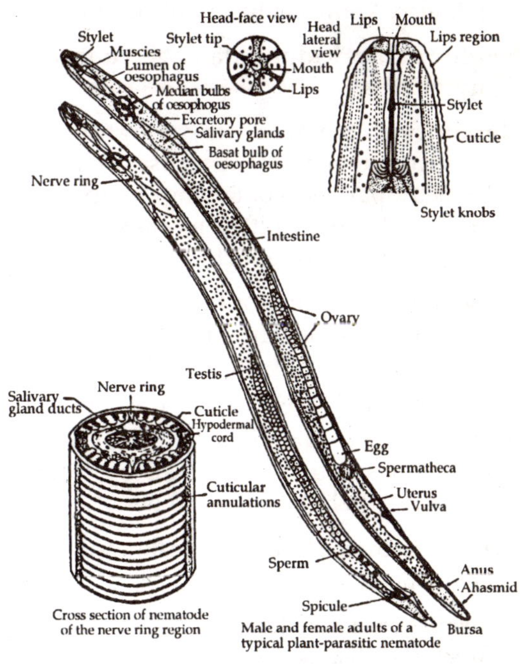

Fig. Figure. Morphological and anatomical characteristics of typical male and female plant parasitic nematodes.

While nematodes cause serious plant disease problems in other areas of the country, nematode diseases are rare in Montana. Exceptions are damaging nematode diseases in alfalfa and sugar beets in some parts of the state. To determine whether nematodes are the cause of a plant disease, roots, soil, and sometimes stems and branches must be examined, usually in a laboratory.

Nematode infections can cause symptoms on both below ground and aboveground parts of plants. Symptoms include root knots, root galls, root lesions, excessive root branching, stunted stem growth, and rapid browning of needles.

Neoropteris (seed fern)seed fern Any of a number of extinct seed-bearing plants with fern-like leaves.

Neotype A specimen selected as the type subsequent to the original description in cases in which the primary types are definitely known to have been destroyed.

Nephropathy Kidney damage caused by mycotoxins.

Nepovirus Member of a group of multicomponent plant viruses with two isometric particles containing two species of linear RNA, transmitted mechanically and by soil-inhabiting nematodes.

Nerve ring The center of the nervous system of nematodes that encircles the esophagus; composed largely of nerve fibers and associated ganglia.

Neurasthenia Nervous exhaustion. A functional neurosis marked by intense nervous irritability and weakness.

Neuromyasthenia Muscular weakness, usually of emotional origin.

Neuropsychiatric Relating to organic and functional diseases of the nervous system.

Neurotoxin A chemical capable of damaging the structure or function of the nervous system (e.g., the brain, spinal cord, nerves or sensory organs). The nervous system controls movement, thought, vision, respiration, behavior, and numerous other physiological processes.

Neurotransmitters Substances produced in neurons that promote or inhibit the conduction of nerve impulses, such as epinephrine, norepinephrine, dopamine, serotonin, and gamma-aminobutyrate.

Neutralism A relationship between two organisms where neither is helped nor harmed.

Nigrospora sp Nigrospora sp. Nigrospora is reported to be allergenic. Morphological characteristics include shiny black conidia, 1-celled, egg-shaped to flattened-spherical, produced singly, and often have an equatorial colorless line or germ slit. Nigrospora often appears as white wooly colonies growing fairly rapidly. This fungus can be a plant parasite or saprophytic.

Nitrate tolerant symbiosis The ability in some host strain combinations to continue to nodulate and fix nitrogen at levels of soil nitrogen that would normally be inhibitory to these activities. In the case of soybean this has been reported most commonly in genotypes from Korea; it may also be due to mutations affecting the regulatory process.

Nitrification Formation of nitrates and nitrites from ammonia by soil micro-organisms.

Nitrogen cycle The cycling of the element nitrogen from gaseous forms to various inorganic forms, such as ammonium, nitrates and nitrites, and organic forms, such as nucleic acids and proteins and back to elemental nitrogen.

Nitrogen fixation The process by which nitrogen gas from the atmosphere is converted into ammonia. The transformation is energy requiring and can be achieved industrially through the Haber Bosch process or by specific microorganisms — all bacteria.

Nitrogen oxides Primary air pollutants produced from internal combustion engines and other industrial processes.

Nitrogen transport Soon after flowering, plants begin to mobilize fixed nitrogen from the leaves to the developing pod. Species and varieties can differ in the efficiency and timing of this process. In some farming systems it may even be desirable that some N remain in the stover for the fertilization of subsequent crops.

No till A cultural system most often used with annual crops, in which the new crop is seeded or planted directly in a field on which the preceding crop plants were cut down, had the tops harvested, or were destroyed by a nonselective herbicide (the old crop is not removed or incorporated into the soil as is common in preparing a plant bed).

Noble rot A condition in which the mould Botrytis grows on overripe grapes. A rich, sweet, expensive wine is made in small quantities from such grapes (Sauternes, Trockenbeerenauslese, Botrytis-wine).

Nocturnal Is active mostly at night.

Node Enlarged portion of a shoot at which leaves or buds arise.

Nodulation The process by which rhizobia induce the formation of galls or nodules on the roots (or occasionally stems) of their specific hosts.

Nodule Small knot or irregular, rounded lump; on leguminous plants, structures on roots that contain nitrogen-fixing bacteria.

Nodulose Having wart-like excrescences.

NOEL (No observable effect level) The dose of a chemical that produces no observable toxicologically significant effects when administered to laboratory test animals.

Noise Random variations or sampling errors in a set of data.

Nomadism Continual movement of humans and animals with no fixed settlement, generally in search of food or water

Nomenclature The naming of Fungi is governed by the International Code of Botanical Nomenclature as adopted by each International Botanical Congress; any proposals to change the Code are published, debated, and voted on at such Congresses. If more than one name has been applied to a fungus, the rules help decide which is the proper one. The rules also allow separate binomials for anamorph and holomorph.

Non infectious disease A disease (or disorder) caused by unfavorable growing conditions and which can't be transmitted from plant to plant.

Non selective herbicide A chemical that is generally toxic to plants, without regard to species. Toxicity may vary with dosage, method of application, etc.

Non target organisms Organisms found with or near those being treated with a chemical or biological control agent; it is important that these agents have as little effect as possible on non-target organisms.

Non toxic Not poisonous; a relative term.

Noncirculative transmission Virus transmission characterized by a very short period of acquisition of the virus by a vector (e.g., an aphid), no latent period before the vector can transmit the virus, and a short period of retention by the vector after acquisition.

Noninfectious disease A disease that is caused by an abiotic agent that cannot be transmitted from one plant to another.

Nonpathogenic Unable to cause disease.

Nonpersistent transmission A type of virus transmission in which the virus is acquired and transmitted by the vector after short feeding times, and is retained by the vector for only a short period of time.

Nonpersistent virus A virus that is carried on the mouthparts of its insect vector and is lost after the vector feeds once or a few times; styletborne virus.

Nonseptate Without cross-walls; coenocytic.

Northern blot An RNA blot. This term originated as lab jargon; not a generally accepted term.

Notifiable disease Diseases of livestock, which must be reported to the police when an outbreak is suspected on a farm. Certain disease and pests must be notified to MAFF. ie Foot & Mouth or Colerado beetle.

Noxious weed A weed specified by law as being especially undesirable, troublesome and difficult to control.

Nozzle A device for metering and dispersing a spray solution.

Nucellus In plants, the watery tissue composing the chief part of the young ovule in the flower and inside the seed during early development. It furnishes nutrients to the young embryo and is digested by the developing endosperm and embryo.

Nuclear cap A mass of RNA and ribosomes near the nucleus in some zoospores.

Nuclease Any enzyme that can cleave the sugar-phosphate backbone of a nucleic acid.

Nucleic acid Genetic material of all living organisms, including DNA (deoxyribonucleic acid) and RNA (ribonucleic acid).

Nucleoid Region in prokaryotes, chloroplasts, and mitochondria where the DNA is concentrated. Unlike a nucleus, it is not bound by a membrane.

Nucleolus Organelle found in the nucleus of eukaryotic cells; composed mainly of ribosomal RNA in the process of being transcribed from from multiple copies of rRNA genes.

Nucleoprotein A compound of nucleic acid and protein.

Nucleotide A single unit of nucleic acid, composed of a phosphate, a five-carbon sugar (either ribose or deoxyribose), and a purine or a pyrimidine.

Nucleus Dense aggregation of proteinaceous matter and nucleic acid in cells, surrounded by a membrane; contains chromosomes and controls heredity.

Nuisance Anything that annoys or disturbs unreasonably; hurts a person's use of their property; or violates the public health, safety or decency.

Nutrient Something that provides nourishment. Food for an animal, plant, fungi, or other organism.

Nutrient broth A liquid basal medium.

Nymph A stage in the life cycle of insects that do not go through complete metamorphosis (cycle of change). Most insects have four stages (1 = egg, 2 = larva, 3 = pupa, 4 = adult), but some only have three stages. This is called incomplete metamorphosis (1 = egg, 2 = nymph, 3 = adult). With incomplete metamorphosis, nymphs hatch from eggs and slowly change into adults. There is no pupa, or resting stage. Most aquatic insects have a nymph stage, since a pupa (coccoon) could not survive underwater. A few land insects, such as grasshoperrs and crickets, have a nymph stage also.

Fig. Node

Nystatin An antifungal antibiotic derived from Streptomyces noursei, used to treat candidiasis.

O

Ob Prefix, inversely or oppositely.

Obclavate Inversely clavate; widest near the base.

Oblate Flattened at the poles.

Obligate One capable of living only as a parasite.

Necessary; obliged. An obligate parasite is an organism that can live only on living tissue.

Obligate anaerobe An organism that can grow only under anaerobic conditions.

Obligate parasite An organism that occurs in intimate association with, and which is wholly dependent for its nutrition on, another living organism.

Obligate saprophyte Organisms that can live only as saprophytes.

Oblong (Of spores) longer than broad (about twice as long or somewhat less), with sides nearly parallel and with ends more or less flattened.

Obovoid Egg-shaped, with the narrow end outward.

Obtuse (Of pileus, cystidia, spores) rounded or blunt.

Occlusion Block or plug that stops flow of liquids (as in vessels).

Ocellus One of two (or one of three) very small, simple eyes on the top of the head of some adult insects. Adults of many species have no ocelli, but almost all adults have a pair of compound eyes.

Ochratoxins Toxins of Aspergillus ochraceus, Penicillium verrucosum, etc.; causes kidney damage in sheep, cattle and pigs.

Odds ratio A comparison of the presence of a risk factor for disease in a sample of diseased subjects and non diseased controls. The number of people with disease who were exposed to a risk factor (Ie) over those with disease who were not exposed (Io) divided by those without disease who were exposed (Ne) over those without who were not exposed (No). Thus OR=(Ie/Io)/(Ne/No)= Ie No / Io Ne.

This measure should be used for case control studies where we retrospectively look at risks in those with and without disease. Also known as exposure odds ratio.

Oedema Blisters produced on leaves and other plant parts under conditions of high moisture and restricted transpiration.

Oestrogenic syndrome A condition in which the mycotoxin, zearalenone, a steroid, causes severe sexual dysfunction in pigs (vaginal prolapse, shrivelled testes).

Oid Suffix, resembling, e.g. balansioid, resembling the genus Balansia.

Oidium 1. A spermatium formed on hyphal branches, especially in heterothallic hymenomycetes;

2. a flat-ended asexual spore formed by breaking up of a hypha into cells.

Oleiferous hypha A hypha that does not contain latex but often contains resinous substances, and often reacts with acid-aldehydes.

Olfactory Having to do with the sense of smell.

Oligogenic resistance Resistance conferred by a few genes.

Omnivore An animal that eats both other animals and vegetable foods.

Onchiostylet In nematodes: A stylet developed from a special cell in the anterior part of the esophagus from which it moves into place during each molt.

Ontjom An Indonesian fermented food, prepared from peanut press cake, which has been surface inoculated with the Monilia anamorph of Neurospora sitophila (Sordariales: Ascomycetes).

Ontogeny Development of the individual.

Oocyst Synonymous with oogonium.

Oocyte Female germ cell.

Oogamy A style of reproduction involving female gametes which are much larger than the male gametes, are non-motile, and may be called 'eggs.' Oogamy is exhibited by the Oomycota and by the Monoblepharidales of the Chytridiomycota.

Oogenesis Development of the oogonium after fertilisation.

Oogonium Single-celled female gametangium giving rise to one or more 'eggs'.

Oomycetes A class of aquatic and terrestrial fungi (subdivision Mastigomycotina) that typically produce oogonia and zoosporangia in which form zoospores having one anteriorly-directed tinsel flagellum and one posteriorly-directed whiplash flagellum.

Oomycota Phylum of chromistan fungi with biflagellate, heterokont zoospores; oogamous, with non-motile gametes; have cellulose walls, and diploid vegetative thalli; hence, Oomycetes.

Ooplasm In Oomycota, the protoplasm, at the centre of the oogonium, which becomes the oosphere.

Ooplast The spherical, translucent to granular inclusion of the oospores of Oomycota.

Oosphere A large, naked, nonmotile, usually spherical, female gamete.

Oospore A sexual spore produced by a the union of two morphologically different gametangia (oogonium and antheridium).

Thick-walled, sexually-derived resting spore of oomycete fungi.

Ooze 1. to be extruded as a viscous fluid through an aperture in a plant structure (e.g. stoma, lenticel, hydathode, wound etc.).

2. a viscous fluid so extruded.

Ooze, bacterial A mixture of bacteria, secreted mucopolysaccharide etc. so extruded.

Open reading frame Portion of a nucleic acid molecule that is translated into a protein.

Operculaoperculum A lid. Thus, for example, the asci of some pezizomycotines have a sort of lid at the distal end which opens and releases the spores at maturity.

Operculate Describes sporangia (esp. some unitunicate asci), having a special lid or cap (operculum) that opens to permit spore discharge.

Operculum A flap or lid-like covering over the opening of an ascus or sporangium.

Opisthodelphic Of nematodes: Having the uterus (or uteri) directed posteriorly.

Opportunistic Fungi which are normally saprobic, but occasionally act as pathogens when condition unusually favourable for infection arise.

Opposite Leaves attached precisely opposite each other on a stem.

Order One of the primary divisions of the classification system. Consists of several to many families. An order is the major division under class.

Ordered tetrad The linear arrangement in the ascus of spores resulting from meiosis.

Organelles Specialized sub-cellular structures which perform specific functions in eukaryotic cells, e.g., mitochondria, nucleoli, ribosomes: these are now believed to have been prokaryotic organisms which entered a symbiotic union with the ancestral eukaryotic cells.

Organic A material (e.g. pesticide) whose molecules contain carbon and hydrogen atoms. Also may refer to plants or animals which are grown without the use of synthetic fertilizers or pesticides.

Organic farming Basically farming without the use of manufactured chemicals, but now conforming to very detailed production standards

Organic matter Plant or animal material capable of undergoing breakdown and resynthesis (in the soil).

Organic pesticide 1. A pesticide that is an organic chemical, meaning that the molecule consists mainly of a carbon skeleton plus other elements such as hydrogen, oxygen, nitrogen, sulfur, phosphorus or chlorine. By this definition, nearly all pesticides are organic, while only a few are inorganic, such as copper sulfate, cryolite, lead arsenate, mercurous chloride and phosphine.

2. A pesticide that is claimed to be environmentally friendly. By this definition, very few pesticides are organic. Claims that a pesticide is environmentally friendly should be viewed with scepticism.

Organic-phosphorous insecticide A synthetic compound derived from phosphoric acid. Organic-phosphorus insecticides are primarily contact killers with relatively short-lived effects. They are decomposed by water, pH extremes, and high temperatures and micro-organisms. Examples are malathion, diazinon, chlorpyrifos, aspon, phorate, isofenphos, trichlorfon, dimethoate and fenthion.

Organism Any lifeform. It may be a plant, animal, fungi, protist, or moneran.

Organo tin (RSnX) Organic fungicides incorporating tin.

Organoleptic qualities Attributes affecting the senses (taste, smell).

Organomercurial (RHgX) Organic fungicides incorporating mercury.

Organophosphate insecticide A group of organic compounds identified by the presence of phosphorus atoms in the molecule, frequently linked with oxygen, sulphur, or both. Organophosphate insecticides 28 Northwest Coalition for Alternatives to Pesticides are cholinesterase inhibitors. Examples parathion, malathion, chlorpyrifos, diazinon.

Orifice An opening in a nozzle tip, duster, or granular applicator through which the spray, dust or granules flow.

Orthotropic Of a basidiospore, where the longitudinal axis of the sterigma corresponds to that of the basidiospore; the basidiospores are not forcibly ejected.

Osmolality Osmotic concentration. An indicator of fluid balance in the bodies tissues.

Osmophilic Being able to grow under conditions of high osmotic pressure.

Osmotic pressure Most biological membranes are impermeable to many of the solutes found in the cell. If this were not so, all of the valuable biomolecules in the cell would simply diffuse out and be lost. However, the same membranes are often more or less permeable to water. Since there are many solutes trapped in the cell, the "concentration" of water is lower in the cell than outside, i.e., there are more water molecules per unit volume outside the cell than inside. Diffusion of water through the membrane works both ways and is completely random. However, water "concentration" is higher outside the cell. That is, there are more water molecules in contact with the membrane on the outside than on the inside. So, there will be a net flux of water into the cytoplasm until the concentrations equalize. The osmotic potential is the measure of the net tendency of water to enter the cell. Real cells can't usually behave in this fashion, since the cell will expand and ultimately burst. The problem is handled in numerous different ways, depending on the cell type. In fungal cells (and various others), the cell membrane is confined within a semi-rigid cell wall. Water enters the cell only until the elastic reaction force of the cell wall equals the outward force caused by water molecules crowding into the cell. At steady state, the cell then maintains a rather high internal pressure, referred to as turgor pressure. Turgor pressure serves as a sort of internal hydrostatic skeleton which helps to support — and even move — fungal structures.

Osmotolerant Capable of growing under conditions of high osmotic pressure, as some yeasts and moulds on concentrated sugar solutions.

Osmotrophic Acquiring food by absorption, as fungi do.

Ostiole 1. The cavity (often lined with periphyses) ending in a pore, in the papilla or neck of a perithecium;

2. any pore by which spores are freed from an ascigerous or pycnidial fruit-body.

Otheca A collection of eggs enclosed in a covering of secretions that hardens to varying degrees and protects the eggs.

Outerbark Exterior, nonliving portion of bark in woody plants.

Outgroup In a cladistic analysis, any taxon used to help resolve the polarity of characters, and which is hypothesized to be less closely related to each of the taxa under consideration than any are to each other.

Ovary The female reproductive structure of organisms; in plants, enlarged basal portion of a pistil, containing the ovules and developing into the fruit.

Ovate Shaped like a section through the long axis of an egg.

Overseason To survive or persist from one planting season to the next.

Overseed To seed onto an existing turf, usually with temporary turf grass, to provide green, active grass growth during dormancy of the original turf (usually a warm-season turf grass).

Overwinter When an animal survives the Winter and lives until the following Spring. Many insects die when the weather gets cold, but some overwinter.

Overwintering inoculum State in which an organism survives the winter (mb)

Ovicide An acaricide or insecticide that kills the eggs of mites or insects.

Oviduct A short, usually tubular, thick-walled part of the female reproductive system between the ovary and spermatheca.

Oviparous Producing eggs that hatch after expulsion from the body.

Oviposit To deposit or lay eggs with an ovipositor.

Oviposition The act of depositing the eggs.

Ovipositor A long needle-like tube on the abdomens of some female insects. The ovipositor is used to inject eggs into soil or plant stems. Species that have ovipositors include crickets and crane flies.

Ovoid Egg-shaped in three dimensions.

Ovoviviparous Producing eggs that hatch within the body.

Ovule Unfertilized seed; the ovule contains the megasporangium with the megagametophyte, surrounded by one or two integuments.

Oxidationoxidation In the sense used by chemists, this means the removal of electrons from an outer electron shell. Oxidation often involves the use of oxygen, but oxygen is not necessarily required. As a practical matter, the removed electrons have to go somewhere, so biological oxidation is a matter of coupling the oxidation of one chemical with the reduction of another. So, what does this accomplish.The trick here is that the resulting product(s) may have lower energy than the starting products. The energy released by the reaction is used by the organism for its own nefarious purposes. Typically, in biological systems, oxidation of chemicals from a food item is coupled with reduction of a special-purpose carrier molecule. That carrier can then dock with any number of enzymes, proteins that perform all the chemical chores needed by a living cell. The enzymes then use the energy stored in the carrier by coupling oxidation of the carrier with a reduction reaction which performs the job the enzyme was evolved to accomplish. Examples from real life tend to be complex, so we'll use a simplified and somewhat unrealistic example involving the carrier nicotinamide- adenine dinucleotide (NAD), a common carrier in the metabolic reactions handled by enzymes in mitochondria. In this example, the energy released by oxidation is stored in three different ways, all of which are important to biological systems. 1. the energy is first stored in the form of an excited intermediate state of an enzyme. This typically involves a slight change in the shape of the protein. As the protein drops back to its normal state, the change in shape drives the next reaction.

2. The energy is stored by reducing a carrier molecule, here the coenzyme NAD^+, as discussed above.

3. The reaction produces an ion, here H^+. Over many such reactions inside a membrane system (from an entire cell to a single mitochondrial crista), a gradient develops across the membrane. This can be used to drive other types of reactions as discussed elsewhere.

Oxidative phosphorylation The formation of ATP from ADP and inorganic phosphate that takes place in the electron transport chain of the mitochondrion.

Oxygen A gas that all animals need to live. Some animals have lungs to get oxygen from the air, others have gills to get oxygen from water.

Ozone A secondary air pollutant that is a highly reactive form of oxygen (O_3) formed when the exhaust of automobiles and other internal combustion engines is in the presence of sunlight.

Ozone layer A layer of ozone found in the stratosphere, where it absorbs harmful ultra violet radiation.

P

Packing tissue Firm, membranous tissue lining the walnut shell and separating the kernel halves.

Paecilomyces sp Paecilomyces sp. - Commonly found in soil and dust, less frequently in air. P. variotii can cause paecilomycosis. Linked to wood-trimmers disease and humidifier associated illnesses. Some members of this genus are reported to cause pneumonia. It has also been reported as causative agent of allergic alveolitis. It may produce arsine gas if growing on arsenic substrate, this can occur on wallpapers covered with Paris green. Cultivation - Potato dextrose agar, 24°C.

PAGE Polyacrylamide gel electrophoresis.

Paint fungi Wood-inhabiting basidiomycetes with thin, spread out, indefinite basidiomata; many Corticiaceae and Thelephoraceae: Aphyllophorales.

Palaeozoic The geological era covering the period from 600-225 million years before present (m.y.b.p.); includes the Cambrian, Ordovician, Silurian, Devonian, Carboniferous and Permian periods.

Palisade Series of parallel hyphae having the appearance of a wall or fence – or even a palisade.

Palisade parenchyma Tissue found beneath the upper epidermis of leaves, composed of elongate, tubular cells arranged upright in the manner of posts in a palisade fortification.

Palmate Three or more lobes, leaflets or veins arising from one point.

Palynomorph A fossil spore or pollen type.

PAN A secondary air pollutant formed when the exhaust of automobiles and other internal combustion engines is in the presence of sunlight

Pandemic A widespread and destructive outbreak of disease simultaneously in several countries.

Panicle A branching cluster of flowers held on a stem, such as the flowering parts of most grasses.

Panzootic A widely distributed epizootic, often one affecting more than one host species.

Papilla Nipple-like projection; used to describe the tip of some sporangia and the localized wall thickenings on the inner surface of plant cell walls at sites penetrated by fungi.

Papillate Bearing a papilla.

Papillate projection Bearing a papilla, i.e. a hump or a swelling.

Pappus The modified calyx of flowers in the sunflower family; usually takes the form of bristles, scales, or awns.

Papulospore A small, asexual, indehiscent sclerotium-like structure differentiated into central and sheathing cells.

Papyraceous Parchment- or paper-like.

Paracapillitium In gasteromycetes, A capillitium consisting of thin-walled, hyaline, septate hyphae in contrast to a true capillitium.

Paragynous Of Pythiaceae, Having the antheridium at the side of the oogonium .

Paraphyletic A taxon is said to be paraphyletic if it does not contain all descendants of its last common ancestor. So, for example, "seedless plants" is paraphyletic because its last common ancestor would also be the ancestor of plants with seeds. A clade must contain all descendants of its last common ancestor. It is usually fairly important to stick to a consideration of clades. However, that importance can often be outweighed by the utility of being able to discuss the earliest forms of some group. We normally signal that we are talking about such a paraphyletic group by saying "basal seedless plants" or "paraphyletic seedless plants" or by the use of some other, similar linguistic artifice.

Paraphysis A sterile, hyphal element in a hymenium originating at the base, and growing upwards, usually unbranched and not anastomosed, especially in the Ascomycota.

Paraphysoid 1. In Basidiomycota, a sterile, accessory, hymenial structure.

2. in Ascomycota, interascal tissue stretching and eventually resembling pseudoparaphyses, often only remotely septate, anastomosing and very narrow; trabeculate pseudoparaphyses.

Parasexual cycle A sequence involving heterokaryon formation, diploidization, and haploidization, often resulting in the formation of recombinant nuclei. Unlike the sexual cycle, the parasexual cycle can occur at any point or continuously throughout the life cycle.

Parasexualism A type of recombination, found in certain heterokaryotic fungi, that is based on mitosis rather than meiosis. Genetically distinct haploid nuclei fuse in the heterokaryon. The resulting diploid nuclei multiply by mitotic division, with some crossing-over, and a diploid homokaryon develops. During repeated mitotic divisions, chromosomes may be shed by the diploid nucleus to produce a haploid nucleus.

Parasexuality Genetic recombination during the mitotic cycle, best known in conidial fungi (ascomycetous anamorphs, such as Aspergillus).

Parasite An organism or virus existing in intimate association with a living organism from which it derives an essential part of the material for its existence while conferring no benefit in return (BMS). An organism or virus existing within or attached to or in intimate association with another living organism, from the functioning tissues of which it derives part or all of the material for its nutrition. An organism or virus for which the tissues of another living organism serve as substratum and source of nutrition (APS). The essential characteristic of a parasite is that it is wholly or partly dependent for its nutrition on the functioning tissues of a living organism with which it lives in intimate association; the BMS addendum, that the parasite confers 'no benefit in return,' is difficult to establish in practice, and was presumably added to enable a distinction to be made be tween parasitic and symbiotic relationships. However,there seems no reason why, for example, a mycorrhizal fungus should not be regarded as a parasite having a symbiotic relationship with its host.

The terms 'parasite' and 'pathogen' are not necessarily interchangeable; both may be applicable to a given organism, but 'parasite' indicates its mode of existence, whereas 'pathogen' relates to its ability to cause disease.

1. Any disease causing organism.

2. An organism exhibiting an obligatory dependence on another organism, its host, which is detrimental to the host.

An organism which lives on or in another living organism and obtains part or all of its nutrients from that other living organism.

An organism living on or in another living organism (host) and obtaining its food from the latter.

An organism that lives within or upon another living organism from which it derives nourishment and in which it may cause various degrees of injury.

Parasite, ecologically obligate A parasitic organism which can also grow sapro phytically, but only if it is not in competition with less-specialized parasites or wholly saprophytic fungi and bacteria.

Parasite, facultative One able to live as a saprophyte and to be cultured on labora tory media (not obligate).

Parasite, obligate One capable of living only as a parasite.

Parasitic Deriving nourishment from another living organism (the host).

Parasitic insect An insect that lives in or on the body of another insect.

Parasitic plants Dodder, mistletoe, witchweed, Indian pipes, and beech drops are seed-bearing plants that lack chlorophyll. They gain all or some of their nutrients by parasitizing other seed-bearing plants.

In Montana, dodder forms a pale green, viney growth on garden plants and on crops such as alfalfa. Mistletoe causes witches' broom symptoms on some forest pine trees.

Parasitism Partial or complete nutritional dependence of one organism or virus on the tissues of another living individual.

The above definitions of parasite and parasitism do not exclude such struc tures as embryo sacs and the sporophytes of Bryophyta. If this is desired it will be necessary to qualify the phrase ' other living organism ' by some such phrase as 'in whose development or functioning it is not a necessary or normal part ' (APS).

The phenomenon of the growth of one organism, the parasite, at the expense of another, the host.

Parasitoid An organism that, during its development, lives in or on the body of a single host individual, eventually killing that individual. Larra wasps, Ormia flies, and Steinernema nematodes are examples of parasitoids; note that the wasps and flies feed like this during their larval stages, while in the adult stages they feed on nectar of plants. Many people use the word parasite when they really mean parasitoid; this is unfortunate because it obscures the lethal effect of parasitoids.

Paratype A specimen other than the holotype and its isotypes that the author cited at the time of publication of the original description.

Parboiling The process of boiling food briefly in water.

Parenchyma The most common type of plant cell; thin-walled cells varying in size, shape, and function. Parenchyma cells are usually loosely packed cubed-shaped or elongated cells that contain a large central vacuole and have thin, flexible cell walls. Parenchymal cell walls are enriched in pectins and therefore stain pink with Toluidine Blue. They also have simple pits. Parenchymal cells are found in large numbers in the pith and cortex. They are the least specialized cells in the plant and normally act as a sort of reserve of stem cells. They may also serve a number of functions, including food storage (either water or nutrients), photosynthesis, as well as tissue repair.

Parenchymatous Describes tissues or thalli made up of thin-walled, apparently randomly arranged cells, packed tightly together and thus usually polygonal.

Parenthesome A curved double membrane on either side of a dolipore septum which may be perforate, imperforate or vesiculate.

Parsimony Refers to a rule used to choose among possible cladograms, which states that the cladogram implying the least number of changes in character states is the best.

Part spore The separate segments of an aeciospore once it disarticulates into several pieces, as in Ostropales, Hypocreales and some other Ascomycota.

Parthenocarpy Development of fruit without fertilization and seed.

Parthenogenesis Reproduction by the development of an unfertilized egg.

Partial resistance Resistance in which disease develops more slowly on the host or to a lesser extent, but the host does become diseased.

Partial veil In agarics, a layer of tissue which joins the stipe to the pileal margin during the development of the hymenium, and which ruptures to become an annulus or cortina.

Particle drift Sprays which are carried away from the application area by air movements.

Parts per million A rule of thumb is that one hectare of land area, taken to the depth of the plow layer weighs two million kilograms. ie. 30 ppm equals 60 kilograms per hectare.

Passage An inoculation of a parasite into a host (or a saprobe into a substratum) and its re-isolation. It is often used in the sense of increasing or decreasing the virulence of a pathogen following successive passages, or in the case of a bac terium in selecting a new population.

Passive immunity Immunity which has been acquired through the transfer of maternal or other antibodies. Passive immunization does not induce immunological memory.

Pasteurization To free a material, usually a liquid, of selected harmful microorganisms using heat.

Pathogen An organism or virus able to cause disease in a particular host or range of hosts. The term can be applied to genera, species and sub-specific taxa of fungi, bacteria, etc., and to strains etc. of viruses. It indicates a general disposition within the taxon, but does not preclude the possibility that a particular race, strain, isolate etc. may be unable to cause disease in a given host; such an exception can be distinguished as a non-virulent (or avirulent) race, strain or isolate of the pathogen, distinct from others of high, intermediate or low virulence in respect of the same host. Similarly, the use of the term is not precluded if a particular host does not become diseased because it is tolerant of the pathogen.

Same as a parasite.

Any agent of the environment capable of inciting disease.

An entity that can incite disease.

An entity capable of producing disease.

Pathogenesis That portion of the life cycle of a pathogen during which it becomes, and continues to be, associated with its suscept.

Pathogenesis-related (PR) proteins Proteins, such as antimicrobial proteins and hydrolytic enzymes, that are synthesized in the early events of the plant defense response.

Pathogenic Possessing the attributes of a pathogen (BMS).

Disease-inciting; possessing the ability to induce disease (APS).

Pathogenicity The quality or characteristic of being able to cause disease. The term is frequently given quantitative qualification (high, moderate, low) and used synonymously with virulence. However, it is useful to preserve a distinction between pathogenicity and virulence, in accordance with the recommendation of A. A. Miles that pathogenicity should be regarded as an attribute of a genus, species or some other commonly accepted grouping, irrespective of the fact that particular entities (strains, formae, isolates or individuals) within the group may not be pathogenic in a given host, and virulence as the ability of an individual entity in the group to cause disease under defined conditions. The frequency with which virulence occurs in the group determines whether or not the group is described as ' pathogenic '. Thus, one may refer to a non-viruent strain of a pathogenic species. The degree to which a pathogen debilitates its host.

The capability of a pathogen to cause disease.

An entity's capacity for producing a disease.

Pathogenicity, horizontal Robinson has proposed the use of this term, but defines it only by analogy with horizontal resistance. It can be defined as pathogenicity which is effective against all varieties of the host; or as pathogenicity expressed without differential interactions between races and varieties. The term is applicable only to a characteristic of a genus or species; however, its use is not recommended because the implications of ' horizontal ' are too far removed from the biological situation to which it refers.

Pathogenicity, vertical Pathogenicity which is effective 'against some varieties of the host but not against others; pathogenicity expressed in differential interactions between races and varieties. The term is not recommended.

Pathognomonic Characteristic or indicative of a disease, denoting especially one or more typical symptoms.

Pathology 1. The study of disease.

2. The abnormal condition that constitutes disease.

Pathophysiologic Characteristic derangement of function seen in disease.

Pathotoxin A toxin produced by a pathogenic organism and capable of simulating the disease. H. E. Wheeler & H. H. Luke erected three criteria to test the toxin theory of plant disease,viz.: (i) a reasonable concentration of the toxin induces in the susceptible host all of the symptoms characteristic of the disease. (ii) the pathogen and the toxin exhibit similar host specificity. (iii) a positive correlation exists between toxin production and pathogenicity. When these conditions are met, a toxin may be said to induce disease. However, with many toxin-induced diseases the toxins may be highly labile and produced in very small amounts, and when the substance is produced only in vivo the demonstration of cause and effect may be very difficult.

Pathotype A subdivision of a pathogen species characterized by its pattern of virulence or avirulence to a series of differential host varieties.

Pathovar A subdivision of a plant pathogenic bacterial species defined by host range; pathovar for bacteria is equivalent to forma specialis for fungi.

Patulin A mycotoxin from Aspergillus clavatus, Penicillium patulum, P. expansum; antibacterial and antifungal; toxic to plants and animals (carcinogenic to mice) and the cause of neurotoxicosis in cattle.

PCR (acronym for polymerase chain reaction) A technique used to amplify the number of copies of a specific region of DNA in order to produce enough of the DNA for use in various applications such as identification and cloning.

Peach leaf curl A fungal disease which causes the leaves of peach or nectarine trees to grow distorted and discolored, falling off prematurely.

Peat-based inoculants Because it is readily available, and can absorb large quantities of inoculant culture, peat has long been used as a carrier for inoculant bacteria. Not all peats are equally effective, and some are acid and must be limed before use.

Pectin A methylated polymer of galacturonic acid found in the middle lamella and the primary cell wall of plants; jelly-forming substance found in fruit.

Pectinase The enzyme that breaks down pectic substances to simple carbohydrates.

Pectinpectin Yet another sugar polymer. In plants, pectin occurs on cell walls of parenchyma and related cell types where it apparently strengthens the cell wall and assists in cell-to-cell adhesion. It is frequently extracted and used as a thickener for preserves.

Pedicel A stalk or stem supporting an organ or other structure, for example: the second segment of the insect antenna. In Hymenoptera (ants, bees, wasps) the one or two segments of the base of the abdomen, particularly in ants, which are much reduced.

Pediculocide A pesticide intended to kill lice.

Peduncle Stalk or main stem of an inflorescence; part of an inflorescence, or a fructification.

Peg roots Primary roots.

Pellet A dry formulation of herbicide and other components in discrete particles usually larger than ten cubic millimeters.

Pellicle The covering (skin) that encloses the kernel; it is white during development but becomes brown at maturity.

Pellis In a basidioma, the outer layers of a pileus (pileipellis) or stipe (stipitipellis), not belonging to the veil.

Peloderan In nematodes: Caudal alae that meet posterior to the tail tip.

Penetration Initial invasion of a host by a pathogen.

Penetration The initial invasion of a host by a pathogen.

Penetration peg The specialized, narrow, hyphal strand on the underside of an appressorium that penetrates host cells.

Penicillin Antibiotic produced by certain species of Penicillium and effective against Gram-positive bacteria.

Penicillium brevicompactum Commonly found in water damaged carpet, wallpaper, and some types of insulation. Penicillium is reported to be allergenic (skin) and it may cause hypersensitivity pneumonitis and allergic alveolitis in susceptible individuals. It can cause other infections such as keratitis, penicilliosis, and otomycosis. Penicillium brevicompactum can produce the mycotoxin mycophenolic acid.

Penicillium chrysogenum This species, formerly known as Penicillium notatum, was one of the first known producers of penicillin. It is commonly found and can act as a food source for some types of dust mites. Penicillium chrysogenum is often found growing as a dark green colony and can produce the mycotoxins roquefortine C, chrysogine, and meleagrin.

Penicillium sp A wide number of organisms belong to this genus. Identification to species is difficult. Often found in aerosol samples. Commonly found in soil, food, cellulose, paint, grains, and compost piles. It is commonly found in carpet, wallpaper, and in interior fiberglass duct insulation. Although this fungus is less allergy-provoking than the other molds, Penicillium is reported to be allergenic (skin) and it may cause hypersensitivity pneumonitis and allergic alveolitis in susceptible individuals. It can cause other infections such as keratitis, penicilliosis, and otomycosis. Some species can produce mycotoxins including 1. Ochratoxin which is damaging to the kidneys and liver and is also a suspected carcinogen; there is also evidence that impairs the immune system.

2. Citrinin that can cause renal damage, vasodilatation, and bronchial constriction.

3. Gliotoxin which is an immunosuppressive toxin, and

4. Patulin that is believed to cause hemorrhaging in the brain and lungs and is usually associated with apple and grape spoilage. It can also cause extrinsic asthma. P. camemberti has been responsible for inducing occupational allergies among those who work with soft white cheeses on which the fungus grows. P. chrysogenum has been found on building materials, including paints, chip boards, and wallpaper. Cultivation - Potato dextrose agar or Malt extract agar, 24°C.

Penitrem A A mycotoxin produced by various Penicillium spp., incl. P. cyclopium, which affects the nervous system, causing tremors.

Per capita rate A rate which is proportional to the number of individuals in a population.

Percurrent 1. Producing new growth in the direction of the longitudinal axis, as in a conidial germ tube emerging through the scar left by the abscission of a conidium or a proliferation growing through the tip of the conidiogenous cell.

2. extending throughout the entire length, as of the columella of a gasteromycete basidioma.

Perennial A plant that does not die when the weather gets cold. Sometimes the leaves and stems above ground die, but the plant sends up new ones from its roots in the Spring.

Perfect Sexual; capable of sexual reproduction.

Perfect flower Flower possessing both stamens and pistils.

Perfect state The teleomorph of a fungus; characterised by the production of sexual spores (ascospores, basidiospores, etc.)

Pericarp Fruit wall.

Periclinal Curving in the direction of, or parallel to, the surface or the circumference.

Periclinal thickening In Coeloanamorphoses, The zone of thicker wall material surrounding the protoplasmic channel at the apex of a phialide.

Pericycle Layer or layers of cells between the phloem and the endodermis of roots, giving rise to branch roots.

Periderm Several layers of corky cells located on the outside of the epidermis of a potato tuber and containing high amounts of suberin.

Peridermium A blister-like aecium as in the form genus Peridermium.

Peridioles The 'eggs' in bird's-nest fungi (Nidulariales); disc-shaped bodies within which the basidiospores are borne, and which are splashed out of the cup-shaped basidioma by raindrops.

Peridium Outer covering of various sporangia in the Zygomycetes, and of the gleba in many Gasteromycetes.

Peridiumperidium The outer membrane of a mitospore. A more familiar example of a peridium may be the outer sac visible on fungal puffballs. The term is not restricted to strictly cleistothecial (completely enclosed) ascocarps (spore organs). Any outer membrane which encloses spores, even if not completely, is a peridium.

Perinatal Between the 28th week of pregnancy and the end of the first week of life.

Perineal pattern Fingerprint-like pattern formed by cuticular striae surrounding the vulva and anus of the mature Meloidogyne female.

Period prevalence The number of existing cases of an illness during a period or interval, divided by the average population.

Periphysis A hair-like projection inside, or near, the ostiole of a perithecium, pycnidium or pycnium, unbranched and not anastomosing.

Periphysoid In an ascoma, a short hypha originating above the level of the developing asci but not reaching the base of the cavity.

Periplasm Peripheral cytoplasm; in Oomycota, the outer, non-functioning protoplasm of an oogonium or antheridium.

Perisporium The frequently fugacious spore wall layer external to the exosporium, enveloping the whole spore and limited by the hardly visible ectosporium. On the disappearance of the perisporium and the ectosporium, the exosporium becomes the outer spore wall; myxosporium.

Peristome A combination of Greek roots roughly meaning "stuff around the mouth." In mosses, this means the ring of peristome teeth around the orifice of the spore capsule that control spore dispersal.

Perithecial ascoma Walled ascoma usually having an ostiole through which spores can be expelled at maturity; the asci are arranged in a hymenium; characteristic of many ascomycetes with unitunicate-inoperculate asci.

Perithecial Relating to a perithecium.

Perithecioid Like a perithecium.

Perithecium Flask-shaped or subglobose, thin-walled fungus fruiting body (ascocarp) containing asci and ascospores; spores are expelled or released through a pore (ostiole) at the apex.

Peritrichate Having flagella all over the outside of the cell.

Peritrichous Of bacterial flagella: Distributed more or less uniformly over the cell surface.

Permanent pasture An established plant community in which the dominant species are perennial grasses, there are few or no shrubs and trees are absent

Permease Transport protein or carrier molecule that assists in the movement of substances across cellular membranes; not permanently altered in the process.

Permitted tolerance 1. The maximum content of toxicant allowed (e.g. of lead or arsenic) in foodstuffs for human consumption.

2. the maximum content of diseased plants allowed in a consignment moving in international or interstate trade.

Peronosporales Specialized forms of the Oomycetes, including aquatic and terrestrial species; many species in this order are plant pathogens (damping-off fungi, downy mildews, and white rusts); unlike the true fungi, they lack chitin in their cell walls.

Peroxyacetal nitrates (PAN) A secondary air pollutant formed when the exhaust of automobiles and other internal combustion engines is in the presence of sunlight.

Persistence (of viruses) The length of time a virus is retained by a vector after being acquired from an infected plant. Persistence is a crucial feature which has been used to establish three categories with which other characteristics of transmission are associated. Viruses are referred to as non-persistent, semipersistent or persistent according to whether they are retained by feeding vectors for hours, days or weeks respectively.

Persistent herbicide A herbicide which, when applied at the recommended rate, will harm susceptible crops planted in normal rotation after harvesting the treated crop, or which interferes with regrowth of native vegetation in non-crop sites for an extended period of time.

Persistent transmission A type of virus transmission in which the virus is acquired and transmitted by the vector after relatively long feeding times and remains transmissible for a prolonged period while in association with its vector.

Persistent virus A virus that systemically infects its insect vector and usually is transmitted for the remainder of the vector's life.

Perthophyte An organism feeding on dead tissues of living hosts.

Pest An organism that people find bothersome. Mosquitos, ticks, dandelions, crabgrass, and yellow jackets are often called pests because they annoy people.

Pest resurgence The rapid rebound of a pest population after it has been controlled.

Pesticide A substance or agent used to kill pests. Comes in many different formulations and types. There are residual, and non-residual (contact) pesticides. Pesticides are also described as insecticides, herbicides, fungicides, nematicides, and rodenticides. depending upon the class of pests they are being used to control.

Pesticide resistance Genetic qualities of a pest population that enable them to resist the effects of certain types of pesticides that are toxic to other members of that species.

Petal Portion of a flower surrounding the stamen and pistil; sometimes colorful or showy.

Petiole The slender stem that supports the blade of a foliage leaf.

PH (potential hydrogen) A symbol for hydrogen ion concentration in a solution; pH values run from 0 to 14 on a logarithmic scale, the lower the value, the greater the concentration of hydrogen ions, and the more acidic a solution is; pH 7 is 'neutral', pH less than 7 is acidic, pH greater than 7 is alkaline or basic.

PH soil A numerical measure of the acidity or hydrogen ion activity of a soil. A pH of 7 is neutral; above 7 is basic (alkaline); below 7 is acidic.

Phaeosphaeria sporespore "Biologically speaking, a fungal spore is a microscopic reproductive unit one or multicelled, used by fungi and other organisms on dispersal of new individuals. Spores contains some food reserve, usually oil or glycogen, they can be produced by meiosis or not. Even if spores are mostly dispersive units, some act as resting structures over unfavorable conditions as intense cold or prorogated [sic] drought. The great majority of spores posses firm cell walls. The most recognizable characteristics used to identify spores are color, size, shape, septation and surface characteristics (e.g. ornamentation). However all of them should be very variable even for a genus alone." MycoSpora - generalities. "By wet weight spores generally contain 25% protein and 20% fat, and they have a low water content relative to vegetative mycelium. Cell walls of spores are generally not fibrillar, but they are multi-layered and often contain melanin and have ornamentations. Spores contain all normal mycelial organelles. Respiratory reserves include lipids, glycogen, phospholipids and polysaccharides that can include sugar alcohols like Trehalose). Respiration rates in spores are only 1-4% those of vegetative mycelium, but obviously the more reserves a spore has, the longer it will survive."

Phage (bacteriophage) Used as a noun with a singular verb this word usually means phage species (singular); used with a plural verb it usually means phage particles; used in the plural it usually means phage species (plural).

A virus parasitic on bacteria. The action of a phage may be highly specific (i.e. acting on only one kind' of bacterium) or broad spectrum, acting on any.

Phage resistant Resistant to infection, usually owing to failure of phage to adsorb to the bacterial cell.

Phage typing A method of distinguishing between different bacterial species or types within a species by testing reaction to a number of selected phages.

Phagotrophic Feeding by ingestion; engulfing food (no fungus does this).

Phagotype A subdivision of a species which is distinguished by its sensitivity to a particular bacteriophage or by a distinctive pattern of sensitivity to a set of specific bacteriophages.

Phaneroplasmodium The typical, robust plasmodium of Myxomycetes, with a clear network of veins ending in a fan-shaped sheet with a delicate margin, and a very granular protoplasm.

Phase (in general use) Some defined period, condition, activity etc. This word has, and should continue to have, many different uses; it should not be given a restricted meaning in plant pathology. Hence its use in bacteriology for a well-defined stage of a naturally occurring alternating variation.

Phaseoliform Of a spore, Curved; in the shape of a french bean.

Phasmid In nematodes: A pore-like structure located in the lateral field of the posterior region of nematodes belonging to the class Secernentea. Function is believed to be sensory. Sometimes called precaudal glands.

Phellem Cork; a protective tissue composed of nonliving cells with suberized walls produced by the phellogen and accumulating on the surface of stems and roots.

Phelloderm Tissue formed by and internal to the phellogen; resembles the cortical parenchyma in morphology.

Phellogen Cork cambium; lateral meristem forming the periderm, a protective tissue in stems and roots; phellem (cork) is produced toward the surface, phelloderm toward the inside.

Phellum The suberized tissue produced by the cork cambium in the bark.

Phenological synchrony The coordinated development of parasites and their hosts based on degree-days and other environmental factors so that parasites can successfully infect their hosts, such as the production and release of ascospores of Venturia inaequalis, the apple scab pathogen, from fallen leaves from the previous season, when new apple leaves are expanding.

Phenology The seasonal life history of a plant, insect or animal.

Phenotype The observable characters of an organism constitute its phenotype and are the product of the interaction between the hereditary potential (genotype) and the internal and external surroundings.

Phenoxy herbicides These hormonal type compounds have a phenol ring to which is attached at least one chlorine atom. They mimic auxin, a plant growth hormone, and stimulate plants to literally grow themselves to death. Examples of phenoxy herbicides: 2,4- D, dicamba (Banvel), mecoprop.

Pheromone Chemical signals emitted by insects and other organisms that enable them to communicate with others of the same species. Often used by insects to attract mates. Pheromones are used as attractants in traps, or applied in large amounts over a crop to confuse insects so they fail to find a mate and reproduce.

Fig. Petiole

Phialide A conidiogenous cell which develops one or more open ends from which a basipetal succession of conidia develops without an increase or decrease in length of the conidiogenous cell itself.

Phialidic Of conidiogenesis, Enteroblastic conidiogenesis where a conidium is delimited by a new wall which is not derived from existing walls.

Phialoconidium A phialidic conidium.

Phloem Conducting tissue of vascular plants. Phloem differs from xylem both structurally and functionally. Functionally, phloem transports solutes both up and down the plant, while xylem functions as a one-way transport system for moving water from the roots to the leaves (or equivalent). Phloem carries sugars and other direct or indirect products of photosynthesis to the stem and roots, generally during the day. Phloem also carries ions and soil nutrients up from the roots to the stem and leaves, generally at night. Xylem makes considerable use of the cell walls of dead or enucleated cells. Phloem uses living cells. Xylem uses capillary action and suction, like a straw, to transport water. Phloem uses osmotic pressure to move solutes. A high concentration of solute is loaded inside cells of the phloem at a source, such as a leaf where sugars are produced. This creates a diffusion gradient that draws water into the cells. The resulting pressure causes a flow to occur. If the dissolved sugars or other chemicals carried along with the sugars are removed from the phloem at another place in the tree for use (a sink such as a root or fruit), the decline in concentration of sugar causes water to move out of the phloem cells. Because water is moving in at a source and out at a sink, there is a mass flow of water and substances in the phloem. In trees, phloem is shed annually with the bark. Water and Chemical Movement Beneath the Bark

Phloem feeding An organism that withdraws nutrients from the food-conducting tissue of a plant's vascular system.

Phloem necrosis Death of phloem cells, often visible, caused by infection by systemic phloem pathogens such as phytoplasmas, spiroplasmas, and viruses.

Phoma sp Phoma sp. This fungus is reported to be a common indoor air allergen and it is also commonly found on various plant parts and soil. Phoma species are reported to grow extensively on painted walls, particularly in humid places such as showers. Some species can be pathogenic to humans, causing either systematic or subcutaneous diseases (phaeohyphomycosis). Morphological characteristics of this fungus include the production of dark colonies resulting from microscopic dark fruiting structures called pycnidia which harbor one-celled spores. It has been reported to produce pink or purple spots on painted walls and grow on paint, cement, and rubber. Cultivation - Potato dextrose agar or Malt extract agar, 24°C.

Phosphorylation A reaction in which phosphate is added to a compound, e.g., the formation of ATP from ADP and inorganic phosphate.

Photobiont The photosynthetic partner in a lichen symbiosis, either a green alga or a cyanobacterium.

Photochemical Chemical reactions that require sunlight.

Photodecomposition The breakdown of a substance by radiant energy (e.g., sunshine).

Photolytic Pertaining to the chemical decomposition attributed to radiant energy (sunlight).

Photoperiod The number of hours of daylight during a 24-hour period. Photoperiod is often an influential factor in animal (insect) or plant biology.

Photoperiodism The effect of day and night length on plant flowering. Some plants are long day, requiring14 16 hours of sunlight per day to flower. Others are short day requiring only 8 9 hours. Others are day neutral and unaffected by day length. Photoperiosim also effects the breeding cycles of farm livestock

Photosynthate The products of photosynthesis, used to support growth, respiration, and fruit production.

Photosynthesis The process where a plant creates oxygen from carbon dioxide. Photosynthesis is very important, since plants make oxygen for animals (including people) to breathe.

Photosynthetic Having the ability to convert light energy to chemical energy; able to produce carbohydrates from carbon dioxide in the presence of chlorophyll.

Phototrophic Obtaining energy for growth etc. from inorganic substances, utilizing light.

Phototropism Growth toward a light source, as in many dung-inhabiting fungi, e.g., Pilobolus (Zygomycetes), Sordaria, Ascobolus (Ascomycetes).

Phragmobasidiomycetes Basidiomycetes in which the basidia are subdivided by primary septa.

Phragmobasidium A basidium with three transverse septa, characteristic of the Phragmobasidiomycetes and also the Teliomycetes.

Phragmospore An asexual spore with two to many transverse septa; with a length:breadth ratio not exceeding 15:1; if elongated, with only a single axis, and that axis not curved through more than 180 degrees; any protuberances, other than setulae, not more than 1/4 the length of the spore body.

Phragmosporous Having transversely multiseptate spores.

Phthalimides The protectant fungicides: Captan, Captafol/Difolatan, Folpet/ Phaltan.

Phycobiont The algal partner in a symbiotic relationship, as in a lichen.

Phycomycetes A group of fungi that may consist of one cell or have filaments (hyphae) with few or no cross walls and that reproduce sexually by union of two sex cells.

Phyllid The leaf of a bryophyte. By many more exacting definitions, this does not constitute a leaf, hence the purported need for this apparently pointless addition to your vocabulary.

Phyllody The replacement of floral parts by leaf-like structures.

Change from a normal flower to leafy structures. Characteristic of certain phytoplasma infections.

Phylloplane The leaf surface; the non-parasitic flora and mycota of the leaf surface.

Phylloplane-competent Used to describe epiphytic microorganisms adapted to living on the leaf surface.

Phyllotaxy Arrangement of leaves on a stem in relation to one another.

Phylogenetic Relating to the evolution of a taxon or, more specifically, the particular evolutionary context of a taxon. Thus, a phylogenetic definition is one which defines a taxon by its evolutionary path, e.g., Archosauria = crocs + rocs. This is shorthand for "Archosauria is defined as the last common ancestor of crocodiles and birds and all descendants of that ancestor." This is discussed in connection with the Fungi on the opening page.

Phylogeny The evolutionary relationships among organisms; the patterns of lineage branching produced by the true evolutionary history of the organisms being considered.

Phylospace A portmanteau word used to describe the placement of organisms in space, time and phylogeny. It is an evolving concept, so to speak.

Phylum The second largest group that scientists classify living things into. Each

Kingdom (Animal, Plant, Fungi, Protist, Moneran) is split into phyla. Example: The Chordate Phylum (animals with backbones) is in the Animal Kingdom.

Physical controls Manual removal or mechanical killing of pests (e.g., vacuuming, or using a hoe to remove weeds,) the use of barriers (e.g., screens, nets, caulking,) the setting of traps, or the use of heat or cold to destroy pests (e.g., flamers, freezing or cold storage of food or woolens.)

Physiogenic disease A disease produced by some unfavorable physical or environmental factors (e.g., light, temperature, water, soil nutrients, chemical, physical or mechanical injury).

Physiologic race A taxon of parasites (particularly fungi) characterized by specialization to different cultivars of one host species. The 6th International Botanical Congress, Amsterdam, 1935, recommended that physiologic race be substituted for physiologic form. Other synonyms are physiological form (or race), biological form, and frequently race).

Physiologic specialization The existence of a number of races or forms of one species of pathogen based on their pathogenicity to different cultivars of a host.

Physiological disorder A disorder caused by factors other than a pathogen; abiotic disorder.

Physiology The study of processes, activities and phenomena related to life.

Physiotype A population of a pathogen in which all individuals have a particular character of physiology (but not of pathogenicity) in common which would appear to be the preferred usage.

Phytiatry The treatment of plant diseases; the term usually implies the use of chemical methods for preventing or eliminating infection.

Phyto (prefix) Plant.

Phytoalexin Substances which inhibit the growth of certain microorganisms and which are produced in higher plants in response to a number of chemical, physical and biological stimuli.

The term as originally used by K. O. Muller (warding-off ' substances) described non-specified anti-fungal substances produced in the host as a response to fungal infection. The argument has been advanced that since anti-microbial substances can be found in low concentrations in healthy plants and can be induced by wounding, heat treatment and poisons in addition to infection, the criteria of Muller are not met and the term 'pseudo phytoalexin' should instead be used. Moreover, substances produced in response to infection which can also be produced in vitro metabolism of plant constituents by microorganisms should be termed 'false phytoalexins'. However, the original definition of the term seems to be no longer tenable in the light of subsequent research, and the prerequisite absence of phytoalexins from healthy tissues or their non-induction by agencies other than infection constitute an unnecessary restriction, provided that they are present at biologically inactive concentrations and increase concomitantly upon infection. The terms 'pseudo phytoalexin' and 'false phytoalexin' therefore should be rejected, and 'phytoalexin' used as defined above.

Phytobiont The plant partner in a mutualistic symbiosis, e.g. a mycorrhiza.

Phytokinin Synonymous with cytokinin and now obsolete.

Phytoncide A non-specific term applied to any substance which confers resistance by preventing the growth of the attacking organism. The term phytoncide was first introduced by B. Tokin (Bacteriocides of plant origin for anti bacterial substances in plants but was later extended by other

Russian workers to include substances of plant origin active against other organisms, both animals and plants. Phytoncides by definition include phytoalexins, and preformed and post-infectional substances conferring resistance to plant pathogens, in addition to those which deter or destroy insect pests. In the plant pathological sense, the term is too all-embracing to be useful and the individual categories of chemical resistance so far known are adequately covered by more specific terms.

Phytopathogen An organism or virus able to induce disease in plants.

Phytopathogenic Term applicable to a microorganism that can incite disease in plants.

Phytopathology The study of plant diseases.

Phytophagous Feeding in or on plants.

Phytoplasma Microorganisms found in phloem tissue that resemble mycoplasmas in all respects except that they cannot yet be grown on artificial nutrient media. Formerly known as mycoplasmalike organism (MLO).

Phytoplasmas Phytoplasmas (formerly called mycoplasma-like organisms) are similar to bacteria, but they have no cell wall. They reproduce like bacteria and usually reside in the phloem (food-conducting vessels) of an infected plant. Leafhopper insects transmit phytoplasmas.

Symptoms of phytoplasma diseases include dwarfed leaves, yellowing or reddening of leaves, excessive proliferation of shoots (witches' brooms), green or sterile flowers, and very short internodes.

Phytosanitary certificate Official document that indicates that plant material has been inspected and found to be free of certain pathogens.

Phytosanitation Measures requiring removal or destruction of infected or infested plant material likely to be a source of pathogens or pests.

Phytotoxicity The ability of a material such as a pesticide or fertilizer to cause injury to plants.

Phytotoxin 1. A toxin produced by a microorganism and active against a plant or against plant cells/tissues.

2. A toxin produced by a plant.

Piercing-sucking mouth parts Mouth parts adapted for penetrating and sucking juices from tissues.

Pigment A coloring matter in animals and plants especially in a cell or tissue.

Pileate Possessing a cap or pileus.

Pileipellis The superficial layer of cells covering the upper side of the cap of an agaric.

Pileus The expanded caplike portion of some basidiocarps or ascocarps that supports the hymenium.

Pilose Hairy.

Pinhead square A square approximately 1/8 inch (3mm) or less in length.

Pinnae Smaller "mini-leaves" on a leaflet.

Pinnaepinna First order of subdivision of a compound leaf or fern frond.

Pinnate Describes a leaf made up of smaller leaflets; usually on both sides of a stem. Each leaflet is called a "pinna" (more than one called "pinnae").

Pinnulet As you have undoubtedly guessed, the third order of segments of a compound leaf or fern frond.

Pioneer A plant that is one of the first to take over an area. For example, after a fire burns all the trees and wildlife in an area (or after people bulldoze), certain weeds and other plants will"pioneer trees."

Pistil Female structures of a flower, usually consisting of an ovary and one or more stigmas and stiles.

Pit The thin, remnant primary wall of tracheids, i.e., primary wall which has not been covered by lignified secondary wall. Pits are involved in translocation of fluids from the xylem.

Pit membrane In many sources, the term pit refers to the absence of secondary wall, i.e. the "hole" in the secondary wall, while the primary wall is the pit membrane. The distinction becomes important in detailed xylem anatomy because vessels may have pits but usually lack pit membranes. In that case the pit is simply a hole in the vessel wall.

Pith A central unspecialized parenchymous ground tissue in monocot roots and dicot stems.

Pithomyces sp Pithomyces sp. This fungus is found mainly growing on decaying plants, specially grasses. It produces a mycotoxin called sporidesmin (a piperazinedione) known to be pathogenic in animals causing facial eczema and liver damage. Spores are produced at the apex of short side branches of vegetative filaments, dark brown, 2- to several celled. The most common isolated species is P. chartarum and its spores have both longitudinal and transverse septa. Cultivation - Corn meal agar, 24°C.

Pituitary Also known as the hypophysis, a gland at the base of the brain with two functionally distinct lobes involved in regulating growth, metabolism, and maturation.

Plage A smooth, paler or unpigmented spot on the surface of a spore.

Plankton A collection of microscopic organisms, including algae, animals, and protists, that float in huge numbers in water. Plankton is usually in largest numbers near the surface. Many animals feed on plankton, including young fish. Zooplankton is plankton that is mostly tiny animals, rather than plants.

Planocyte A motile cell.

Planogamete A motile gamete; zoogamete.

Planospore A zoospore.

Planozygote A motile zygote.

Plant An organism in the Plant Kingdom. Plants do not have the ability to move like animals, but they are able to make their own food by pulling water and nutrients from the soil, and by using light. Plants provide food for animals and fungi. Includes trees, shrubs, vines, weeds, wildflowers, ferns, mosses, grasses, and others.

Plant disease control Almost all control methods aim to prevent plant diseases rather than cure them after they have become diseased. The five basic principles of plant disease control are exclusion, avoidance, eradication, protection, and resistance.

Plant disease defination Plants and plant products are tremendously important for human survival since they provide food, clothing, furniture, a stable environment, and often housing. Plants, whether cultivated or wild, generally grow well when the soil provides them with sufficient nutrients and moisture, sufficient light reaches their leaves, and the temperature stays within a "normal" range. However, like people, plants can get sick. Agents similar to those that cause disease in people also can cause diseases in plants.

The broadest definition of plant disease includes anything that damages plant health. This definition can include such diverse factors as pathogens, insufficient nitrogen, air pollution, lawnmower damage, and deer damage.

A stricter definition usually includes any persistent irritation resulting in plant damage and characteristic symptoms. This definition includes such factors as pathogens, insufficient nitrogen, and air pollution.. However, it excludes factors such as lawnmower injury to trees and lightning injury since this damage presumably is a one-time occurrence.

A very strict definition includes only infectious organisms (pathogens) that multiply and spread to other nearby plants. Most pathogens are microscopic and include bacteria, fungi, nematodes, viruses, mollicutes, protozoa, and parasitic plants. These are called biotic diseases. Plant pathologists usually do not consider organisms such as insects, deer, rodents, and birds to be pathogens.

Plant diseases history Certain diseases have had far-reaching impacts on our society. Perhaps foremost among these is Phytophthora late blight which caused the potato famine in Ireland. As a direct result of this disease, hundreds of thousands of people died and approximately one and a half million people immigrated to the United States.

Downy mildew of grape almost ruined the French wine industry in the late 1800's. The success of Bordeaux mixture in controlling this disease gave great encouragement to agriculturists and stimulated the study of the nature and control of plant diseases.

Two forest diseases which caused huge economic losses in America are Chestnut blight and Dutch elm disease (DED). Both were introduced accidentally to the United States in the early 1900's. Chestnut blight completely destroyed the most valuable timber trees in the Applalacians. Dutch elm disease continues in its destruction of American elm shade trees. Every year, hundreds of thousands of elm trees in towns across the country die from DED.

These examples are prominent because they caused so much damage. However, plant diseases cause variable amounts of damage every year. Often the severity of damage depends on weather patterns.

Plant growth regulator A substance that alters the expected growth, flowering or reproduction rate of plants. Fertilizers and other plant nutrients are excluded from this definition.

Plant pathology The branch of science for plant disease. The content of plant pathology is not always clearly defined. It is centred on disease in plants caused by pathogenic fungi, bacteria (including actinomycetes), mycoplasmas and viruses; disease caused by nematodes is frequently included in plant pathology, but damage caused by insects and other animals is normally excluded. Disorders do not strictly fall within its purview. Synonymous with phytopathology, but plant pathology is the preferred usage in Britain.

Plant-growth regulator A substance used for controlling or modifying plant-growth processes without appreciable phytotoxic effect at the dosage applied.

Plaque A restricted, clear area in a lawn of confluent bacterial growth in culture, resulting from lysis of the bacteria in that area by a phage.

Plasma The fluid portion of the blood, rich in soluble proteins with a wide range of functions.

Plasmalemma Or plasma membrane; outer boundary of the protoplast, next to the cell wall; consists of a single membrane; also called cell membrane and ectoplast.

Plasmid In many types of prokaryotic and eukaryotic cell: a linear or covalently closed circular molecule of DNA, (distinct from chromosomal DNA, mtDNA, ctDNA, or kDNA and commonly dispensable to the cell), that can replicate autonomously (i.e., independently of other replicons).

Plasmodesma A fine protoplasmic thread connecting two protoplasts and passing through the wall separating the two protoplasts.

Plasmodiocarp in Myxomycetes, an elongated, branched, usually sessile fruit-body, still showing signs of its plasmodial origin.

Plasmodiophoromycetes A fungus-like group of organisms, sometimes called the Plasmodiophoromycota or the endoparasitic slime molds, characterized by the production of zoospores and plasmodia that are restricted to the cells of their host.

Plasmodium a multinucleate mass of protoplasm, motile in Myxomycetes, non-motile in Plasmodiophoromycetes.

Plasmogamy Fusion or mixing of the cytoplasm of two cells; follows anastomosis and often precedes karyogamy.

Plasmolysis The shrinking and separation of the cytoplasm from the cell wall due to exosmosis of water from the protoplast.

Plastids Organelles in the cells of certain groups of eukaryotes manufacture and store food; plastids are bounded by a double membrane, e.g., chloroplasts.

Plectenchyma A dense tissue formed by hyphae becoming twisted and fused together.

Plectomycete In general, a group of primitive or reduced ascomycetous forms that have an angiocarpous fructification without an ostiole, the entire interior of which is irregularly penetrated by ascogenous hyphae, with the result that the generally spherical asci, without accompanying paraphyses or other threads, lie scattered irregularly in a pseudoparenchymatous tissue composed of the ascogenous hyphae.

Pleiosporous With many spores.

Pleomorphic Able to assume various shapes (and perhaps sizes); occurrence of more than one distinct form in the life cycle of an organism.

Pleomorphism 1. In fungi: Having more than one independent form or spore stage in the life cycle.

2. In general: An inherent variability in size and shape (e.g., among the cells in a pure culture or clone of a given organism).

Plerome The plant tissues inside the cortex.

Plerotic Of an oospore, Filling the oogonium.

Plesiomorphic A plesiomorphic character of a taxon is one which it shares with an ancestral group. It is not a unique, derived character (synapomorphy or apomorphy), but a primitive character which tells us nothing about its evolutionary position.

Plesiomorphy A primitive character state for the taxa under consideration.

Plesionecrosis A symptom exhibited by tissues not yet dead but in the process of dying (e.g., wilting).

Pleurogenous Of conidia, Formed along the sides.

Plicate Folded into pleats like a fan.

Ploidy The number of (complete) sets of chromosomes in a cell, e.g. haploid, diploid, polyploid.

Plowpan A compacted layer beneath the plow layer produced by pressure exerted on the soil during plowing.

Plurivorous Utilising a number of hosts or substrata.

Poaching Wet soils which are trampled and 'churned up' by cattle, they become muddy and it often leads to a breakdown in soil structure.

Pod A type of fruit that contains several seeds in it. Pods will split open when ripe and release the seeds.

Podetia Upright, cylindrical structures in lichens (esp. Cladonia) which often bear apothecial ascomata at their apex.

Point sampling A type of grid sampling in which a sample is taken at the center of each grid location.

Poison A substance that causes harm to an organism when the organism ingests (eats) it. Poisons can vary in strength; sometimes a poison just makes the organism sick, other times it may kill it. Many species of mushrooms and plants are poisonous to humans. Poison should not be confused with venom.

Poisson distribution The distribution which arises when parasites are distributed at random amongst hosts.

Polar At one end or pole of the cell (e.g., a flagellum, spore inclusion, germ tube, etc.).

Polarilocular Of an ascospore, Having two cells separated by a central perforated septum.

Polarity of characters The states of characters used in a cladistic analysis, either original or derived.Original characters are those acquired by an ancestor deeper in the phylogeny than the most recent common ancestor of the taxa under consideration. Derived characters are those acquired by the most recent common ancestor of the taxa under consideration.

Pollen A fine powder made by flowers. With flowering plants, pollen must get from one flower to another flower for a plant to make fruit. Most plants depend on bees or other insects to "pollinate" flowers.

Pollinate To transfer pollen from the flower of one plant to another plant of the same species. This allows the plant to grow fruits and seeds so it can grow new plants. Bees and other insects help most plants spread by pollinating them.

Pollination The transfer of pollen from the pollen organ to the ovule; for example in flowering plants from stamens to the stigmatic surface of the carpel.

Pollinator The agent of pollen transfer, usually bees.

Pollinizer The producer of pollen; the variety used as a source of pollen for cross-pollination.

Polycentric With a number of centres of growth and development and with more than one reproductive organ.

Polyclonal antibody A preparation containing antibodies against more than one epitope of an antigen.

Polycyclic Of a disease or pathogen: Producing many generations of inoculum and many cycles of infection during a single growing season.

Polyetic Of plant disease epidemics: Continuing from one growing season to the next.

Polygenic Pertaining to, or governed by, many genes.

Polygenic resistance Resistance conferred by several genes.

Polygyne Having many queens in the same colony. For example, some ant species have several queens in the same colony, others tolerate only one, and are monogyne.

Polymer A compound made by linking many identical smaller molecules (monomers).

Polymerase chain reaction The selective amplification of DNA by repeated cycles of (a) heat denaturation of the DNA, (b) annealing of two oligonucleotide primers that flank the DNA segment to be amplified and (c) the extension of the annealed primers with the heat insensitive Tag DNA polymerase.

Polymorphic Having more than one form, or more than one size. For example, the workers of some ant species are polymorphic.

Polymorphism The existence of several asexual spore stages in the life cycle of an organism.

Polynucleate Having more than one nucleus per cell.

Polypeptides Organic compounds made up of amino-acids linked by peptide bonds.

Polyphialides A phialide with more than one open end.

Polyphyletic Term applied to a group of organisms which does not include the most recent common ancestor of those organisms; the ancestor does not possess the character shared by members of the group.

Polyploid Having three or more complete sets of chromosome.

Polypore The shelf or bracket-fungi; hymenomycetes living on dead (or sometimes living) trees and often producing perennial basidiomata in which the hymenium lines annual layers of corky, vertically oriented tubes.

Polyprotein A protein translated from an entire viral genome which is then cleaved by proteases (protein-degrading enzymes) into the active protein products.

Polysaccharide A polymer made up of many linked monosaccharides, e.g., cellulose, a polymer of glucose.

Polystelic Having more than one stele.

Pome fruit A simple fleshy fruit, the outer portion of which is formed by the floral parts that surround the ovary.

Pond A usually mostly still body of water, smaller than a lake. Usually fed by a small stream.

Pores Small openings of tubes where the spores are developed, under the caps of certain mushrooms such as polypores and boletes.

Poroid Having pores.

Porospore A conidium produced by an extension of the inner wall of the conidiogenous cell and extrusion through a pore in the wall of the conidiophore.

Positive sense RNA RNA that can serve directly as messenger RNA.

Positron emission tomography (PET scan) Imaging technique that relies on the detection of gamma rays emitted from tissues after administration of a natural biochemical substance into which positron-emitting isotopes have been incorporated.

Post emergence herbicide A herbicide that is applied after emergence of the crop or weed.

Post emergence treatment Treatment made after the crop plants emerge.

Posterior Hind or hindmost; opposite of anterior.

Postnatal Subsequent to (and within 1 year of) childbirth.

Potash A chemical element necessary in the soil for the development of plants.

Potexvirus (siglum of potato virus X.) Member of a group of plant viruses that infect a wide range of hosts, including monocots and dicots; individual members infect only a narrow host range, typically causing mosaic and ringspot symptoms. Type member: potato virus X (PVX).

Potyvirus (siglum of potato virus Y) Member of a large group of plant viruses with flexuous particles containing a single molecule of linear RNA, most of which are transmitted by aphids in a noncirculative manner.

Powdery mildew A white, powdery, superficial ascomycetous fungus that is an obligate parasite and generally is found only on one or a few closely related species of plants.

PPB (parts per billion) One ppb is equivalent to one microgram of compound per liter of solution.

PPM (parts per million) One ppm is equivalent to one milligram of compound per liter of solution.

PPT (parts per trillion) One ppt is equivalent to one nanogram of compound per liter of solution.

Pr toxin A toxin produced by Penicillium roquefortii fortunately not during the ripening of blue cheeses.

Pre patent period The time from infection until when a female starts to produce eggs in helminth infections, equivalent to latent period in micro-parasitic infections.

Pre plant herbicide A herbicide that is applied before the crop is planted.

Precambrian Geological era from the earliest days of the earth until 600 million years ago, at the end of which the earth's atmosphere is believed to have attained a level of oxygen capable of supporting multicellular, eukaryotic organisms.

Precipitin An antibody that causes precipitation of soluble antigens.

Precipitin test A serological test in which the reaction between soluble antigen and antibody is apparent from the formation of a visible precipitate. The test may be done in agar, or in tubes (tube precipitin test) or by using droplets on slides (slide precipitin test) or petri dishes (microprecipitin test). The detection of the precipitate is facilitated by incorporating polystyrene latex particles (latex test), red blood corpuscles (haemagglutination test) or by using crude sap (chloroplast agglutination test).

Precision agriculture Using the best available technologies to tailor soil and crop management to fit specific conditions found in a small unit area within a larger management unit such as a field, orchard block, etc.

Precursor A substance that precedes and is the source of another substance.

Predaceous Preying upon other organisms, as in the nematode-exploiting fungi.

Predaceous fungi Specialized fungi of several phyla which trap or infect nematodes, rotifers, collembola, amoebae, tardigrades, etc., exploiting them either as principal diet or as a nitrogen supplement.

Predation A relationship where one organism eats another for food.

Predator An organism that, during its development, consumes more than one prey individual. Tiger beetle larvae are examples of predators.

Predatory or predaceous insect An insect that feeds on other insects.

Predispose To make prone to infection and disease.

Predisposition The weakening of an organism by some factor(s) of either the physical or biotic environment so as to render the organism more susceptible to a pathogen.

Pre-emergence herbicide A herbicide that is applied before emergence of the crop or weed.

Pre-emergence treatment Treatment made after a crop is planted but before it emerges.

Preference in nodulation Cultivars in some plant species select for particular rhizobia from among the numerous strains present in their environment. They are said to have a preference in nodulation for these strains.

Premia payment paid on the number of ewes qualifying for the sheep quota and calculated on the basis of the actual market price set against the EC price guide.

Premunity A term defined by Z. Klement & R. N. Goodman as an acquired non-specific immunity manifested on plants pretreated with one bacterium that immunises or protects the plant from infection by another pathogen. This is a particular form of acquired immunity and is subject to the same objection, viz. that any degree of infection renders the use of the term invalid since immune is defined as exempt from infection.

Prepenetration period The period between the time that a vector is placed on a plant and the time that the host tissues are penetrated.

Prepupa The active, but non-feeding, last instar larva. This stage occurs when the cuticle is separated from thc hypodermis but has not yet been cast off. Occurs in thrips and female scales.

Prevalence The proportion of the host population infected (or with some marker of past or present infection) at a particular time.

Prevalence models Prevalence models are compartmental models dividing the host population into, for example, susceptible, latent, infectious and immune individuals.

Prevention select a suitable location for planting Some plants like sunny locations; others prefer more shade. Plants also differ in their tolerance to soil types, pH, moisture, wind, and other factors. Avoid extremely wet soils because they may favor many root disorders.

Practice rotation in your garden plot. If possible, change the location of the entire garden occasionally, to help prevent the buildup of disease-causing organisms. Even when a garden is maintained in the same location, it is important to rotate specific crops to a different area of the garden each year. An example of a poor rotation is tomato followed by eggplant followed by potato since all are susceptible to similar pathogens. A better rotation is tomato, followed by corn, beans, or lettuce.

Select disease-resistant varieties where they are available and adapted to your growing conditions. Lists of resistant plants can be found in many texts and seed catalogues.

Plant resistant cultivars when 1. minimum maintenance is desired (home grounds, rights of way, low-budget operations);

2. inoculum is always present;

3. environmental conditions usually favor disease at some time; or

4. adequate chemical controls are not available.

Using susceptible cultivars is acceptable when 1. inoculum is rarely present,

2. environmental conditions rarely favor disease development,

3. or intensive management practices are being used. Non-host species should be used when the pathogen has a very wide host range and cannot be eliminated from a site and when no resistant cultivars of the disease-prone plant are available.

Use only disease-free seed and transplants. Buy seeds and plants from a reputable source. Many disease-causing organisms are carried to your garden on seeds and transplants.

Maintain soil fertility through careful use of fertilizers, soil amendments, costs, etc. Plants that are too weak or too vigorous as a result of improper fertilization will be more subject to diseases.

Do not overcrowd your plants. The high humidity that results beneath the plant canopy may favor certain fungal diseases. Thin your plants to permit free air movement.

Water properly. Try to maintain an even water supply and avoid dry-wet fluctuations. Drip systems and soaker hoses are better than overhead sprinklers.

Control insect pests. Some insects spread diseases to other plants if they remain in the garden.

Destroy and remove diseased plants as soon as you observe them. They may serve as a source of spread to others plants if they remain in the garden. Do not compost diseased plant parts.

Spade under or compost crop refuse as soon as possible after harvesting the garden. This not only recycles some good organic matter to the soil, but also removes the possibility of disease-causing organisms remaining overwinter in the garden.

Be alert to leaf diseases that occur annually on certain plants and take care of them early, before the damage becomes so severe that the plant is lost.

Prey An animal that is eaten by another animal for food.

Primary First or most important. A "primary food source" means a food that an animal eats most.

Primary bloom The first production of flowers on a potato plant, occurring after 8 to 12 leaves have been formed on the mainstem and generally coinciding with the beginning of the tuber growth phase.

Primary cycle Of plant disease: the first infection cycle to begin in a given season; usually occurring only once per season.

Primary growth Growth in length, controlled by the apical meristem

Primary host The principal host of heteroecious rust fungi, which bears the sexual phase (the teliospores).

Primary infection The first infection of a plant, usually in the spring by an overwintering sexual pathogen propagule.

The first infection of a plant by the overwintering or oversummering pathogen.

Primary inoculum The overwintering or oversummering pathogen, or its spores, that cause primary infection.

Inoculum, usually from an overwintering source, that initiates disease in the field, as opposed to inoculum that spreads disease during the season.

Primary leaf First true leaf that emerges on a plant following the cotyledons.

Primary pinna Leaflet of first subdivision of a fern frond or compound leaf.

Primary pollutant Air pollutant that is released directly into the atmosphere and is harmful to plants, e.g. SO_2 or NOx.

Primary production Elaboration of complex, energy-rich organic compounds by photosynthetic organisms. Fungi and animals are ultimately totally dependent on this process.

Primary root Root that develops directly from the radicle of an embryo rather than from a crown or node.

Primary septum A septum formed in association with nuclear division separating cells and having a pore which may, in Basidiomycota, be modified as a dolipore or which may, in Ascomycota, be associated with Woronin bodies; characteristic of higher fungi.

Primary xylemprimary xylem Xylem is found in leaves and young stems or in young roots it is referred to as primary xylem. Primary xylem is xylem derived from apical meristem, rather than cambium. It contains fiber cells and tracheids (pipe-like) elements which progressively develop secondary, lignified walls until little of the primary wall is left except at pits. These pits operate to permit fluids to move between tracheids. The precise centrifugal series of secondary wall structures, and the shape and pattern of mature pits are useful in taxonomic and forensic work.

Primer Small fragment of nucleic acid with a free 3'-hydroxyl group necessary for initiation of DNA, and, sometimes, RNA synthesis; often specific fragments chosen for use in polymerase chain reaction (PCR) for rapid identification of pathogens.

Primitive Describes a character state that is present in the common ancestor of a clade. A primitive character state is inferred to be the original condition of that character within the clade under consideration. For example, "presence of hair" is a primitive character state for all mammals, whereas the "hairlessness" of whales is a derived state for one subclade within the Mammalia.

Probasidium The primary basidial cell in which karyogamy (or enlargement of a single nucleus) occurs prior to nuclear division. It may be thin- or thicker-walled and may either remain wholly or partly at the base of the mature basidium or be obliterated as the basidium develops.

Probing Used of aphids and some other stylet-bearing vectors for the brief exploratory penetration of superficial cells. This should be distinguished from feeding which usually involves penetration to the vascular tissues.

Proboscis A mouthpart of an insect, usually long and skinny. Mosquitos have a proboscis to suck blood; butterflies and moths have a coiled up proboscis which they unravel to suck nectar from flowers.

Procorpus In nematodes: Cylindrical portion of the corpus anterior to the metacorpus.

Prodelphic In nematodes: Having uteri parallel and anteriorly directed at the origin.

Progametangium In Zygomycetes, A hypha which forms a gametangium and suspensor cell.

Proinhibitin A plant defense compound that is already present before invasion by a pathogen.

Prokaryotic Lacking a membrane-bound nucleus and paired chromosomes; also lacking cytoplasmic organelles such as ribosomes, mitochondria and golgi apparatus: describes the bacteria and the cyanobacteria.

Prolapse A condition in which an organ becomes displaced .

Proleg Any process or appendage that serves the purpose of a leg; specifically, the fleshy, unjointed, ventral abdominal projections of caterpillars and certain sawfly larvae.

Prolepsis A hyperplastic symptom of disease in which organs appear before the natural time.

Proliferation A rapid and repeated production of new cells, tissues, or organs; specifically, a hyperplastic symptom of plant disease in which organs continue to develop after they have reached the point beyond which they normally do not grow.

Promoter A region of DNA, usually upstream of a coding sequence, that binds RNA polymerase and directs the enzyme to the correct transcriptional start site.

Promycelium In rust and smut fungi, a germ tube issuing from the teliospore and bearing the basidiospores.

Pronotum The upper or dorsal surface of the prothorax.

Propagative transmission Pathogen transmission characterized by a long period of acquisition of the pathogen (usually a mollicute, e.g. phytoplasma or spiroplasma, and sometimes a virus) by a vector (typically an insect), a latent period before the vector is able to transmit the pathogen, and retention of the pathogen by the vector for a long period because the pathogen reproduces or replicates in the vector.

Propagative virus. A circulative virus that replicates in its insect vector. Such a virus is said to be propagatively transmitted (e.g., potato yellow dwarf virus).

Propagule That form or part of an organism by which it may be dispersed or reproduced.

The part of an organism that may be disseminated and reproduce the organism.

The part of an organism that may be spread so as to reproduce the organism.

Prophage The form in which phages are perpetuated in lysogenic bacteria.

Prosenchyma A dense tissue formed by hyphae becoming twisted and fused together, where the hyphal elements remain distinguishable.

Prospective study A study in which people are initially enrolled and then followed up at subsequent times.

Protease A generic term for an enzyme that cleaves a polypeptide chain.

Protectant A fungicide that can give protection from a pathogen by killing its spores or germ tubes while they are still outside the host, but cannot cure an existing infection .

Protectant fungicide A fungicide that remains on the surface where it is applied; no after-infection activity .

Protection Protection involves treating a healthy plant before it becomes diseased. There are both chemical and biological means of protection.

Chemical protection is one of the most widely-used means of control. Some fungicides (such as copper and sulfur products) are allowed for use under several "organic" growing guidelines. There are many fungicides on the market, but few can be obtained easily by the home gardener. A knowledge of the disease cycle and host susceptibility is needed to obtain good control using fungicides. Proper timing, coverage, and selection of fungicides also is needed.

Interest in using biological organisms to control diseases has increased in recent years. With biological control, one organism is used to attack or inhibit the activity of another organism. One of the most successful biological protections has been the use of a bacterium to protect against a bacterial disease known as crown gall. In this case, the roots of a seedling or nursery plant are dipped into a suspension of the biocontrol bacterium before planting.

Protective coverings Any cloth, screen, plastic or other material placed over growing plants to prevent damage by pests or harsh weather.

Protein A long chain of amino acids (usually more than 100) joined by peptide bonds.

Prothorax The first segment of the thorax, it bears the anterior or first pair of legs but no wings.

Protonema First growth from a germinating spore, especially, the thread-like process from a germinating moss (Bryophyta) spore.

Protoplasm Fluid living content of a cell, the cytoplasm and nucleoplasm.

Protoplasmodium A small (less than 1 mm in diameter) primitive plasmodium with highly granular protoplasm, no vein-like strands and no advancing margin; known in only a few genera of myxomycetes.

Protoplast The membrane-bounded living contents of a cell from which the wall has been removed enzymically to facilitate transformation.

Protostele A type of stele with a solid core of primary xylem.

Prototroph A strain of microorganism whose nutritional requirements do not exceed those of the corresponding wild-type strain.

Prototrophic 1. of an isolate, strain etc. of a microorganism having no nutritional requirements in addition to those of the wild type.

2. requiring only inorganic nutrients.

It is recommended that this term should be used only in the sense of in which it is complementary to auxotrophic.

Prototunicate Describes a kind of ascus that is basically unitunicate, but whose wall lyses at or before maturity; such asci may develop in a hymenium or may be distributed randomly in the interior of the ascoma.

Prototunicate ascusprototunicate "Some asci have no active spore-shooting mechanism. These asci are usually more or less spherical, and are found in cleistothecial (occasionally perithecial), and hypogeous ascomata. Sometimes the wall of this kind of ascus dissolves at maturity and releases the **ascospores**, which can then ooze, rather than be shot, out of the ascoma; or they may wait inside until it decays or is ruptured. These asci are often called prototunicate. ... [T]hey are found in several otherwise rather different orders, [and] it seems likely that they represent a secondary condition, [which has] evolved several times from unitunicate asci." Ascomycetes and anamorphs. The term may be used simply to mean spores without any mechanism for ballistospory. Such spores typically have a fragile membrane which disintegrates spontaneously at maturity or breaks open with very slight mechanical stress.

Protozoan A class of minute animals whose entire body consists only of a single cell.

Protozoa is a plural word, the singular is protozoan. There are large numbers of species. Some protozoa are entomopathogens.

Protuberate Of conidia, With short projections not qualifying as appendages.

Proximal Nearest to the point of attachment .

Pruina A frost-like or flour-like surface covering.

Pruning Any removal or cutting of wood from a tree or vine.

Pseudo Prefix, indicating close or deceptive resemblance to the following element of the word.

Pseudocapillitium In myxomycetes, Empty plasmodial strands from which the cytoplasm has been withdrawn, found in some sporocarps, especially aethalia.

Pseudocoel In nematodes: Body cavity containing a fluid in which the various internal organs are suspended.

Pseudoextinction The apparent disappearance of a taxon. In cases of pseudoextinction, this disappearance is not due to the death of all members, but the evolution of novel features in one or more lineages, so that the new clades are not recognized as belonging to the paraphyletic ancestral group, whose members have ceased to exist. The Dinosauria, if defined so as to exclude the birds, is an example of a group that has undergone pseudoextinction.

Pseudohypha One of the short chains of cells resembling hyphae found in some yeasts.

Pseudomonopodial Branching a type of branching characterized by the presence of a main branch which has outgrown the other of the dichotomous pair, resulting in a main axis with smaller lateral branches. A type of branching where the apical meristem appears to divide to form two branches, one of which is dominant resulting in an upright main axis with distinct side branches.

Pseudoparaphyses Specialized hyphae found in some bitunicate ascomycetes; they grow down from the roof of the pseudothecial ascoma and dissolve a space for the developing asci.

Pseudoparaphysis In Ascomycota, a hypha originating in the top of the ascoma and growing down to the base where it becomes attached and may eventually become apically free, often regularly septate, branched and anastomosing and broader than paraphysoids.

Pseudoparenchyma Thin-walled, usually angular, randomly-arranged cells in fungi, which are tightly packed. Often used to form walls in specialized structures such as ascocarps. They are similar in appearance to the parenchyma cells of plants and have some of the same functions, i.e. as a population of undifferentiated cells competent to form various different types of specialized tissues.

Pseudoparenchymatous Describes 'tissue' made up of very densely packed fungal hyphae which in mass have come to resemble a perithecial ascoma.

Pseudoplasmodium A structure resembling a multinucleate plasmodium that has retained its cell membrane boundaries; an aggregation of amoebae, especially that forming the vegetative stage preceding sporocarp formation in the cellular slime moulds, and moving as a unit.

Pseudopodium A temporary cytoplasmic protrusion of an amoeboid cell used for locomotion or phagocytotic feeding.

Pseudosclerotium A compacted mass of intermixed hyphae and substratum.

Pseudostem Of gasteromycetes, A stem consisting of spongy tissue in which the hyphae are not orientated parallel to the stipe axis.

Pseudothecial ascoma Ascoma containing bitunicate asci; often resembling a perithecial ascoma, though developing differently.

Pseudothecium Perithecium-like fruiting body containing asci and ascospores dispersed rather than in an organized hymenium; an ascostroma with a single locule or cavity and containing bitunicate asci.

PSI pounds per square inch (a measure of pressure).

Psilocin and psilocybin Hallucinogenic indoles found in many species of Psilocybe, and some species of Panaeolus, Gymnopilus and Conocybe.

Psychedelic Mind-altering, hallucinogenic, psychoactive.

Psychrotolerant Growing at temperatures below 10ºC.

Pteridophyte A paraphyletic group including (roughly) all Rhyniophyta except Spermatophytata, i.e. stem rhyniophytes, lycophytes, ferns and horsetails.

Pubescent Having soft, short hairs.

Puccinia pathway The region through which rust urediniospores move from southern areas through all grain-producing areas of the United States to Canada each season.

Puffball A gasteromycete basidioma in which the basidiospore mass (gleba) is enclosed by a papery peridium at maturity; an ostiole allows compression or wind suction to disperse basidiospores in the Lycoperdales and Tulostomatales; the basidiomata of the Sclerodermales have no ostiole.

Puffing A phenomenon in which thousands of asci in an apothecial ascoma discharge their ascospores simultaneously, producing a visible cloud of spores.

Pulleywheel occlusion The structure blocking the septal pore in hyphae of the rust fungi (Uredinales).

Pulmonary Relating to the lungs.

Pulses (grain legumes) Leguminous plants or thier seeds, chiefly those plants with large seeds used for food.

Pulverulent Appearing as though covered with a fine powder.

Pulvinate Cushion-like in form.

Punctate Having surface dot(s), pore(s), etc.

Punctuated equilibrium A model of evolution in which change occurs in relatively rapid bursts, followed by longer periods of stasis.

Punctum A very small spot or depression that appears like a dot.

Punky Of a basidioma, soft and tough.

Pupa The third stage of life for insects (1 = egg, 2 = larva, 3 = pupa, 4 = adult). An insect larva changes into a pupa in order to make a final change into an adult insect. This is a resting stage where the insect usually doesn't move around or eat. With moths, the pupa is called a "coccoon." With butterflies, it is a "chrysalis." For most insects, we just call it a "pupa."

Puparium Some fly pupae are protected by a puparium. This is the dried, hardened, cast skin of the larva

Pupate To change from the larval stage to the pupal stage.

Pure live seed (PLS) Percentage of the content of a seed lot that is pure and viable.

Pustule A blister-like spot, on a leaf, stem, fruit etc., from which erupts a fruiting structure of a fungus.

Fig. Pustule

A local elevation of the epidermis that may rupture to expose the causal agent (e.g., rust, smut, white rust, etc.).

PV A subdivision of a plant pathogenic bacterial species defined by host range; pathovar for bacteria is equivalent to forma specialis for fungi.

Pychnoxylic Dense wood that contains little parenchyma; typical of Archeopteris, and conifers.

Pycnial stage (pycnium) A flaskshaped spore stage of the rust fungi; oozes out spores in a sticky matrix.

Pycnidial conidioma A flask-shaped sporoma produced by coelomycetous anamorphs.

Pycnidioid Resembling a pycnidium in shape.

Pycnidiospore Conidium produced within a pycnidium.

Pycnidium An open-pored, flaskshaped fruiting structure in which asexual spores called conidia are produced.

The asexual, globose or flask-shaped fruiting body of fungi-producing conidia.

Pycniospore A specialized spore, produced in a pycnium by rust fungi, that functions as male gamete. Synonym: spermatium.

Pycnospore An obsolete term for a pycnidiospore or a pycniospore.

Pycnothyrium A superficial, shield-shaped, flattened or hemispherical conidioma, comprised of radiating cells, sometimes with only an upper wall, sometimes also with a basal wall.

Pygmism The state of being dwarfed or reduced in size.

Pyramiding The addition, through plant breeding or genetic engineering, of several resistance genes into a single plant cultivar.

Pyreniform Shaped like a nut.

Pyrenoid A starch storage structure found in the chloroplasts of green algae.

Pyrenolichens Lichens in which the mycobiont produces perithecial or pseudothecial ascomata.

Pyrenomycetes Class of Ascomycotina, traditionally based on taxa with perithecioid ascomata that are ascohymenial in ontogeny and have unitunicate asci, often with an apical annulus.

Pyrethroids These are synthetic compounds produced to duplicate or improve more or less successfully on the biological activity of the active principles of the pyrethrum plant. Pyrethrum is a natural botanical insecticide the active principles of which are extracted from the flowers of the pyrethrum plant, and are know collectively as "pyrethrins."

Pyriform Pear-shaped.

Q

Quadrant One of four equal parts into which a field is divided for monitoring.

Quadriseriate In four ranks or rows.

Quantitative resistance Resistance reactions that have no distinct classes but vary continuously from resistant to susceptible, the result of few to many genes the individual effects of which may be small and difficult to detect.

Quarantine All operations associated with prevention of dissemination of unwanted organisms within and between territories.

Control of import and export of plants to prevent spread of diseases and pests.

Regulation forbidding sale or shipment of plants or plant parts, usually to prevent disease, insect, nematode, or weed invasion of an area.

Quiescence A period during which an infection is present but not active within a host: for example the period between an acute attack of chickenpox (varicella) and a subsequent recrudescence of shingles (zoster); not the same as latency.

Quiescent Dormant or inactive.

Quorum sensing The ability of bacteria to interact with each other through a variety of mechanisms; allows a population of bacteria to behave more like a multicellular organism.

R

Race 1. A genetically, and as a rule geographically, distinct mating group within a species.

2. physiologic race.

A genetically and often geographically distinct mating group within a species; also a group of pathogens that infect a given set of plant varieties.

A strain of a pathogen characterized by the limitation of its host range to certain species and varieties of plants.

Race non-specific resistance Resistance which is effective against all. biotypes of the pathogen.

Raceme Type of inflorescence in which the spikelets are borne on short stems (pedicels) attached directly to the main axis.

Rachis In fungi: A conidiophore elongating to one side of a terminally produced spore, often resulting in a zig-zag-shaped structure.

Radiate Diverging or spreading outwards from a central point.

Radiation Event of rapid cladogenesis, believed to occur under conditions where a new feature permits a lineage to move into a new niche or new habitat, and is then called an adaptive radiation.

Radicle Part of the plant embryo that develops into the primary root.

Radioimmunoassay Aan immunoassay in which radioactive tracers are introduced into the substance to be analyzed.

Radionuclide scans Any of a variety of medical imaging methods that rely on atomic isotopes that decay and emit radiation.

Random sampling Removing and/or examining a subset of a larger collection of objects by selecting the objects to be sampled without any pattern or pre-judgment (i.e., closing one's eyes, pointing toward an area, then opening the eyes and selecting that plant or plant part closest to the pointing fingertip).

Range Of a plant pathogen: The geographical region or regions in which it is known to occur.

Rank In traditional taxonomy, taxa are ranked according to their level of inclusiveness. Thus a genus contains one or more species, a family includes one or more genera, and so on.

Rapd Random amplified polymorphic DNA. PCR is used with short primers to amplify parts of target DNA that can then be compared in gel electrophoresis. Useful in finding species-specific patterns and in dealing with very small amounts of material.

Raptor Larger bird of prey. Raptors include hawks, eagles, ospreys, and owls.

Rasping Sucking mouth parts fitted with structures for roughening plant tissues, and also provided with modification of structures for sucking juices.

Rat tail bloom A secondary bloom in Bartlett pears that results when terminal buds form and open on the current season's growth.

Rate and dosage These terms are synonymous. "Rate" is the preferred term. Usually refers to the amount of active-ingredient material (such as 2,4-D acid equivalent) applied to a unit area (such as one acre), regardless of percentage of chemical in the carrier.

Rate The number of events happening divided by the length of time over which they happen. A rate of change is the amount of change happening in a interval divided by the length of the interval; for small intervals the rate of change might be given by a simple rule (a differential equation).

Ray Tissue that extends radially in the secondary xylem and phloem of a woody plant.

RDNA The DNA which codes for ribosomal RNA. It that term isn't familiar. However, there are some significant differences. In contrast to prokaryotes, the four RNAs contained in eukaryotic ribosomes are coded by two types of genetic units which are generally not linked. Each type occurs in its own randomly repeated clusters. The larger unit, the rDNA, is transcribed by RNA polymerase I as a single precursor containing the small subunit (18S) rRNA, 5.8S rRNA and the large subunit (28S) rRNA, each bracketed with spacer sequences. The second type of unit codes for 5S rRNA and is transcribed by RNA polymerase III. Peyretaillade et al.

Receptacle Any organ-bearing or hymenium-bearing structure; in Phallales, the stipe and pileus or the clathrate structure which supports the hymenium.

Receptive hyphae Special hyphae arising from spermagonia of rust fungi; the spermatia from other spermagonia fuse with them and initiate the dikaryon.

Receptor A site that recognizes and binds an elicitor; any organ or molecular site that is sensitive to a distinct (specific) signal molecule.

Recessive Describes a phenotypic trait that is expressed in diploid organisms only if both parents contribute the trait to the progeny .

Recombinant DNA DNA molecules in which sequences, not normally contiguous, have been placed next to each other by in vitro methods.

Recrudescence Reappearance of disease in a host whose infection has been quiescent.

Rectum In nematodes: Posterior gut of the female. A narrow, dorsoventrally flattened tube that is lined with cuticle and separated from the intestine by a sphincter muscle.

Recuperative potential The ability of a turf grass to recover from injury through vegetative growth.

Recurrent selection A form of plant breeding in which the best plants from one evaluation are used as parents and intercrossed to generate plants for the next round of evaluations.

RED (reregistration eligibility decision) EPA's conclusion about whether a pesticide's uses are eligible for reregistation, and if so, what tests and data must be completed before a reregistration decision can be made.

Reentry interval The EPA-designated period of time following the application of certain pesticides after which workers may return to the treated area without protective clothing.

Re-establishment Rebuilding a lawn by complete removal of any existing turf, followed by site preparation and planting.

Reflexed Bent back.

Refugia An area, untreated with pesticides, provided to preserve susceptible populations of pests.

Regenerate The ability to grow back part of a body that was lost. Crayfish can regenerate a lost leg, and earthworms can regenerate lost segments.

Registered pesticide A pesticide product that has been approved by the EPA for the uses listed on the label. Pesticide registration decisions are based primarily on EPA's evaluation of test data provided by the chemical manufacturers. Most EPA testing requirements focus on the active ingredients of the pesticide products only. EPA uses a risk/benefit analysis when deciding whether to approve a particular pesticide. This means that even if animal test results indicate that a pesticide is highly toxic, carcinogenic, teratogenic, or poses other health or environmental risks, the chemical may still be registered if certain economic, social, environmental or human health benefits are deemed to be greater than the risks.

Registration Formal licensing with a government-approved body of a new pesticide formulation that allows it to be distributed and sold. Registration normally requires data that demonstrates no unreasonable adverse health or environmental effects when applied according to the directions on the label.

Regrowth bud The buds on alfalfa crowns that become new stems.

Regurgitate Throw up. Some animals regurgitate their food to get rid of parts they can't digest, such as bones and hair. Some birds swallow food and then regurgitate it for their young to eat.

Reindeer 'moss' The fruticose lichen, Cladonia rangiferina, and similar species.

Relatedness Two clades are more closely related when they share a more recent common ancestor between them than they do with any other clade.

Relation 1. A specific association and interaction of one organism with another, as in host-parasite relations'.

2. a relationship that may be assumed when two organisms are compared and are found to have some or all characters in common.

With microorganisms the degree of similarity between two or more strains is determined by their common characters (i.e. multivariate association), but only in a limited number of cases can the relatedness or relationship be proved by the transfer of genetic material from one individual to another. This use of 'relation' is to be deprecated.

Relationship The suffix -ship implies a genetic connexion and the word is often wrongly used to indicate an association. In microbiology most relationships are only assumed, few (such as that between Escherichia, Shigella and Salmonella) are known. There are different degrees of relationship or relatedness; when genetic material can be transferred to produce forms that bear and reproduce the characters of both parents the degree of relatedness is close; there are other cases where transfer of perhaps non-nuclear material shows a less close relatedness.

American usage does not distinguish between relation and relationship.

Relative humidity Water vapour in air compared with the amount of water vapour held at the same temperature when saturated

Relative risk (RR) The proportion of diseased people amongst those exposed to the relevant risk factor divided by the proportion of diseased people amongst those not exposed to the risk factor. This should be used in those cohort studies where those with and without disease are followed to observe which individuals become diseased.

REMI (restriction enzyme mediated integration) A method of transformation that generates tagged mutations.

Remote sensing Examination of a site from a distance (i.e., collecting data about a farm field from an airplane or satellite).

Renal Relating to the kidneys.

Reniform Kidney-shaped.

Renovation, turf Turf improvement involving replanting into existing live and/or dead vegetation.

Repeating Of spores, giving rise to the same type of mycelium as that on which they developed.

Repellent A compound that keeps insects, rodents, birds, or other pests away from plants, domestic animals, buildings, or other treated areas.

Repent Prostrate; creeping.

Replant effect Severe reduction in rate of root and shoot growth of a second planting of a given perennial species following the same or a closely related species on the same site. B. M. Savory proposed the term specific replant disease,to distinguish the condition from one caused by a 'soil sickness' affecting unrelated plant species. In accordance with our recommendations for the use of disease and disorder, we suggest that 'specific replant disease' should be used when the effect is attributable to the action of a parasitic organism; 'specific replant disorder' when attributable to a non-parasitic factor, and 'specific replant effect' when the cause is undetermined.

Replicase An enzyme involved in the replication (copying) of nucleic acid.

Replication The process by which a DNA or RNA molecule makes an exact copy of itself; repetition of an experiment or procedure at the same time and place (one of several identical experiments, procedures, or samples); the process by which a virus particle induces the host cell to reproduce the virus.

Replicon Any DNA sequence or molecule that possesses a replication origin and is therefore potentially capable of being replicated in a suitable cell.

Reproduce The way animals, plants, or other organisms make new versions of themselves.

Reproductive bud The buds on alfalfa stems that become flowers.

Reproductive ratio 1. Basic reproductive ratio, Ro, a dimensionless parameter which encapsulates the biological details of different transmission mechanisms. For microparasites, Ro, is defined as the average number of secondary cases of infection to which one primary case gives rise throughout its infectious period if introduced into a defined population consisting solely of susceptible individuals. For macroparasites, Ro, is the average number of female offspring (or just offspring in the case of hermaphroditic species) produced throughout the lifetime of a mature female parasite, which themselves achieve reproductive maturity in the absence of density-dependent constraints on the parasite establishment, survival or reproduction. Also known as the basic reproduction rate, number.

2. Effective reproductive ratio, R, The number of secondary cases (microparasites) or female offspring (macroparasites) produced in a host population not consisting entirely of susceptible individuals (microparasites) or within which density dependent constraints limit parasite population growth (macroparasites). Under conditions of stable endemic infection, R=1.

Reptile Cold-blooded, scaly animals from the Chordate Phylum. This class of animals breathes air and lives mostly on land. It includes turtles, snakes, and lizards.

Reregistration A reassessment by EPA of previously registered pesticides according to current scientific standards. This process, mandated by Congress, is underway for approximately 20,000 pesticide products registered prior to 1972.

Reservoir The site where a pest population or quantity of inoculum can survive in the absence of a host crop, and from which a new crop may be invaded.

Residual The property of a pesticide for persisting after application in amounts sufficient to kill pests for several days to several weeks or even longer.

Residual pesticide A pesticide that continues to be effective on a treated surface or area for an extended period following application.

Residue management Management of rice straw and stubble after harvest.

Residue The quantity of a pesticide, its degradation products, and/or its metabolites remaining on or in a plant, animal tissues, soil, or surfaces after spraying or any other method of pesticide application.

Resin Sticky to brittle plant product derived from essential oils; often a defense compound of conifers.

Resinosis An abnormal exudation of resin from conifers.

Resin flow through bark or from wounds or cankers on conifers.

Resistance 1. The ability of an organism to withstand or oppose the operation of or to lessen or overcome the effects of an injurious or pathogenic factor (APS).

2. the ability of the host to suppress or retard the activity of a pathogenic organism or virus (APS).

The first definition is appropriate to the resistance of a fungus or bacterium to a plant-protective chemical as well as to that of a host against a parasite. The term may properly be given quantitative qualification, e.g. high, moderate or low, according to the effectiveness of the protective mechanism.

Host resistance and pathogen virulence are complementary aspects of a host-pathogen interaction, and each must be assessed in relation to the other: thus if strain ' X ' of a pathogen induces severe disease in a host and strain ' Y ' induces negligible disease, the host can be said to possess low resistance to ' X ' and high resistance to ' Y '. Conversely, if the hosts are genotypically identical it may be said that strain ' X ' is virulent and strain ' Y ' non-virulent.

The resistance of a pathogen to a chemical protectant is assessed in relation to the dosage of the material.

In expressions such as rust-resistant, scab-resistant, wilt-resistant etc., the name of the disease is used loosely as a substitute for the name of the pathogen, but in some instances a distinction must be made between resistance to the pathogen and resistance to the disease: the two may not be synonymous if, for example, reduced disease (i.e. reduced symptomatic evidence of disorder) results from increased tolerance of a given level of infection rather than to reduced infection.

The phenomena of resistance have been studied by descriptive pathologists, geneticists, physiologists, epidemiologists etc., and each group has described resistance in terms that relate to its own point of view. These sets of terms concern the same master-set of phenomena but they do not correspond precisely: thus resistance may be described in terms relating to its mode of inheritance. e.g. oligogenic or polygenic major gene or minor gene or to specificity in host-parasite relations,e.g. race-specific or race non-specific vertical or horizontal or to the presence or absence of host reactions, e.g. active or passive.

The sum of the qualities of the host and causal agent that retard the activities of the causal agent. 1. The reduction, due to genetic selection, of susceptibility of a parasite or its vector to chemotherapy.

2. The ability of a host to resist a pathogen.

Resistance, acquired Non-inherited resistance conferred by some predisposing treatment.

Resistance, active Resistance resulting from host reactions occurring in response to the presence of the pathogen or its metabolites, e.g. formation of phytoalexins, thickening of cell walls, suberization etc.

Resistance, adult plant Resistance detected at post seedling stages of development. There is no implication as to its specificity or mode of inheritance. Synonymous with resistance, mature plant.

Resistance, field Resistance which is observed under natural infection in field conditions but is not detected under the experimental conditions of resistance tests; it is expressed at low inoculum potentials and is likely to be markedly sensitive to environmental factors. Horizontal resistance is commonly of this type, but the terms are not synonymous because horizontal resistance refers specifically to the absence of differential interactions between varieties of the host and races of the pathogen. The term field resistance does not carry any implication either as to the mechanism of resistance or to the manner in which it is genetically controlled: it is often polygenic but field resistance is not synonymous with polygenic resistance.

Resistance, general (generalised) Resistance to more than one disease. Sometimes used synonymously with race non-specific resistance but this use is not recommended.

Resistance, horizontal Resistance which is evenly spread against all races of the pathogen. Resistance in which there is no differential interaction in an analysis of variance on a series of pathotypes and a corresponding series of pathodemes, The expressions 'vertical' and 'horizontal' resistance derive from a graphical representation by van der Plank of the relationships between races of a pathogen and varieties of its host (pathotypes and pathodemes respectively, according to Robinson) . Horizontal resistance is generally quantitative in effect, e.g. with increased horizontal resistance rates of infection are reduced, incubation times are increased, reproduction rates of the pathogen are reduced and the spread of an epidemic is consequently slowed down. Although horizontal resistance is commonly polygenic it is not necessarily so, and the terms are not synonymous. Field resistance and horizontal resistance also are not synonymous, even though field resistance is often of this type. However, the term is equivalent to race non-specific resistance and the latter term is preferred because it is descriptive of the condition to which it refers.

Resistance, intermediate Synonymous with moderate resistance.

Resistance, juvenile Resistance detected at early stages of host development.

Resistance, major gene Resistance controlled by genes which have a large observable effect. Not synonymous with oligogenic resistance. The classification of genes into major and minor determinants is arbitrary and depends on subjective judgement of the relative size of the effects observed.

Resistance, mature plant Synonymous with resistance, adult plant.

Resistance, mature tissue Resistance which develops when tissues are mature; it may be expressed at any stage of host growth.

Resistance, minor gene Resistance controlled by genes which have small observable effects.

Resistance, monogenic (digenic etc.) Resistance controlled by one (two etc.) genes. Resistance controlled by a few genes is termed oligogenic.

Resistance, multigenic Synonymous with polygenic but not recommended because of its mixed etymological derivation.

Resistance, oligogenic Resistance controlled by few genes the phenotypic effects of which may be large or small, i.e. major or minor genes.

Oligogenic resistance is not synonymous with major gene or vertical resistance.

Resistance, passive Resistance which does not entail any defensive reactions of the host to the presence of the pathogen. Passive resistance may result, for example, from physical or chemical attributes of the host which prevent infection, or from an absence of qualities which stimulate or facilitate infection.

Resistance, polygenic Resistance controlled by polygenes, i.e. a series of genes controlling a quantitative character. Not synonymous with adult plant, field, minor gene or horizontal resistance.

Resistance, qualitative Resistance that is expressed in large and discontinuous increments. It is not substantially influenced by environmental factors within normal limits. It is commonly associated with oligogenic inheritance. The term should not be used as an alternative to oligogenic or major gene resistance because it refers to the manifestation of resistance and not to its inheritance. It is not synonymous with vertical resistance. The term should only be used when there is insufficient information to describe the form of resistance more precisely.

Resistance, race nnon specific Resistance to all races of a pathogen. The term is essentially synonymous with horizontal resistance and is preferred because it is more obviously descriptive of the condition to which it refers. This form of resistance is commonly, but not necessarily, polygenic and the two terms are not interchangeable.

Resistance, race specific Resistance to some races of the pathogen, but not to others. Usually related to cases of oligogenic resistance where there is a gene-for-gene relationship between host and pathogen. The term is preferable to the synonymous vertical resistance because it is descriptive of the condition to which it refers. It should not be used as a synonym for oligogenic resistance, for the latter is not necessarily race specific; the terms relate to different attributes of resistance.

Resistance, seedling Resistance detectable at the seedling stage. There is no implicate to its specificity, mechanism or mode of inheritance.

Resistance, stable Resistance which retains its usefulness for a long period. Durable resistance may be preferred. Care should be used with both terms since they refer to the genetic equilibrium of host-pathogen interaction and not the mechanism involved. Not synonymous with non-specific resistance.

Resistance, vertical Resistance to some races of a pathogen but not to others. Resistance in which significant interactions are detectable in an analysis of variance on a series of pathotypes and a corresponding series of pathodemes.

The use of 'horizontal' and 'vertical' in this context derives from a graphicalre-presentation by van der Plank of the relationships between races of a pathogen and varieties of its host. Vertical resistance is often expressed as a qualitative modification of disease development, e.g. as a hypersensitive reaction to infection. Vertical resistance is not synonymous with oligogenic, major gene or active resistance because it refers specifically to the interactions between races and varieties and not to the mode of inheritance or the mechanism of resistance. However, vertical resistance is exactly equivalent to the more familiar and obviously descriptive race specific resistance, and the latter is recommended in preference to the former.

Resistant Possessing qualities that hinder the development of a given pathogen. The same considerations apply to the use of this term as to resistance.

Resistant species Pests that survive relatively high rates of pesticide application.

Respiration The oxidative breakdown and release of energy from from fuel molecules from fuel molecules with by reaction with oxygen in aerobic cells

Response The change produced in an organism by a stimulus.

Resting spore A sexual or other thick-walled spore of a fungus that is resistant to extremes in temperature and moisture and which often germinates only after a period of time from its formation.

Restricted use pesticide A pesticide which may only be purchased and used by certified applicators. This designation is given by EPA to those pesticides deemed to pose the highest hazard to applicators, Learning About Pesticides at School 29 the public, or the environment.

Restriction endonuclease An endonuclease that binds to double-stranded DNA at a specific nucleotide sequence and then, if both strands of the DNA lack appropriate modification at that sequence, cleaves the DNA either at the recognition sequence or at another site in the DNA molecule.

Resupinate Describes basidiomata that lie flat or spread out in a thin layer on the substrate, with a superficial hymenium: 'paint fungi' of families Thelephoraceae, Corticiaceae (Aphyllophorales: Holobasidiomycetes).

Resurgence The rebounding of a pest population to higher than pre-treatment levels following a pesticide application. This is due to the effect of the pesticide in destroying or reducing populations of natural predators, parisitoids or pathogens that would otherwise control the pest. Any pest organisms that escape the pesticide's effects (or that arrive after the application) are then able to flourish unchecked.

Retention period The time following acquisition feeding during which a vector is capable of transmitting virus.

Reticulate With a net-like pattern; net-like.

Reticulation Joining of separate lineages on a phylogenetic tree, generally through hybridization or through lateral gene transfer. Fairly common in certain land plant clades; reticulation is thought to be rare among metazoans.

Retina A light-sensitive area, as in the subsporangial vesicle of Pilobolus.

Retroculture A culture resulting from the re-isolation of an organism from an experimentally infected host or inoculated substrate. A term rarely used in plant pathological literature and not to be recommended.

Retrogressive A mode of blastic conidiogenesis in which the short cylindrical segments of the conidiogenous cell differentiate into conidia; the conidiogenous cells becomes shorter during conidiation.

Retrorse In a backward or downward direction.

Retrospective study A study in which people are enrolled and then have their history of risks, infections or disease measured.

Retrovirus A family of RNA viruses that have the unique characteristic of producing an enzyme that makes a DNA copy of its genetic information from an RNA template (the opposite of what normally takes place). The most widely recognized of these viruses is HIV, the causative agent in AIDS. Another virus from this family (HTLV-1) has been associated with T cell leukemia. Initial reports of an association of an HTLV-II-like retrovirus with CFS could not be confirmed in subsequent studies.

Reverse transcriptase An enzyme so called because it operates in a manner which reverses normal transcription. That is, a reverse transcriptase uses RNA as a template to create a complementary single-stranded DNA molecule.

Revolute Of the margin of the pileus: rolled back or up.

RFLP (restriction fragment length polymorphism) Inherited differences in sites for restriction enzymes (e.g., caused by base changes in the target site) that result in differences in the lengths of the fragments produced by cleavage with the relevant restriction enzyme. RFLPs are used for genetic mapping to link the genome directly to a conventional genetic marker.

Rhabdions In nematodes: Plates in the cuticular lining of the stoma that make up the walls of the various divisions of the stoma.

Rhexolytic Secession of conidia involving the circumscissile rupture or breakdown of the longitudinal wall of the cell below the conidium.

Rhinocladiella sp Rhinocladiella sp. Naturally found in soil and woody plant materials as a saprophyte. Reported among the principal fungi causing chromoblastomycosis, a disease characterized by a chronic localized infection of the skin and subcutaneous tissue that follows the traumatic implantation of the causal agent. The lesions are verrucoid, ulcerated, and crusted, and may be flat or raised 1-3 cm. The mycosis usually remains localized with extensive keloid formation. Forms of the disease include Verrucous dermatitis, Brain abscess syndrome, Single or multiple cysts, Local or systemic lesions. Morphological characteristics - Conidiophores simple, or branched in some species. Conidia apical on new growing points of conidiophore, subhyaline to dark, mostly 1-celled, ovoid to oblong-ellipsoid, dry. Cultivation - Potato dextrose agar or Malt extract agar, 24°C.

Rhizobia The common name for several genera of bacteria which have the ability to infect the root of legumes and to produce root nodules. Each species of rhizobia can infect some but not all legumes.

Rhizobium Genus of bacteria that lives symbiotically with roots of leguminous plants; during the symbiosis, atmospheric nitrogen gas is converted into a form useable by the plant.

Rhizobium nodules Galls on roots caused by Rhizobium spp.

Rhizoid 1. A root-like structure;

2. filamentous extensions of a chytrid or other thallus used for attachment and assimilation (rhizomycelium); (3) in Basidiomycota, a rather large and distinct hypha at the base of a stipe.

Rhizome A (usually) underground stem that is horizontally oriented; rhizomes may appear like roots, but have a definite node and internode architecture.

Rhizomorph A macroscopic, typically rope-like strand of compacted tissue formed by certain higher fungi. Rhizomorphs often are enduring structures that can remain dormant under adverse conditions.

Rhizomucor sp Rhizomucor sp. Belongs to the Zygomycetes (mucorales) and it is reported to be allergenic and often linked to occupational allergy. It may cause mucorosis in immune compromised individuals. It occupies a similar biological niche to Mucor sp.. The sites of infection are the lung, nasal sinus, brain, eye, and skin. Infection may be multiple sites. It may also cause zygomycosis (rhino-facial-cranial area, lungs, gastrointestinal tract, and skin). This disease is associated with the acidotic diabetes, malnourished children, severely burned patients, and other diseases such as leukemia and lymphoma, immunosuppressive therapy, or use of cytotoxins and corticosteroids. The fungi show a proclivity for vessel invasion resulting in embolization and necrosis of surrounding tissue. Morphological characteristics of this fungus are the production of sporangiospores, which are diminute, rounded thick walled bodies resistant to heat and drought. They are produced in large numbers in globular envelopes (sporangia) at the tip of special hyphae (sporangiophores). Identification is based on the way the sporangia are formed. Cultivation - Potato dextrose agar, 24°C.

Rhizomycelium Branching, anucleate or sparsely nucleate, rhizoidal filaments of variable width forming part of the thallus in some fungi.

Rhizoplane The root surface.

Rhizoplast A fibrillar structure in a zoospore connecting the kinetosomes (at its proximal face) with the nuclear envelope.

Rhizopus sp Rhizopus sp. Fungus found throughout the environment. It has been reported to be allergenic and it is often linked to occupational allergy. It may cause mucorosis in immune compromised individuals. It may also cause zygomycosis (rhino-facial-cranial area, lungs, gastrointestinal tract, and skin). This disease is associated with the acidotic diabetes, malnourished children, severely burned patients, and other diseases such as leukemia and lymphoma, immunosuppressive therapy, or use of cytotoxins and corticosteroids. The fungi show a propensity for vessel invasion resulting in embolization and necrosis of surrounding tissue. Cultivation - Potato dextrose agar, 24°C.

Rhizosphere An environment regarded, variously, as (a) that region of the soil modified as a result of the uptake and deposition of substances by a growing root, (b) the root itself, together with that volume of soil which it influences, (c) the root surface together with that region of the surrounding soil in which the microbial population is affected by the presence of a root.

Rhizosphere-competent Used to describe microorganisms adapted to living in the rhizosphere of a plant.

Ribonucleic acid Several nucleic acids composed of repeating units of ribose (a sugar), a phosphate group, and a purine (adenine or guanine) or a pyrimidine (uracil or cytosine) base; transcribed from DNA and involved in translation to proteins.

Ribosome A subcellular protoplasmic particle, made up of one or more RNA molecules and several proteins, involved in protein synthesis.

Ribozyme An RNA molecule with catalytic properties, typically in association with a cofactor such as a neucleotide or metal ion.

Rickettsia A single-celled animal and human diseasecausing organism with a partial cell wall that has not been grown in culture.

Rimose Cracked, usually in a radial manner.

Ringshake Peripheral cracks in woody tissues of stems. The pattern of damage is concentric with the annual rings.

Ringspot A term denoting a spot (usually of leaves) surrounded by a ring or rings which may be of chlorotic, necrotic or abnormaly dark green tissue.

A circular area of chlorosis with a green center.

Symptom of a disease characterized by yellowish or dead (necrotic) rings with green tissue inside them, as in certain virus diseases.

Ringworm Superficial mycoses caused by keratinophilic Hyphomycetes.

Risk assessment A quantitative or qualitative evaluation of the environmental and/or human health risks resulting from exposure to a pesticide or other substance. Considers proposed or existing uses, exposure potential and toxicity factors.

Risk factors Factors whose presence is associated with an increased probability that disease will develop later.

Risk/benefit analysis An approach that attempts to weigh the risks (e.g., to human health or the environment) of use of a certain pesticide against the benefits (e.g., social, economic) of its use.

River A body of moving water, usually larger than a stream. Rivers flow until they empty into a larger river or other body of water, such as a bay or ocean.

Rivulose Marked with lines shaped like rivulets.

RNA Several nucleic acids composed of repeating units of ribose (a sugar), a phosphate group, and a purine (adenine or guanine) or a pyrimidine (uracil or cytosine) base; transcribed from DNA and involved in translation to proteins.

RNA blotting A technique for transferring RNA from an agarose gel to a nitrocellulose filter on which it can be hybridized to a complementary DNA.

Rock tripe Rock-inhabiting lichens of the genus Umbilicaria.

Rodenticide A pesticide that is used to kill rats, mice and other rodents.

Roestelium In the Uredinales an aecium with a cornute peridium, thin at the sides, usually rupturing by longitudinal slits, and made up of characteristically marked and imbricated cells. The name is taken from the form-genus Roestelia.

Rogue To remove and destroy undesired individual plants from a planting on the basis of disease infection, not being true-to-type, insect infestation, or other reason.

Rolling Mechanical crushing of potato vines to hasten vine death, sometimes used synonymously with vine-killing.

Roost A perch where an animal, especially a bird, rests or sleeps.

Root The underground part of a plant that pulls in water and nutrients from the soil. Roots also help keep soil in place so it doesn't wash away (erosion).

Root cap A group of cells on a root that protects the growing tip.

Root crown The uppermost portion of the root system where the major roots join together at the base of the stem.

Root disturbance One of the most common and yet least recognized causes of tree problems is root disturbance, often associated with some sort of construction activity such as laying of pipe or cable, street widening, placing a new building in an older site or turning residential areas into parking lots or patios. Even shallow cultivation in the root zone of trees can damage important roots growing close to the surface. Transplanting established trees with wide, spreading root systems can also result in decline through loss of a substantial part of the root system.

Piling only a few inches soil under the dripline of established trees can suffocate the roots resulting in tree death. This commonly occurs as a result of building construction or establishment of landscaping berms.

Root exudates The various compounds that leak from growing and expanding sections of roots as well as from broken cells at exit points of lateral roots.

Root graft The fusion of roots of two adjacent plants so that their water and food conducting (vascular) systems become joined.

Root hair Threadlike, single-celled outgrowths from a root epidermal cell.

Root rot Decay of roots in a living plant; often caused by pathogenic fungi.

Root zone The upper six to eight inches of soil, in which most of the turf-grass roots are concentrated.

Rootlet A small root, or a branch of a larger root. Vine rootlets are often above ground and used to grab onto tree trunks and other structures so the vine can climb. Many birds use these rootlets as nest material.

Rootstock Portion of the stem (trunk) and associated root system into which a bud or scion is inserted in grafting; fleshy overwintering part of a herbaceous perennial plant with buds and eyes.

Roseola Also known as roseola infantum, exanthem subitem, and pseudorubella. An acute disease of infants or very young children caused by HHV-6 and characterized by high fever and a skin rash.

Rosette A group of leaves, making a circle, reaching out from a central spot. Rosettes usually are mostly flat and close to the ground. Biennial plants grow a rosette thier first year.

Rosetted bloom A flower whose petals have been tied together with silk by a pink bollworm larva.

Rosetting Severe reduction of internode growth in a vertical axis without comparable reduction in the size of the leaves.

Rostrum A beak-like process.

Rot Disintegration of tissue as a result of the action of invading fungi or bacteria; a disease so characterised.

Softening, discoloration, and often disintegration of succulent plant tissue as a result of fungal or bacterial infection.

Rot, brown Rotting of wood in trees invaded by cellulose-destroying fungi, leaving brown lignin residue. Also applied to diseases caused by Monilinia (Sclerotinia) fructigena in fruits of apple, pear etc.

Rot, collar Rotting of the stem or main axis at or about the level of the soil surface; a disease in which this is a typical symptom.

Rot, dry 1. Rotting that proceeds at a rate which allows drying of the lesion to keep pace with lysis.

2. A term applied specifically to rotting of apparently dry timber by Serpula lacrimans.

Rot, foot 1. In a seedling with hypogeal cotyledons, rotting of the axis from immediately above the seed.

2. generally, rotting involving the lower part of the stem-root axis, but not the distal part of the roots.

Rot, neck Rotting of the ' neck ' of a bulb or mangold.

Rot, root Rotting of any part of the root system (not confined to the foot).

Rot, soft Rotting of tissue (usually parenchymatous, e.g. storage organs, fruit etc.) by action of the pathogen on the middle lamella of cell walls; cells are separated but retain their identity for a period.

Rot, wet Tissue rapidly and completely disintegrated, with release of free water from the lysed cells. Probably differs from soft rot only in the rate at which cells lose their capacity to retain water.

Rot, white Rotting of wood in trees invaded by lignin-destroying fungi,leaving white cellulose residue.

Rotation A cropping system in which crops are grown in a field in a fixed annual sequence. A rotation reduces the build up of diseases and pests, aids weed control, improves soil fertility, spreads the risk of crop failure and allows even distribution of labour requirements.

Rotational grazing The practice of of imposing a regular sequence of grazing and rest on the area to be grazed

Rotting The disintegration of tissues by the action of microorganisms.

Roundworm Nematode.

Routine test dilution (RTD) The highest dilution of phage that will produce confluent lysis on the homologous bacterial type.

Row covers Any fabric or protective covering placed over rows of plants to protect them from pest damage or harsh climate.

RRNA (ribosomal RNA) RNA molecules forming part of the ribosomal structure.

Rubella Also known as German measles, an acute disease marked by skin rash and swollen lymph nodes, but generally without fever. It is caused by an RNA virus of the togavirus family.

Rue leaf Any leaf produced after the seed leaves (cotyledons).

Rugose A rough appearance of leaves in which veins are sunken and interveinal tissue raised, caused by certain virus infections.

Ruminant An animal thet chews the cud and has a complex digestive system with a four part stomach enabling bacteria to break down food. Ruminants lack upper incisor teeth and their complex stomach allows them to store and digest large amounts of bulky and fibrous food

Rumposome A membranous organelle appressed to a lipid globule in zoospores of Chytridiales and Monoblepharidales to which the flagellar rootlet system microtubules run from the kinetosome.

Run off 1. The process of shedding spray from plant surfaces during and immediately after application, when droplets coalesce to form a continuous film and the surplus liquid runs off the surface. Commonly used in the expression ' spray to run-off '.

2. the quantity of spray which runs off unit area of plant surface.

Runner A slender horizontal stem that grows close to the soil surface.

Runoff Pesticide material that is carried away from an area by the flow of surface water. Also used to describe the rate of application to a surface "spray to a rate of runoff."

Russet Scorched or burnt appearance of plant surfaces, especially leaves or pods; roughened surface of fruit or vegetables.

Rust Formerly a name for diseases characterised by the formation of rust coloured pustules on stems and leaves. Now properly 1. a disease caused by a fungus of the order Uredinales;

2. a fungus of the order Uredinales. White rust is an acceptable name for a rust disease of chrysanthemum caused by Puccinia horiana, but it should not be applied to the white blister diseases caused by species of Albugo (Peronosporales).

Rust fungus Fungus belonging to the Uredinales.

S

Saccate Pouch or sac-like.

Saccate pollen Pollen with a saccus or sacci; characteristic of many conifers.

Saccus A wing-like or bladder-like extension on a pollen grain.

Safener A chemical that when used in combination with a pesticide reduces its effects on non-target organisms. Only herbicide safeners have so far been produced, and these protect crops from injury by herbicides but do not prevent the herbicide from killing weeds.

Sagenogen A specialised organelle on the cell surface of Labyrinthulomycetes that produces the ectoplasmic net; bothrosome; sagenogenetosome.

Salinity An excess of soluble salts in the soil; turf-grass growth and quality are impaired.

Saliva Watery body fluid in the mouth of an animal; spit. Sometimes certain animals have chemicals in their saliva which helps them eat.

Salt bridge A reference to an electrode of an EC meter.

Salt toxicity Salt applied to streets, driveways, or sidewalks in winter is absorbed by tree roots during the growing season. This results in varying degrees of browning of the leaf edges, especially during drought periods. Gradual dieback may occur in some salt damaged trees; some trees may be killed.

Saltant A discontinuous variation of unknown origin, e.g. a mutant unconfirmed by genetic evidence. Use of this term should be avoided.

Saltation A mutation occurring in the asexual state of fungal growth, especially one occurring in culture.

Samara Fruit of certain trees, such as maples and elms. Samaras have a seed and a "wing." Sometimes they are called "keys" because their shape is similar to the shape of a key. The wing helps the seed travel by wind, since it is wide and lightweight and catches the air.

Sampling Removing and/or examining a portion of an entire set (i.e., examining three leaves per plant on 20 plants in a 10-acre field).

Sanitation Destruction or removal of infected and infested plants or plant parts; decontamination of tools, equipment, containers, work space, hands, etc.

Sap Liquid from a plant. Sap can be in a trunk, stem, leaf, or fruit. Some saps are poisonous.

Sap transmission Transmission, usually of viruses, by rubbing sap from an infected plant onto a healthy plant to cause infection.

Saprobe An organism that obtains its nutrients from non-living organic matter (commonly dead and decaying plant or animal matter) by absorbing soluble organic compounds. (Also saprotroph.)

Saprogenesis Survival; that phase of the life cycle of a pathogen during which it is not actively causing disease in a living suscept.

Saprophyte An organism that derives its nourishment from dead organic matter.

a plant (organism) using dead organic material as food (BMS). If the dead material is part of a living host the term perthophyte is applicable, as in the case of those organisms which kill host tissues in advance and obtain nutrients from the dead cells, e.g. Pythium debaryanum, Botrytis cinerea. Saprobe is sometimes used in the sense of saprophyte for fungi, protozoa etc., saprophyte being restricted to higher plants.

An organism living on dead organic matter.

Saprotroph An organism obtaining its nutrients from dead organic matter; saprobe.

Saprotrophic Utilising non-living materials as nutrients.

Sapwood Physiologically active zone of wood contiguous to cambium .

Sarcody A hyperplastic symptom in which swellings occur above and below portions of organs that are tightly encircled.

Sarcoidosis A systemic disease involving the lungs, lymph nodes, skin, liver, spleen, eyes, phalangial bones, and parotid glands, characterized by granular nodules. Its cause is not known.

Satellite virus A defective virus requiring a helper virus to provide functions necessary for replication. It may code for its own coat protein or various other products.

Scab A discrete superficial lesion involving localized severe roughening or pitting; more commonly, abnormal thickening of the surface layers, with or without the development of cork layers; localized hyperplasia of the surface tissues; a disease characterised by lesions of one of these types.

Crustlike disease lesion.

Scabrid Rough with delicate and irregular projections.

Scabrous Rough with short, rigid projections.

Scald 1. A term for lesions (or diseases involving lesions) that suggest to the eye the apparent consequences of scalding with hot water. Such lesions are mainly bleached and may be partly translucent: they are not chlorotic in the usual sense of that description.

2. a term applied to a non-parasitic disorder of apple.

3. American common name for leaf blotch of barley, caused by Rhynchosporium secalis.

Scald, sun A superficial form of damage to fruits that results from the action of intense sunlight, especially when the surface is wet with dew etc., e.g. of grape, apple and peach.

Scale insects A large group of insects that are minute and highly specialized. The females are wingless and usually legless, while the males have only a single pair of wings or none at all. Adult male scales look like gnats, but can usually be recognized by the absence of mouthparts. The first immature stage is called a crawler and moves about to find a spot to feed. Later stages lose their legs and antennae (or they are greatly reduced) and lose the ability to move. These insects are one of the more destructive groups found on ornamental plants.

Scales These are flat unicellular outgrowths of the body wall in various shapes that are common to many groups of insects. Some example are: on butterflies and moths, scales are flattened plates that often contain pigments that lend color to the insect; in scale insects, scales are coverings made of waxy substances for the protection of eggs, nymphs, and adults.

Scalp To remove an excessive quantity of functioning green leaves at any one mowing; results in a shabby, brown appearance caused by exposing crowns, stolons, dead leaves, and even bare soil.

Scat Animal poop. Many scats can be identified by thier shape, size, and color.

Scavenger An animal that eats dead or decaying material.

Scedosporium sp Scedosporium sp. This fungus is found in soils, decaying plants matter or dung. Listed among the principal fungi that cause Phaeohyphomycosis. This disease consists of a group of mycotic infections characterized by the presence of dematiaceous septate hyphae and sometimes yeast or combination of both. Species of Scedosporium mostly affect people with compromised immune systems, but healthy people may also become infected. The hyphae may be short to elongate, distorted or swollen, regularly shaped or any combination of the above. Infections of the eyes and skin by the black fungi could also be included in this disease. . Cultivation - Potato dextrose agar, 24°C.

Schizolytic Secession of conidia involving a splitting of the delimiting septum so that one half of the crosswall becomes the base of the seceding conidium and the other half covers the apex of the conidiogenous cells.

Schizophrenia The most common type of psychosis, characterized by extensive withdrawal of the individual's interest from other people and the outside world and the investment of it in his/her own self.

School A small or large group of fish swimming together. Fish can work together to trap prey or protect themselves from predators.

Scientific name Every organism (animal, plant, fungus, protist, moneran) is given a scientific name. This way, scientists and other people from around the world can talk about the same thing. Many countries also give an organism a "common name." For example, the species Rana catesbeiana is called "Bullfrog" in America.

Scion Portion of a shoot used for grafting onto the root stock .

Sclerenchyma Tissue composed of cells with walls thickened with lignin; sclerenchyma tissue functions primarily in strengthening and support

Sclerotium 1. A firm, frequently rounded, mass of hyphae with or without the addition of host tissue, normally having no spores in or on it.

2. in myxomycetes, the firm, resting condition of a plasmodium.

Sclerotized In nematodes: Hardened refractive regions.

Scoalesce Grow together into one body or spot.

Scolecospore An asexual spore with or without septa; with a length:breadth ratio greater than 15:1; with a single axis, not curved through more than 180 degrees; any protuberances, other than setulae, not more than 1/4 the length of the spore body.

Scolecosporous Having long, thin (filiform) spores.

Scopulariopsis sp Scopulariopsis sp. This fungus is ubiquitous, and can be found on a wide variety of materials including old carpets and water-damaged wallpaper. Exposures from Scopulariopsis brevicaulis have been associated with cases of occupational allergy in the tobacco industry. It can decompose arsenic compounds found on building materials with an arsenic substrate, such as, some types of wallpaper and paints.

Scorch Any symptom, such as a lesion or a system of lesions that suggests the action of flame or fire on the affected part, especially as affecting extended organs such as green leaves or petals. On a given host, scorch symptoms may be caused by a variety of agencies and observations as to shape, distribution, age of organs affected and colour(s), provide clues as to the nature of such agency.

Appearing as if tissues were burned by heat; usually affecting marginal portions of leaves.

Burning" of plant tissue from infection, lack or excess of some nutrient, or weather conditions.

Scouting Systematic, regular monitoring of a crop or ornamental planting or landscape.

Scrapie Long established disease of sheep that causes intense irritation: cause unknown, thought to be neither bacteria or virus.

Screening host A plant in which one or more of the viruses of a mixture can be separated from the others.

Scum The layer of algae on the soil surface of thinned turfs; drying can produce a somewhat impervious layer that can impair subsequent shoot emergence.

Scutate Like a round plate or shield.

Scutellum In nematodes: An enlarged, shield-like phasmid.

Secession Of conidia, separation of conidia from the conidiogenous locus or neighbouring conidium by schizolysis or rhexolysis.

Secondary bloom A second production of flowers on a potato plant, occurring at the end of the mainstem of an indeterminate cultivar; secondary bloom may occur on a determinate cultivar at leaf axils along the mainstem.

Secondary cycle Of plant disease: any cycle initiated by inoculum generated during the same season.

Secondary growth Growth in width initiated and maintained by the vascular cambium and cork cambium. The vascular cambium surrounds the core of the stem, branch or root. It creates xylem medially and phloem radially. If present, the cork cambium, a second, outer layer of cambium, creates a new outer layer (typically bark) radially.

Secondary infection Any infection caused by inoculum produced as a result of a primary or a subsequent infection; an infection caused by secondary inoculum.

Infection resulting from the spread of infectious material produced after a primary infection.

Secondary inoculum Inoculum produced by infections that took place earlier during the same growing season.

Secondary metabolite A compound produced in microbes (e.g., mycotoxins, syringomycins) or plants (e.g., caffeine or nicotine) that is not necessary for normal growth and development.

Secondary organism An organism that multiplies in already diseased tissue but is not the primary pathogen.

Secondary outbreak The increase of a nontarget pest to harmful levels following a pesticide application, caused by destruction of natural enemies that normally control the nontarget pest.

Secondary parasitoids Hyperparasitoids or parasitoids that parasitize other parasitoids.

Secondary pest A pest that resurges following disruption of control by a natural enemy.

Secondary pollutant Air pollutant that must be chemically produced from other air pollutants, e.g. ozone (O_3) that is a product of a photochemical reaction of exhaust products from combustion engines in the atmosphere.

Secondary roots The network of fine roots that develops from the primary roots of a strawberry plant and picks up water and nutrients from the soil; white roots.

Secondary spread The spread of a pathogen within a field after the initial or primary infection.

Secondary stems Stems formed by stolons that emerge from the soil.

Secondary symptom A symptom of virus infection appearing after the first (primary) symptoms.

Sector A well-defined segment of an in vitro colony of a fungus differing from the rest of the colony, e.g. in pigmentation, growth rate, sporulation etc.

Sedentary Remaining in a fixed location.

Sedges A group of grasslike, herbaceous plants that, unlike grasses, have unjointed stems. Stems are usually solid and often triangular in cross section.

Fig. Sheath

Seed The part of a plant that can make another plant. Usually a plant makes flowers, then once the flowers have been pollinated, they turn into fruit. The seeds are inside the fruit. Seeds can be spread by wind, water, or animals. Then, when they are in a new place with water and nutrients, they can grow into a new plant. Sometimes an animal eats some fruit, then poops the seeds out someplace, and a new plant grows there!

Seed blend A combination of seeds of two or more cultivars of the same turf-grass species.

Seed dressing The process of cleaning seed. The term has been used to refer to the coating of seed with fungicide but this usage is deprecated.

Seed leaves The first leaf in grasses or the first two leaves in broadleaf plants. On seedlings, seed leaves are the cotyledons.

Seed piece Portion of a potato tuber containing at least one eye that is planted to produce a new potato plant.

Seed plants A monophyletic clade of plants that reproduces by seeds; megagametophyte is retained on the parent sporophyte and enclosed in an integument; microgametophyte is transferred to the megagametophyte.

Seed size Seed size in Phaseolus vulgaris is one of several seed characteristics influencing consumer acceptance. Thus Guatemalans prefer small-seeded black beans while Colombians favor large-seeded red beans. In general beans from the Mesoamerican center of domestication are small-seeded; those having an origin in the Andes of Ecuador and Peru are large-seeded.

Seed sterilization A method used to remove contaminant organisms from the surface of seeds. Laundry bleach is commonly used, although, where the seed must be scarified as well to break dormancy, concentrated sulfuric acid is also effective.

Seed treatment An application of a biological agent, chemical substance or physical treatment to seed, usually to protect the seed or plant from pathogens or to stimulate germination of the seed or growth of the plant.

Seedborne Carried on or in a seed.

Seedcotton Harvested lint that is still attached to seeds; i.e., the lint before ginning.

Seedhead Floral development; in the case of grasses, usually a fruiting cluster or spike.

Seedling diseases - Seedlings fail to emerge, or fall over and die (damping-off caused by fungi such as Rhizoctonia, Pythium, and Fusarium)

Dead areas on cotyledons or stems (fungi, bacteria).

Segment A ring or subdivision of the body or of an appendage between areas of flexibility associated with muscle attachments.

Seir model A class of compartmental prevalence models, with compartments Susceptible, Latent (Exposed), Infectious and Recovered. Takes its name from a common notation. In the notation used in the course, this would be an XHYZ model.

Selaginella reproductive cycleheterospory A condition in which an organism produces two different types and sizes of spores, viz. microspores and megaspores. As a reproductive strategy, heterospory involves the following steps: 1. The mature diploid plant (sporophyte) produces "male" and "female" spores. Typically, the plant produces numerous small "male" spores (microspores) and fewer, but larger "female" spores (megaspores).

2. The megaspores are usually retained on the sporophyte, where a haploid female gametophyte develops inside the spore.

3. The microspores may or may not be retained on the sporophyte. They develop into sperm-producing, haploid gametophytes.

4. The sperm are released into the environment, and dispersed between plants.

5. The sperm fertilize female gametophytes grown from megaspores, producing a diploid zygote (fertilized cell).

6. The zygote develops into an embryo and finally a new diploid sporophyte. Heterospory evolved a number of times among different ancient plants. Lycopsids, ferns, sphenopsids and progymnosperms all developed heterospory. Many of these plants also evolved into large trees. The obvious advantage for heterosporous trees is that their great height would enhance the dispersal of windborne microspores. Competition for light in the shady Archaeopteris forests, may have also have encouraged heterospory, since the macrospores, like seeds, included food reserves to assist early plant growth. Heterospory is generally regarded as the precursor to the evolution of seeds. Microspore dispersal and macrospore viability in pteridosperms thus become pollen dispersal and ovule viability in seed-bearing plants. The change in sexual reproduction from moisture-bound gametophytes to seeds-bearing sporophytes enable vascular plants to colonize drier habitats and upland environments, and also provided the embryo with a protective case against hungry invertebrate herbivores.

Selection Process which favors one feature of organisms in a population over another feature found in the population. This occurs through differential reproduction—those with the favored feature produce more offspring than those with the other feature, such that they become a greater percentage of the population in the next generation.

Selective herbicide A chemical that is more toxic to some plant species than to others.

Selective medium A culture medium containing substances that specifically inhibit or prevent the growth of some species of microorganisms.

Selective pesticide Pesticides that are toxic primarily to the target pest (and perhaps a few related species), leaving most other organisms, including natural enemies, unharmed.

Selectivity The ability of a pesticide to kill some pests but not others.

Self fertilization A process in which sexual reproduction occurs as a result of the fusion of sex cells produced by the same individual.

Self fruitful The ability to set fruit with pollen from the same flower or tree.

Self incompatible In some plant species, embryos initiated by fertilization of the ovary with pollen from the same plant do not develop. All seed then results from cross pollination between plants.

SEM (scanning electron microscope) Objects are coated with a thin layer of a metal in a vacuum, and examined in an electron beam.

Seminicolous Growing on seeds.

Senescence Nodules formed on the root of a host have a finite life span, usually 50-60 days. Plants can have several "crops" of nodules in a single growing season, and in leguminous trees current nodules may be at some distance from the stem. The breakdown of nodules over time is called senescence.

Senescent The growing old and dying back of plant tissue.

Sensitive Reacting with severe symptoms to the attack of a given pathogen. Similarly sensitivity; the inability of the affected organism to endure the operation of an injurious or pathogenic factor or the activity of a pathogenic organism or virus without more or less strong reaction, evidenced by varying degrees of symptom expression and damage.

The first definition refers solely to a host-parasite relation, whereas the second is equally relevant, for example, to the response of both plants and microorganisms to toxic chemicals.

Sensitive is the antonym of tolerant and when qualified quantitatively both words need to be used with caution, giving due consideration to the 'dosage' of the damaging factor. In relation to the action of an antibiotic or toxic chemical the expression 'very sensitive' generally implies that a given damaging effect (e.g. LD50) is brought about by a very low dosage of the material, and the meaning is unlikely to be misinterpreted. However, the same expression applied to a host plant invaded by a pathogen may mean that a severe effect has been caused by a very small amount of the pathogen (a usage analogous to that for a toxic chemical), but it is sometimes used to describe a situation in which severe symptoms have occurred as a result of a massive infection of the plant, i.e. the expression is used as a substitute for 'of low resistance'. This is not a valid usage: different grades of sensitivity should be related to a constant grade of infection.

In a non-quantitative use, in the sense of the BMS definition, sensitive is a preferred alternative to susceptible.

Sensitivity The ability of a test to work on people you know have the infection. More precisely TP/(TP+FN), where TP is the number of true positives and FN is the number of false negatives.

Sensitization The development of a hypersensitive or allergic reaction upon reexposure to a substance. The reaction may be immediate or delayed and may be of short-term or chronic duration.

Sensu lato In a broad sense.

Sensu stricto In a narrow sense.

Sentinel surveillance A monitoring method that employs a surrogate indicator for a public health problem, allowing estimation of the magnitude of the problem in the general population.

Sepal A whorl of a flower; sepals often resemble reduced leaves and function in the protection of the bud; sepals may be modified to function more like a petal.

Sepedonium sp Sepedonium sp. This fungus is mainly found as parasite of mushrooms, but it is also isolated from soil. Most easily recognized by the spores, which are colorless to yellow, spiny, round, 1-celled, and produced singly at the ends of short filaments. Sometimes spores of Sepedonium can be very similar to those of the human pathogen Histoplasma capsulatum and species of Morticrella. . Cultivation - Potato dextrose agar, 24°C.

Septa Bulkheads occurring at regular intervals in most eumycotan fungi; they strengthen hypha and define the compartments, but are perforate and in many cases allow passage of cytoplasm and nuclei.

Septal pore A possible synapomorphy of the Ascomycota. The hyphae of ascomycetes are partitioned off into compartments by septal walls. The septae are bridged by pores which provide controlled cytoplasmic continuity throughout the hypha.

Septate Having cross walls, or septa, that divide hyphae or spores into a number of separate cells.

Septum An dividing wall between adjacent cells of a hypha or spore.

Sequelae Morbid conditions following as a consequence of a disease.

Sequence The order of nucleotides in RNA or DNA or of amino acids in a polypeptide.

Sequential sampling A sampling method in which the number of samples is not fixed in advance.

Sequestrate Describes fungal fruit bodies which have evolved from exposed hymenia and forcibly discharged spores to a closed or even hypogeous habit in which the spores are retained in the fruit body until it decays or is eaten by an animal vector. Many sequestrate taxa can be clearly recognized a s being derived from specific spore-shooting ancestors, e.g. Rhizopogon from Suillus (Boletaceae).

Serial transmission The transmission of a virus by a test vector, after a single acquisition access period, to two or more test plants in succession.

Sericeous Silky.

Serological testing Bacteria injected into an animal such as a rabbit, induce the production of substances termed antibodies. These then react with the bacteria causing them to be precipitated. Since antibodies only react with the organism that caused them to be formed, or one very similar to it, serological reactions are also of value in identifying particular bacteria, for example the rhizobia.

Serology Literally the science of serum reactions, but used in microbiology for the study of the reactions between microorganisms or viruses and antibodies in the serum of immunized animals (including man). Serological reactions are reactions between antigen(s) and antibody(ies), and the different manifestations or kinds of reaction (agglutination, precipitation, complement fixation etc.) depend on the physical conditions in which the tests are carried out.

On the basis of serological tests, differences between apparently similar organisms or viruses may be found, and the subdivisions formed are called serotypes or strains.

The study of antigen-antibody reactions. More generally, the use of serotype data to infer an individual's history of infection.

Seropositive An individual whose serotype suggests that they have experienced infection in the past.

Seroprevalence The proportion of a population who are seropositive.

Serotype 1. (of bacteria) 'A subdivision (infrasubspecific) of a species or subspecies distinguished from other strains of the same species on the basis of its antigenic structure' (Bact. Code, Rec. 8a (2) annot.)

2. (of plant viruses), a group of viruses sharing only a few of its antigens in common with another group,

The range of antibodies which an individual possesses, usually based on sampling from blood serum or saliva.

Different strains of a pathogen can sometimes be distinguished by the different antibodies they induce in a host, or with which they can be made to react in vitro; thus the word serotype has also come to be applied to a particular strain ('the virulent serotype'). This is the more common clinical usage.

The range of antibodies used to define a serotype obviously depends on those available to the researcher. Sometimes, as for measles, the presence of a known antibody within the serum of an individual correlates extremely well with the clinical observation that that individual is protected against any further infections. But sometimes, as for malaria, there is as yet no definite relationship between a given serotype and the presence of a functional immunity, which may make the use of the word serotype unhelpful when trying to distinguish between different parasites for the purposes of understanding transmission.

Serrate With a notched or toothed edge, like a saw.

Serum The fluid fraction of coagulated (clotted) blood.

Sessile Used in reference to a leaf, leaflet, flower, floret, fruit, ascocarp, basidiocarp, etc., without a stalk, petiole, pedicel, stipe or stem; (of nematodes) permanently attached; not capable of moving about.

Set aside Set aside was introduced as part of a programme for tackling the over production of cereals within the EU. To help compensate farmers for their loss of income the Arable Area Payments Scheme was introduced. Under this scheme, farmers can claim support payments based on the size of the area used to grow cereals, linseeds, oilseeds and protein crops such as peas, beans and lupins. However, to qualify for these payments, all but the smallest producers must 'set aside' part of their arable land, taking it out of production.

Set stocking Grazing systems in which stock remain in one field or paddock for a long period.

Seta This incredibly over-used anatomical term is simply Latin for a bristle or a stiff hair. It is used in a variety of contexts, not only for things that look like bristles (or, to be sure, stiff hairs), but also for things like the stalk of the sporophyte in mosses, which looks nothing like a bristle (or, for that matter, a stiff hair).

Setose Bristly; beset with bristles.

Setula A fine, hair-like appendage which is extracellular in origin.

Severity values A means to quantify accumulating opportunities for pathogen infection (e.g. infection periods) to a pre-determined threshold that requires a disease management activity (e.g. a fungicide application); used in disease prediction or forecasting.

Sexual Participating in or produced as a result of a union of nuclei in which meiosis takes place.

Sexual dimorphism A pronounced difference in the morphologies of the two sexes within a species.

Sexual reproduction Reproduction involving fusion of two haploid nuclei (karyogamy) to form a diploid nucleus followed by meiosis (reduction division) back to haploid nuclei at some point in the life cycle, resulting in genetic recombination.

Sexual spore Spore produced during the sexual cycle.

Sexual stage Reproductive stage of the life cycle of an organism.

Sexual state The state in the life cycle of a fungus in which spores are produced after sexual fusion. Synonym: perfect state.

Sexually compatible Able to be cross-mated or cross-fertile.

Sheath The part of a grass leaf that encloses the stem below the collar region.

Shelf life The maximum period of time a pesticide can remain in storage before losing some of its effectiveness.

Shelter A place where an animal lives and is protected. Shelter is one of four parts (along with food, water, and space) that make up an animal's habitat.

Shii take Lentinus edodes; a domesticated edible agaric used in Japanese and Chinese cuisine; the fungus is grown on logs of Fagaceae (oak, beech).

Shock symptoms The severe, often necrotic symptoms produced on the first new growth following infection with some viruses. Also called acute symptoms.

Shoot density The relative number of shoots per unit area.

Short cells Cells of a dimorphic plant root exodermis which remain unsuberized long enough to permit ingress by VAM fungi.

Short day Term applied to strawberry cultivars that require a period of time with day length shorter than a minimum (about 14 hours) for the induction of flower buds; June-bearing.

Shothole A symptom of disease on leaves; necrotic lesions of limited size fall out from the leaf lamina leaving holes that are approximately circular.

Disease symptom characterized by the dropping out of small, round fragments of leaves, making them look as if riddled by shot.

Small holes in a leaf caused by feeding activity and giving the appearance of injury via a shotgun.

Shot hole disease A leaf disease in which the spread of infection is limited when the host sheds the infected area of the leaf.

Shoyu (soy sauce) An Oriental sauce of soybeans and wheat fermented by Aspergillus, yeasts, and bacteria.

Shrub A woody plant of lower height than a tree. Shrubs usually have many thick stems instead of one big trunk. Some plants are considered shrubs if they don't grow to their full size, and trees if they are able to become fully grown. Examples of this are Eastern Redcedar and American Holly.

Side dressing Fertilizer or other material added to the soil around a growing crop.

Siderophores Substances produced and expelled from the cell of some species of bacteria under conditions of iron deficiency. They complex to iron in the soil solution and are then reabsorbed and processed, providing the organisms with an efficient mechanism for obtaining a scarce resource.

Sieve element A tube-shaped living cell in the phloem functioning in the transport of dissolved organic substances in the plant.

Sieve plate Perforated wall area between two phloem sieve cells through which they are connected.

Sieve tube A series of phloem cells forming a long cellular tube through which food materials are transported.

Sigmoid Doubly curved in opposite directions, like the Greek letter sigma.

Sign The actual presence of the causal organism in association with the disease symptoms.

The manifestation of disease by the presence of structures of the causal agent.

Signal word One of three words (Danger, Warning, or Caution) which must appear on a pesticide label. Signal words indicate the level of acute toxicity posed by the pesticide product. Hazard Level I II III IV Signal Word Danger Warning Caution Caution oral LD50 up to 50 mg/kg 50-500 mg/kg 500-5,000 mg/kg >5,000 mg/kg inhalation LC50 up to 0.2 ug/l 0.2-2 ug/l 2-20 ug/l >20 ug/l dermal LD50 up to 200 mg/kg 200-2,000 mg/kg 2,000-20,000 mg/kg >20,000 mg/kg eye effects corrosive, corneal corneal opacity no corneal opacity; no irritation opacity, not reversible within irritation reversible reversible w/i 7 days 7 days; irritation within 7 days persists 7 days skin effects corrosive severe irritation moderate irritation mild or slight at 72 hours at 72 hours irritation

Silt Particles of dirt in ponds, lakes, rivers, streams, and marshes. Silt moves around a lot when it is disturbed. Too much silt in the water is a problem, since it can cover and destroy eggs of fish, amphibians, and other organisms.

Simple tissue Tissue composed of only one cell type

Single gene resistance Resistance conferred by a single gene.

Single photon emission computed tomography (SPECT) scan An imaging technique that measures the emission of photons of a given energy from radioactive tracers introduced into the body.

Sinuate With deep wave-like depressions along the margin; of a lamella, where the lamella is notched at the proximal end at the junction with the stipe.

Sinus The indentations of a lobed pinna or pinnule of a fern or of a dicot leaf.

Sirenin A hormone secreted by female gamete of Allomyces which attracts male gametes.

Sister group The two clades resulting from the splitting of a single lineage.

Site specific Management by accurate assessment of needs and application of amendments to smaller unit areas within a larger management unit such as an orchard, field or grove.

Skeletal hyphae Hyphae that are thick-walled, unbranched or sparsely branched, aseptate, straight or slightly flexuous and have thin-walled apices.

Skeletonizer Insects which consume leaf tissue, often from the lower side of the leaf, leaving the upper epidermis and vascular tissues intact.

Slaframine A mycotoxin produced by Rhizoctonia leguminicola; causing excessive salivation or slobbering in ruminants feeding on certain fungus infected legume forage crops.

Sleep apnea A group of potentially lethal disorders in which breathing recurrently stops during sleep for long enough to cause measurable blood deoxygenation.

Slicing A method of turf cultivation in which rotating flat tines slice intermittently through the turf and soil.

Slime flux A thick liquid exuding from branches and trunks of trees and associated with bacteria and fungi, and often also with gas production within the tree causing it to split.

Slime mold A member of a category of eukaryotic organisms that typically have some fungal-like attributes and some animal-like attributes.

Slit-trench drain A narrow trench (usually five to ten centimeters wide) backfilled to the surface with a porous material such as sand, gravel, or crushed rock. Used to intercept surface or lateral subsurface drainage water.

Slowly available fertilizer Designates dissolution less than that obtained for completely water-soluble fertilizers. May involve compounds that dissolve slowly, materials that must be decomposed microbiologically, or soluble compounds coated with substances highly impermeable to water. Used interchangeably with delayed release, controlled release, controlled availability, slow-acting and metered release.

Sluglike Larvae of sawflies and of some moths which resemble slugs.

Slurry A semi fluid mixture of faeces and urine which may also include rain water and yard washings.

Smith period A period of 2 days during which the temperature does not fall below 10C and the relative humidity is 90% or higher for at least 11 hours on each of the days, such conditions being favourable for infection of potato leaves by Phytophthora infestans.

Smut Ubiquitous, type I allergen. They are parasitic plant pathogens that require a living host. Most often found on corn, grass, weeds, flowering plants and other fungi; usually disseminated by wind. Indistinguishable from myxomycetes under 600x microscopy.

Smut balls Teliospores and accompanying sterile cells of flag smut fungi.

Smut fungi Fungus belonging to the Ustilaginomycetes.

Smut spore A dark, thick-walled resting spore of a smut fungus; may germinate to produce a promycelium; often improperly termed a chlamydospore.

Snout beetle A weevil an insect with a long proboscis or snout bearing chewing mouthparts.

Sociable When a species of animal likes to be around others of the same species. Since goldfinches like to be around other goldfinches, we call them "sociable."

Sod Plugs, squares or strips of turf grass, with adhering soil which are used in vegetative planting.

Soft rot Rotting of tissue, usually the parenchyma, by the activities of a pathogen on the middle lamella of cell walls. The cells become separated but retain their shape for some time.

Soil The unconsolidated mineral material on the immediate surface of the earth that serves as a natural medium for the growth of land plants.

soil application A herbicide applied primarily to the soil surface rather than to vegetation.

Soil drench Application of a solution or suspension of a chemical to the soil, especially pesticides to control soilborne pathogens.

Soil extract The solution separated from a soil or soil suspension.

Soil inhabitant An organism that maintains its population in soil over a period of time.

Soil invader An organism whose population in soil diminishes in several months to years.

Soil map A map showing the distribution of soil types or other soil mapping units in a relation to the prominent and cultural features of the earth's surface.

Soil modification Alteration of soil characteristics by adding soil amendments; commonly used to improve physical conditions.

Soil pasteurization Process used to free soil of selected harmful microorganisms using heat.

Soil penetration problems Trees planted in small planting holes, especially in soils that are severely compacted, exhibit transplant shock symptoms for unusually long periods. Symptoms include annual scorching and wilting, nutrient deficiency symptoms, tip dieback, slower healing of pruning wounds, and general decline.

Sometimes trees never recover completely and languish their entire lives. Roots can eventually penetrate the interface of the planting hole (where the existing soil meets the prepared planting hole soil). However, when the existing soil is a very compact clay, it may take several years for the new roots to penetrate extensively enough to support "normal" stem and canopy growth.

Soil probe A cylindrical soil-sampling tool with a cutting edge at the lower end.

Soil profile A vertical section of the soil through all its horizontal layers, extending into the parent material.

Soil sterilant A material which renders the soil incapable of supporting plant growth. Sterilization may be temporary or almost permanent.

Soil water deficit (SWD) The amount of water required to restore field capacity

Soilborne Refers to many fungi able to survive in the soil as saprophytes. Also called "soil inhabitant."

Solarization Weevils are a major problem in stored beans and force farmers to sell their produce soon after harvest, lowering prices. In solarization, the beans are covered with plastic and exposed to the heating effects of the sun, killing the weevils.

Solubility The maximum amount of a liquid or solid that will dissolve in a liquid.

Soluble powder A powder formulation that dissolves and forms a solution in water.

Solute Anything dissolved in a solvent. In salt water, the water is the solvent, and the salts are solutes.

Solution A mixture in which a pesticide is dissolved in a liquid. The pesticide is evenly dispersed as individual molecules among the molecules of liquid.

Soma Same as thallus.

Somaclonal variation Variability in clones generated from a single mother plant, leaf, etc. by tissue culture.

Somatic cell hybridization Production of hybrid cells by fusion of two protoplasts with different genetic makeup.

Somatic Pertaining to the vegetative or assimilative body of an organism.

Some Simple, fleshy, indehiscent fruit derived from several carpels, e.g. apple, pear, quince.

Songbird A small bird, usually in the Passeriformes Order (perching birds). Songbirds include well-known birds such as cardinals, sparrows, and bluebirds. Larger birds, such as hawks, owls, ducks, and herons are not considered songbirds. Most songbirds live in forests, parks, fields, yards, or marshes. Most of them also build nests in trees or shrubs.

Sooty mold A fungus of the family Capnodiaceae and of certain other families of the order Dothideales. The organisms grow epiphytically, utilizing honeydew, and form dark, spongy, hyphal mats on the surfaces of certain plants.

Soredia lichen propagules – small aggregations of fungal hyphae around algal cells; formed by the break up of a thallus.

Sori Part of a fern that makes spores. Sori are the equivalent of flowers on other plants.

Sorocarp The minute, usually microscopic, stalked fruit-body of the cellular slime moulds (Acrasiomycetes and Dictyosteliomycetes) formed by migration and aggregation of the cells composing the pseudoplasmodium.

Sorophore Stalk of a sorocarp.

Sorus A small, simple, immersed or erumpent fruiting structure, consisting of a discrete mass of spores (often powdery at maturity) and sporogenous cells, sometimes with associated sterile hyphae or bounding (peridial) structures; found in Uredinales, Ustilaginales, Tilletiales, Albuginaceae (Peronosporales) and some other taxa; the superficially similar structures in some Synchytriaceae (Chytridiales).

Sorussorus A cluster or group of sporangia of a fern; most frequently applied to clusters of fern sporangia attached to the bottom of fern leaves.

Southern blotting Procedure for transferring denatured DNA from an agarose gel to a nitrocellulose filter where it can be hybridized with a complementary nucleic acid.

SP A genus name followed by sp. means that the particular species is undetermined; spp. after a genus name means that several species are being referred to.

Spathulate Spoon shaped.

Spawn Mycelium produced artificially for the purpose of cultivating mushrooms.

Specialist natural enemy A natural enemy with a very restricted diet. The most specialized natural enemies are those that attack only one species (for example, the sphecid wasp Larra analis is known to attack northern mole crickets only). Others may attack several species belonging to only one genus (for example, Ormia depleta is known to attack only two members of the genus

Scapteriscus [tawny and southern mole crickets]). Specialist natural enemies may be useful in classical biological control.

Species The seventh and smallest group that scientists classify living things into. Actually, only one organism is in this group. Each genus is split into species. Also, a particular species of organism must reproduce (mate) with another organism of the same species (Example: A Bullfrog (Rana catesbeiana) must mate with another Bullfrog (Rana catesbeiana)). Each species has a scientific name. The ancient language of Latin is used to give each species a specific name so that everyone around the world, no matter what language they speak, can refer to the same organism. A species has a first and last name. The first name is also the name of the genus it is in. (Example: Rana catesbeiana (Bullfrog) and Rana clamitans (Green Frog) are in the same genus (Rana)).

Species epithet (bot) The second word of a species name (e.g., graminis in the species Puccinia graminis).

Specific gravity The ratio of the density of a substance to the density of pure water; specific gravity of potato tubers is used as a measure of their dry matter content.

Specific resistance Resistance which is effective against some biotypes or races of the pathogen, but not others, usually inherited monogenically and expressed qualitatively.

Specificity in nodulation Specificity in nodulation can affect benefits to inoculation for closely related species. As example rhizobia nodulating white clover will also nodulate subterranean clover, but will not fix nitrogen with this host.

Specificity The ability of a test to fail on people you know don't have the infection. More precisely TN/(TN+FP), where TN is the number of true negatives and FP is the number of false positives.

Spermagonium Structure in which male reproductive cells are produced; in rust fungi, globose or flask-shaped haploid fruiting body composed of receptive hyphae and spermatia (pycniospores).

Spermatheca In nematodes: An enlarged portion of the female gonad between the oviduct and the uterus functioning in the storage of the sperm.

Spermatium 1. A fertilizing sex cell, e.g. an oidium or the pycniospore of the Uredinales;

2. a microconidium in discomycetes and pyrenomycetes;

3. a non-motile gamete.

Spermatization In certain higher fungi: the union of a spermatium with a female reproductive structure.

Spermatocyte In nematodes: A cell giving rise to spermatozoa or spermatozooids.

Sphaerocysts Large, turgid, thin-walled cells, found clustered in the tissues of Russula and Lactarius (Agaricales), and accounting for the peculiar brittle texture of their flesh.

Sphere (suffix) Near; in the region of influence of; the microhabitat around. Used in such combinations as rhizosphere and spermosphere in relation to roots and seeds, respectively.

Spicate Having the form of a spike.

Spicule Copulatory organ of male nematode.

Spider mites These arthropods are not insects, but are in the same order as ticks, both of which are in the class Arachnida. They are very small and require a strong hand lens to see. Most are plant feeders and can do considerable damage. Some are important predators of other mites.

Spike An elongated inflorescence in which the individual flowers are borne tightly against the main stem or rachis.

Spikelet Spike-like appendage comprised of one or more reduced flowers and associated bracts; unit of inflorescence in grasses; a small spike.

Spiking A method of turf cultivation in which solid tines or flat, pointed blades penetrate the turf and soil surface.

Spindle fibres A group of microtubules that extend from the centromere of chromosomes to the poles of the spindle or from pole to pole in a dividing cell.

Spine A multicellular more or less thorn-like process or outgrowth of the outer skin of an insect not separated from it by a joint.

Spinose Spiny.

Spinulose With minute spines.

Spiracle A breathing pore. Each segment of an insect's body has at least two spiracles on the lateral sides through which air enters the tracheae. These paired lateral openings lead from the exterior into the respiratory system of insects. The respiratory system has ducts (tracheae) throughout the body.

Spiroplasma A member of a group of pleomorphic, wall-less prokaryotes occurring as epiphytes or as intracellular or extracellular parasites or pathogens in a range of invertebrates and plants.

Spliceosome A nuclear ribonucleoprotein complex which (usually along with other functions) removes intron transcripts from newly synthesized RNA.

Spontaneous generation, theory of The theory, now known to be invalid, that plants, animals and microorganisms arose suddenly from non-living materials under certain environmental conditions.

Spoon, coring A method of turf cultivation by which curved, hollow, spoonlike tines remove soil cores and leave a hole or cavity in the sod.

Spora The population ('spore flora') of air-borne (or water-borne etc.) particles of plant or animal origin.

Sporangiole A small sporangium without a columella, usually having only a small number (less than 50) of spores; sporangiolum.

Sporangiospore A spore with a nucleus formed by mitosis, an asexual spore. There seem to be an unreasonable number of names for this concept.

Sporangium 1. A structure that produces endogenous, asexual spores by cytoplasmic cleavage;

2. in Myxomycetes, a discrete, sessile or stalked fruit-body, often occurring in large groups which have developed from a single plasmodium.

Spore A general term for a reproductive structure in fungi, bacteria and cryptogams, commonly one-celled, but in fungi frequently a multicelled structure which is in effect a group of one-celled spores because every cell may produce one or more germ tubes.

Spore axis A line that passes through the centre of the spore from one extremity to the other. If there are more than two extremities, there is more than one axis.

Spore print A visible deposit of basidiospores obtained by allowing an agaric to drop spores onto white paper overnight; the colour of this deposit is an important aid to identification.

Spore wall The walls of spores are usually layered. from the inner layer outwards these are referred to as: endosporium, episporium, exosporium or epitunica, perisporium or myxosporium, and ectosporium.

Sporidesmin Hepatotoxic mycotoxin formed by Pithomyces, causes facial eczema in sheep.

Sporidiole A little spore.

Sporidium A basidiospore of a species of Urediniomycetes, Ustomycetes and bunts and related taxa (Tilletiales and Filbasidiales); in the case of smuts and bunts, any spore other than an ustilospore.

Sporiferous vesicle A thin-walled accessory cell distal to, or behind and to one side of, the chlamydospore of some VAM fungi.

Sporocarp Any spore-bearing organ; fruit-body; especially in Acrasiomycetes, Myxomycetes, Endogonales and Glomales.

Sporodochium A conidioma with superficial pulvinate stroma supporting conidiophores or conidiogenous cells on its upper surface and not covered by the substratum.

Sporogenesis Spore formation.

Sporogenous Producing or supporting spores; of yeasts, yeasts with teleomorphs in either the Ascomycota or the Basidiomycota.

Sporoma Any multicellular structure specially developed to produce spores.

Sporophore A spore-bearing or spore-supporting structure in fungi, which may be simple as in sporangiophore or complex as in ascomata and basidiomata; fruit-body.

Sporophyll A spore-bearing leaf; a modified leaf bearing sporangia.

Sporophyte The spore producing, diploid (2N) phase of the life cycle; compare to gametophyte.

Sporothrix sp Sporothrix sp. It can cause sporotrichosis. Usually only in populations which are immune compromised. Morphological characteristics include conidiophores mostly simple, 1-celled or septate, hyaline, bearing a loose cluster of dry conidia at apex; conidia, hyaline, 1-celled, globose to ovoid, born on short, prominent denticles. Mostly found as saprophyte. Cultivation - Sabouraud's agar, 24°C.

Sporotrichosis A lymphatic mycosis caused by Sporothrix schenckii (Hyphomycetes).

Sporotrichum sp Sporotrichum sp. This fungus is reported to be allergenic. Sporotrichum is commonly found on decaying plant matter, wet and rotting wood and in landscaping mulch. It is morphologically similar to the human pathogen Sporothrix.

Sporulate To produce spores.

Sporulation The process of producing spores.

Spot A discrete lesion, usually more or less circular, differing in tint from surrounding tissues, commonly chlorotic or necrotic.

Spot treatment A herbicide applied over one or more small areas of a whole unit; for example, treatment of patches of weeds within a larger field.

Spraing (sprain) Reddish brown spots, rings, or arcs in tuber tissue caused by tobacco rattle virus; corky ringspot.

Spray 1. To apply (water, plant protective chemicals, herbicides etc.) in the form of droplets suspended in air.

2. a solution or suspension (usually in water) dispersed under air or water-pressure as droplets.

The physical characteristics of sprays used for plant protective measures depend, inter alia, upon the nature of the propellent (air or water) and the size of the droplets, and these determine the volume of material applied to the crop. These characteristics are described by the following terms:

Spray deposit The amount of wet pesticide initially deposited per acre of plant or other surface.

Spray drift The movement of airborne spray particles from the spray nozzle beyond the intended contact area.

Spray, coarse A dispersion of drops of mass median diameter over 200.

Spray, concentrate An undiluted commercial preparation.

Spray, fine A dispersion of drops of 5-150 mass median diameter.

Spray, floor A spray applied to over-wintering dead leaves on the ground.

Spray, high volume Over 100 gal/acre (1120 l/ha) on bushes and trees: over 60 gal/acre (670 l/ha) on ground crops.

Spray, low volume 20-50 gal/acre (225-560 l/ha) on bushes and trees: 5-20 gal/acre (55-225 l/ha) on ground crops.

Spray, medium volume 50-100 gal/acre (560-1120 l/ha) on bushes and trees: 20-60 gal/acre (225-670 l/ha) on ground crops.

Spray, mist A dispersion of droplets 5-100 mass median diameter.

Spray, Ultra-low Volume (very low volume)

Less than 20 gal/acre (225 l/ha) on bushes and trees: 5 gal/acre (55 l/ha) on ground crops.

Sprayer, air blast A machine employing air-blast nozzles.

Sprayer, air carrier A machine employing air to propel the spray.

Sprayer, hydraulic A machine depending on liquid energy both to form and to propel the spray.

Sprayer, logarithmic Equipment for applying a spray in concentrations decreasing exponentially with time or distance.

Spreader A substance added to fungicide or bactericide preparations to improve contact between the spray and the sprayed surface; a surfactant.

Sprig A single turf-grass stem (stolon, rhizome or tiller), usually with some attached roots, used in vegetative propagation.

Springwood Early part of the yearly xylem growth ring in woody plants consisting typically of cells that are larger than those formed later in the season (summerwood).

Sprout The new stem formed from the eye of a potato tuber.

Sprout inhibitor A chemical applied to potato vines or to stored tubers to prevent sprouting.

Spur Short woody shoot that is the primary fruiting structure for most fruit trees.

Squamose With scales.

Squamous Covered with or consisting of scales.

Squamule A small scale.

Squamulose Describes lichen thalli which consist of small scales.

Square A cotton flower bud.

SSU RDNA DNA coding for RNA associated with the small ribosomal subunit, i.e. for "18S" RNA.

Stabilizing selection The theorized competitive disadvantage of unnecessary virulence genes; races with excess genes would have decreased fitness relative to races with fewer virulence genes, so a "super-race" would be less likely to appear in multiline crops.

Stable endemicity Where the incidence of infection or disease shows no secular trend for increase or decrease.

Stachybotryotoxin A mycotoxin produced by the hyphomycete, Stachybotrys, growing on hay; implicated in serious poisoning of horses.

Stachybotrys Stachybotrys sp. Stachybotrys thrives on water damaged cellulose rich materials such as sheet rock, paper, ceiling tiles, cellulose containing insulation backing and wallpaper. The presence of this fungus in buildings is significant because of the mold's ability to produce mycotoxins, which are extremely toxic, such as Satratoxin H. Exposure to these toxins can occur through inhalation, ingestion or dermal exposure. Symptoms include dermatitis, cough, rhinitis, nose bleeds, a burning sensation in the mouth and nasal passage, cold and flu symptoms, headache, general malaise, and fever. Inhalation of conidia may also induce pathological changes (pneumomy-cotoxicoses). Satratoxin H has been reported to be abortogenic in animals and in high doses or chronic low doses it can be lethal. S. chartarum produces other macrocyclic and trichoverroid trichothecenes and, like Memnoniella echinata, produces phenylspirod-rimanes, which are immunosuppressive. Stachybotrys typically appears as a sooty black fungus occasionally accompanied by a thick mass of white mycelia. As a general rule, air sampling for Stachybotrys yields unpredictable results mainly due to the fact that this fungus is usually accompanied by other fungi such as Aspergillus and Penicillium that normally are better aerosolized than Stachybotrys. Bulk or surface sampling of suspect materials can be analyzed in a laboratory for identification by light microscopy. Cultivation - Corn meal agar, 24°C.

Stadium The period of time between two successive molts.

Stage Frequently a synonym of state but, for fungi at least, better reserved for circumstances in which the succession of two or more states is in a fixed order or obligately alternate. One of the successive principal divisions in the life cycle of an insect, e.g., egg, nymph, larva, prepupa, pupa, adult.

Staghead Defoliated, dead or dying major branches in the crown of a tree, usually resulting from inadequate water uptake or translocation.

Staling Accumulation of metabolites which slow or stop growth of the organism which produces them.

Staling effect Slowing-down of growth of a fungus in pure culture as a result of accumulation of self-inhibiting metabolites, adverse pH etc.

Stalk 1. A short plant stem which leaves, flowers, or fruit grow on.

2. The main trunk of a mushroom.

Stamen The "male" structures of a flower, consisting of a supporting filament and the anther, a sac which produces and holds the microgametophytes (pollen).

Staminate flower A male flower.

Stand The number of established individual shoots per unit area.

Stand decline The gradual (over a period of 3-5 years) debilitation of the plants in an alfalfa field caused by the combined effects of pests and unfavorable environmental conditions.

Stand establishment in an alfalfa field The period of plant establishment between planting and first cutting.

Starter doses Where soil N levels are low, small doses of nitrogen fertilizer are sometimes applied at the beginning of the season to ensure vigorous early plant growth and nodule formation. The "starter dose" used should be less than that causing inhibition of nodulation and nitrogen fixation.

Stasis A period of little or no discernible change in a lineage.

State 1. (of fungi) One condition of an organism that differs in morphology or in physiological activity or both at different periods in its existence. (States with an obligate succession are termed stages).

2. (of bacteria) ' the name given to the rough, smooth, mucoid and similar variants which arise in cultures of many species of bacteria.

Statismospore A basidiospore that is not forcibly discharged.

Statute A law passed by a legislature.

Staurospore An asexual spore with or without septa; more than one axis; with axes not curved through more than 180 degrees; protuberances, other than setulae, present and greater than 1/4 the length of the spore body.

Staurosporous Possessing staurospores.

STD Sexually transmitted disease.

Steinernematid nematode (steinernematidae) A family of nematodes. All known species of steinernematid nematodes are entomopathogenic. They have three major developmental phases: egg, juvenile, and adult. One of the juvenile stages is capable of attacking (infecting) an insect host.

Stele The central cylinder inside the cortex of the roots and stems of vascular plants; contains the vascular or conducting tissue.

Stellate Star-shaped.

Stem The upright pillarlike part of the mushroom, which supports the cap. Also called the "stalk" or "stipe."

Stem group All the taxa in a clade preceding a major cladogenesis event. They are often difficult to recognize because they may not possess synapomorpies found in the crown group.

Stem pitting A symptom of some viral diseases characterized by depressions on the stem of the plant.

Stem, branch, and trunk disorders - Cankers and complete or partial death of woody stems or branches (fungi and bacteria)

Sticky ooze from trunks and branches (bacteria, mechanical injury, stress, boring insects, sapsuckers (birds))

Large conks and bracket-like fruiting structures on trunks and branches (wood-rotting fungi)

Galls or swellings on lower trunk and/or branches (crown gall bacterium, white pine blister rust)

Witches' brooms or excessive branching (fungi, mistletoes, phytoplasmas).

Extreme distortion, enlargement, and flattening of stems or branches (physiological/genetic condition known as fasciation).

Stemmother In aphids, the form that hatches from the overwintering egg and gives rise to the series of summer generations which reproduce without fertilization.

Stemphylium sp Stemphylium sp. - Reported to be allergenic. Isolated from dead plants and cellulose materials; it can be parasitic or saprophytic. Morphological characteristics - Dark conidia, with cross and longitudinal septa, variable in shape, frequently globose, broadly ellipsoid, or ovoid, often constricted at major septum. Cultivation - Potato dextrose agar or V-8 juice agar, 24°C.

Sterigma A short apical outgrowth of a basidium, with a minute, pointed extremity on which a basidiospore is produced and from which it is forcibly discharged.

Sterilant Any agent or chemical that destroys all living organisms in a substance such as soil.

Sterile 1. Unable to reproduce sexually.

2. To be free from living microorganisms.

Sterile fungus A fungus that is not known to produce any kind of spores.

Sterilization The process whereby all microorganisms and their propagules are killed by exposure to heat, radiation, or chemicals, or removed by filtration.

Sterilized Free from living microorganisms.

Sterol inhibitors Systemic but non-selective fungicides, e.g., bitertanol; triadimefon (Bayleton); fenarimol (Rubigan, Bloc); triforine (Funginox, Saprol); etaconazole (Vangard); triarimol (Trimidal); prochloraz (Sportak); fenapanil which inhibit sterol formation, and thus membrane formation.

Stichobasidium A cylindrical holobasidium with nuclear spindles longitudinal and situated about the middle of the basidium.

Sticker Substance used in inoculation to ensure that the rhizobia adhere to the seed during planting. They range from milk or sugar solutions used in simple seed inoculation, to stronger adhesives (40% gum arabic, 5% methyl ethyl cellulose) used in pelleting seed.

Sticktight Nut that remains on the tree after harvest (also called mummy); nut with husk firmly adhering to shell.

Stigma The distal part of the pistil, the organ that receives the pollen. Stigmata may be quite varied in shape and structure. They are often button-shaped and studded with hair-like papillae ("dry" stigmata) or coated with a sticky film ("wet" stigmata) to capture pollen.

Stilbaceous Bearing synnemata.

Stilboid Applied to small (to several mm high) stalked fruit-bodies with a swollen head (i.e. stipitate-capitate), which resemble in shape basidiomata of the genus Stilbum; found commonly in some Basidiomycota but also in other divisions.

Stinkhorn Basidioma of some members of the order Phallales (Gasteromycetes).

Stipe The stalk-like portion of some larger ascocarps or basidiocarps; any spore-bearing stalk.

Stipe columella A stipe which supports the receptacle and also penetrates the gleba forming a percurrent columella.

Stipitate Stalked.

Stipple Pigmented spots up to a few millimeters in diameter, often on the upper surface of leaves.

Stipule Small leaflike appendage at the base of a leaf petiole, usually occurring in pairs.

Stochastic model A mathematical model which takes into consideration the presence of some randomness in one or more of its parameters or variables. The predictions of the model therefore do not give a single point estimate but a probability distribution of possible estimates.

Stocking density The number of animals per unit area of land at a point in time

Stocking rate The number of animals per unit area over a given period

Stolon A slender, horizontal stem that grows close to the soil surface; in fungi, a hypha that grows horizontally along the surface.

Stoma Specialised pore in epidermis of plant, through which some fungi infect; in gasteromycetes and some other groups, a small opening of the fruit-body through which spores are discharged.

Stomach poison A poison, sprayed or dusted on the food of insects, which kills the insect when it eats the food.

Stomastoma Tiny pores or opening in the epidermis of leaves; stomata are flanked by two guard cells that regulate opening and closing of the pore and thus regulate gas exchange (respiration) and transpiration.

Stomatopodium A hyphal branch (appressorium) or 'plug' above or in a stoma of a plant.

Stone fruit Fruit with a stony endocarp, e.g. cherry, peach, plum.

Stool A plant from which offsets may be taken or with several stems arising together; a clump of roots or root stocks that may be used.

Store cattle Animals which have been grown slowly so that their skeletal development is not impaired, but muscle tissue is slightly below the animals potential and fatty tissue is not developed.

Strain 1. A group of similar isolates.

2. the descendants of a single isolation in pure culture.

3. a culture of bacteria which corresponds to a cultivated variety (cultivar) of higher plants in having some special economic significance.

4. biological strain.

5. (of plant viruses) (a) ' a virus that is serologically or immunologically related to the type strain of a virus or to other demonstrable strains ' (b) a group of viruses having most of its antigens in common with that of another group (strain)

The descendants of a single isolation in pure culture; an isolate. Also a group of similar isolates; a race. In plant viruses, a group of virus isolates having most of their antigens in common.

Strain diversity Bacteria have a single "chromosome" plus additional smaller pieces of DNA termed plasmids, which can be transferred between organisms. This, plus genetic mutation and DNA rearrangements within the chromosome, means that organisms from the same species can show significant variation.

Stramineous Straw-coloured, yellowish.

Streak A disease characterised by elongate lesions or areas of discolouration, usually of limited length, on leaves with parallel venation or on stems.

An elongated lesion with irregular sides.

Stream A small body of moving water. Also called creek, brook, and spring, depending on location and size. Smaller than a river.

Striae In nematodes: Superficial grooves or clefts on the cuticle; when present, striae may be seen encircling lips or body.

Striate Marked with delicate lines, grooves, or ridges.

Striated disc Flattened, fan-shaped assemblage of microtubules and fibrils extending from the side of the kinetosome in zoospores of the Monoblepharidales.

Strict liability The legal responsibility for damage or injury, even if you are not at fault or negligent.

Strigose Bearing large, bristle-like, coarse hairs.

Stripe A disease characterised by elongate areas of discolouration, of indefinite length, on stems or on leaves with parallel venation.

Strobilus A cone a cone-shaped cluster of spore-bearing leaves (e.g., a pine cone), an aggregation of sporophylls on a common axis.

Structural pest A pest such as a termite or wood rot fungus that destroys structural wood in buildings.

Stub cotton A cotton crop in which the stalks are cut down after harvest but the crown and rootstock are left in the ground to regrow the following season.

Stubble The part of a crop left after harvesting

Stuberization The formation of tubers at the ends of stolons; tuber initiation.

Stunt A wide range of parasitic and nonparasitic agents.

Stunted An unthrifty plant reduced in size and vigor due to unfavorable environmental conditions.

Stunting The reduction in height of a vertical axis, resulting from a progressive reduction in the length of successive internodes or a decrease in their number.

Style A tubular column in the pistil of a flower, through which the pollen is transported from the stigmata to the ovary.

Stylet knob One of the (usually three) basal protuberances of the stylet.

Stylet Slender, tubular mouthparts in plant-parasitic nematodes or aphids.

A small, stiff, needlelike tube inserted into a food source to obtain liquid food.

Stylet-borne transmission A type of virus transmission in which the virus is acquired and transmitted by the vector after short feeding times, and is retained by the vector for only a short period of time.

Styletborne virus A virus that is carried on the mouthparts of its insect vector and is lost after the vector feeds once or a few times; nonpersistent virus.

Stylospore A spore borne on a filament or hypha.

Sub Prefix, under, below, less than; usually in the sense of approximating to the qualified term.

Subacute A zone between acute and chronic, denoting the course of a disease.

Subchronic Intermediate between acute and chronic toxicities; subchronic toxicity studies involve repeated daily exposures of animals to a chemical for part (not exceeding 10%) of a lifespan. In rodents, this period extends up to 90 days of exposure.

Subclinical infection An infection in which symptoms are sufficiently mild or inapparent to escape diagnosis other than by positive confirmation of the ability to transmit the infection or serologically.

Subcrown internode Short culm-like connection between the crown and seed roots of wheat.

Subculture A culture derived from another one.

Suberin Waxy, waterproof chemical in some plant cells, notably cork (in stems) and endodermis cells (in roots). Suberin is an extremely complex and irregular material, like lignin – with which it shares some similarities. Suberin is composed of two physically separated domains: the aliphatic and phenolic. The phenolic domain is rather lignin-like, but with even greater variability, and built on the same basic unit of a di- or tri-hydroxyphenyl group attached to a three-carbon chain, variously oxidized and integrated with the carbohydrates of the cell walls. Perhaps the most common building block is ferulic acid: formally, 3-(2'-methoxy-3'-hydroxyphenyl)-propenoic acid. Distally, the phenolic domain is attached at points by ester Suberin Bernards (2002) linkages to glycerol. The remaining hydroxyls of the glycerol molecule are ester-linked to some strange-looking C-18 to C-30 lipids. These lipids are substituted at C9-10 with one or two hydroxyls, or even with an epoxide link between the two carbons. Finally the position may be oxidized to a carboxylate (alone or esterized to glycerol) or hydroxyl (alone or esterised to ferulic acid). Variations allow for cross-linkage to other suberin molecules via the 9-10 or positions.

Suberization 1. The formation of cork by the action of a cork cambium (phellogen).

2. the deposition of suberin in or on cell walls as in cork cells, endodermis etc.

Because the same term has been used to describe these two distinct processes it is recommended that the process of suberin deposition should be termed suberinization.

Suberized Of cell walls: Hardened by their conversion to cork (suberin).

Subgenomic RNA A piece of viral RNA, shorter than the entire genome of the virus, found in cells infected by the virus and sometimes encapsidated.

Subgleba A sterile, filamentous or chambered tissue which supports the gleba.

Subgrade The surface grade of a turf site before the addition of topsoil.

Subhymenium The generative tissue subtending the hymenium; hymenopodium.

Subhypothallic Mode of fruit-body development in Myxomycetes subclass Myxogastromycetidae where the hypothallus forms as a sheath on top of the plasmodium during fruit-body formation, and tubular sporangial stalks are filled with cells or cell debris.

Subiculum A net-, wool-, or crust-like growth of mycelium under fruit-bodies.

Sublethal infection An infection that does not result in death of a host.

Subphylum Taxon above Class but below Phylum; suffix is -mycotina, e.g. Ascomycotina, Basidiomycotina

Subpseudopodium A fine extension occurring at the leading edge of a pseudopodium.

Subsoiling Deep plowing to break up compacted subsoil.

Subsp Subspecies.

Subspecies An infraspecific population defined on the basis of one or more characters (morphologic for most organisms) that distinguish its members from typical representatives of the species.

Substituted dinitroanilines A class of herbicides widely used for preplant application in cotton.

Substrate 1. A chemical substance acted on by an enzyme.

2. the material forming the growth medium for a microorganism.

The substance or object on which an organism lives and from which it gets nourishment.

Substratum The material on or in which an organism (esp. a microorganism) occurs.

Subulate Slender, more or less cylindrical, and tapering to a point; awl-shaped.

Succulent A plant that is able to store water in its tissues and then withdraw it during times of drought. e.g. a cactus. The water storage tissue may be found in the stem, leaves, or roots depending on the species. Stem succulents, leaf succulents, and root succulents are types of growth forms.

Sucker The short side growth which arises from the root or crown of the plant.

Sucking insects Insects that insert their mouthparts into plant tissues and withdraw nutrients and fluids through stylets.

Sucrose A disaccharide, $C_{12}H_{22}O_{11}$, consisting of glucose and fructose monomers.

Sulcate Grooved, not necessarily longitudinally.

Sulfur dioxide (SO_2) A primary air pollutant produced in industrial processes and coal burning that causes interveinal necrosis on broadleaf plants and tip necrosis on conifers.

Summer planting A system of strawberry culture in which planting occurs in summer and fruit production begins the following spring.

Summerwood Part of the yearly xylem growth ring in woody plants formed late in the growing season and consisting of cells smaller than those of springwood.

Sun checking Cracking or breaking of whole kernels of grain caused by exposure to alternating conditions of dew, sun, and water stress.

Sunscald Bark on the southwest side of tree trunks may be killed by sunscald. Sunscald occurs when bark warms and thaws in the afternoon sun, then refreezes when nighttime temperatures drop rapidly.

Damage is most common on the darker-colored, smooth-barked trees such as mountainash, apple, and maple. Tree wraps can help prevent sunscald. Planting in a site that is shaded in the winter is recommended for susceptible trees.

Superficial Occurring on the surface of the substratum; not enclosed by fungal or host tissue.

Superinfection Attempt to infect a host with a second virus, usually a different strain of the first infecting virus.

Supernatant The material remaining above the pellet after centrifugation of a suspension.

Superparasitism When more parasitoid eggs are deposited into a host than can survive in that host.

Supplements In nematodes: Preanal genital papillae on the ventral side of males; derived from cuticle, but may be provided with glands. Function during copulation. Sometimes called supplementary organs.

Suppress To lower the level of a pest population.

Suppression A hypoplastic symptom characterized by the failure of plant organs or substances to develop.

Suppressive soil A soil in which certain disease(s) fail to develop because of the presence in the soil of microorganisms antagonistic to the pathogen .

Supra Prefix, above; usually emphasising position or situation.

Surface tension Forces on the surface of liquid droplets that keep them from spreading out over treated surfaces.

Surfactant A material that reduces surface tension between two unlike materials, such as oil and water. A spreader or wetting agent used to increase coverage of the surface being sprayed.

Surfactant, anionic A salt of an organic acid, the structure of which determines its activity at the surface of the solution.

Surfactant, cationic A material in which surface activity is determined by the basic component of the compound.

Surfactant, non ionic A material which does not yield ions in solution and is therefore compatible with both anionic and cationic substances.

Suscept 1. An organism affected or capable of being affected by a given disease (APS).

2. a disease-prone host.

The conceptual relationship between pathogen and suscept is analogous to that between parasite and host. A word is needed to fulfil the function served by suscept, but because this latter word is so displeasing many writers prefer to avoid its use.

Any organism that can be attacked by a biotic pathogen.

Any plant that can be attacked by a given pathogen; a host plant.

Susceptible Non-immune i.e. subject to infection. Similarly, susceptibility: the inability of an organism to oppose the operation or to overcome the effects of an injurious or pathogenic factor. Inability of the host to defend itself against or to overcome the effects of invasion by a pathogenic organism or virus (APS). In the BMS definition, susceptible is the antithesis of immune, each has an exact meaning, and both are distinct from resistant. The APS definition is complementary to that for resistance and both terms relate to the same aspect of the host-parasite interaction: low resistance corresponds with high susceptibility, and high resistance with low susceptibility. There seems no necessity to retain both forms, and it is suggested that susceptible should be used only in the sense of non-immune. The expression 'highly susceptible' is often used in the sense of showing severe symptoms. This effect may be a consequence of low resistance, but it is not necessarily so, and the term sensitive is more appropriate.

An individual accessible to or liable to infection by a pathogen.

Lacking the inherent ability to resist disease or attack by a given pathogen; nonimmune.

Susceptible species Pests readily killed by relatively low rates of pesticide application.

Suspension A mixture in which the pesticide in a solid form (tiny particles) is suspended in water.

Suspensor In Zygomycota, the remains of the gametangia that project from the sides of the zygosporangium which has developed between them.

Sustainability from Sustain meaning: To hold up; to bear; to support; to provide for; to maintain; to sanction; to keep going; to keep up; to prolong; to support the life of.

Sustainable agriculture The application of husbandry experience and scientific knowledge of natural processes to create integrated, resource conserving farming systems, based on respect for the people and animals involved which reduce environmental degradation, and which promote agricultural productivity and economic viability in both the short and long term.

Suture Visible seam on hull.

Swarm spore Zoospore; mastigote propagule.

Swath The area covered by one pass of the pesticide application equipment.

Symbiont One member of a symbiotic relationship.

Symbiosis A relationship between two organisms such that each benefits. With legumes and rhizobia, the legume gains by having the microbe supply it an additional source of nitrogen it can use for growth the microbe gains a source of energy for growth and a habitat (the nodule) within which it is protected from outside stress.

Symbiotic relationship (symbiosis) A close relationship between organisms living side-by-side.

Symplastic Entering living cells.

Sympodial Describes blastic conidia formed on a conidiogenous cell by the growth of a succession of apices, each originating below and to one side of the last. The conidiogenous cell usually becomes longer, but in some cases becomes inflated, as a result of repeated spore formation.

Symptom A visible or otherwise detectable abnormality arising from disease. There is a distinction between 'signs' and 'symptoms' in medical terminology, signs being visible (objective) indications of disease, and symptoms subjective; such a distinction is not directly applicable in plant pathology, but some pathologists distinguish abnormalities of the host plant as symptoms and visible structures of the pathogen (pustules, conidiophores etc.) as signs of disease.

A condition of the body reported by an individual when suffering from a disease; here used more loosely to include signs; any evidence used in diagnosis or identification of infected individuals.

Visible or measurable manifestation that an organism is diseased; a change in the organism itself.

The external and internal reactions or alterations of a plant as a result of a disease.

Symptom picture The totality of symptoms detected at one time.

Symptom, primary The first symptom to appear in a disease in which more than one type of symptom may be produced.

Symptom, secondary One following the primary symptom in diseases where more than one type is produced.

Symptomatology The study of symptoms of disease and signs of pathogens for the purpose of diagnosis.

Symptomless carrier A plant that, although infected with a pathogen (usually a virus), produces no obvious symptoms.

Symptoms on storage organs Storage organs include tubers, bulbs, and corms. Symptoms on these organs include:

Discolored or dead areas that go deep into storage organs (fungi, bacteria)

Dry rots (fungi).

Soft rots accompanied by strong, repulsive odors (bacterial pathogens, such as Erwinia spp.) (Frequently, bacterial soft rots move into tissue originally attacked by pathogenic fungi, making diagnosis of the original pathogen difficult.)

Scabby lesions on potato skins (bacteria such as common scab)

Galls on storage organs (fungi, nematodes).

Internal problems (several viruses or bacteria, such as ring rot of potato).

Symptoms, acute 1. (of virus diseases) Symptoms that develop shortly after infection and that are followed on later leaves by less intense persistent symptoms,or by masking of symptoms. Synonymous with shock symptoms.

2. (of diseases caused by fungi or bacteria) severe symptoms, usually leading to death of plants or plant parts.

Symptoms, chronic 1. (of virus diseases) Persistent or recurrent symptoms, which may or may not be preceded by acute symptoms.

2. (of other diseases) symptoms that persist over an extended period without necessarily causing death of the host or its parts.

Although chronic symptoms are often also mild symptoms the two terms are not synonymous: chronic describes the duration or persistence of symptoms, whereas mild is descriptive of their severity.

Symptoms, local (of virus diseases) Symptoms appearing at the site of inoculation.

Symptoms, masked A contradictory term used when symptoms disappear under certain conditions, e.g. symptoms of certain virus diseases with the onset of high temperatures during the summer. Masked infection is to be preferred.

Symptoms, mild Symptoms that are slight in extent or in intensity (or both) usually not culminating in death of major parts of the plant.

Symptoms, systemic (of virus diseases) Symptoms appearing other than at the site of inoculation, usually in leaves produced subsequent to inoculation.

Synanamorph One of two or more different anamorphs which develop simultaneously, or are otherwise known to be associated with a single teleomorph.

Synangium A reproductive unit composed of fused sporangia.

Synapomorphy A unique character which is shared by all basal members of a clade and is derived from their common ancestor, but not shared with ancestral groups. A synapomorphy may be secondarily lost in later descendants. Only a synapomorphy may be used to infer phylogeny.

Synascus The gametangium of the Protomycetales, formed by germination of a thick-walled resting spore, either as a vesicle external to the spore or enclosed by resting spore wall; also used for the gametangium of the Ascosphaerales.

Syncephalastrum sp Syncephalastrum sp. This organism is considered to be primarily non pathogenic in humans and is usually found in soils and in dung. Morphologically it is distinct with conidiophores erect, branched, tips enlarged, bearing a head of rod shaped sporangioles, each producing a row of nearly spherical conidia.

Synchronous Describes blastic conidia initiated simultaneously in a cluster or on a specialized cell.

Syncytium A multinucleate structure in root tissue formed by dissolution of common cell walls induced by secretions of certain sedentary plant-parasitic nematodes, e.g. cyst nematodes.

Syndrome The totality of effects produced in a plant by one disease, whether all at one time or successively, and including effects not directly detectable to the unaided eye.

Synergism When the simultaneous action of separate chemicals together have a greater total effect than the sum of their individual effects. Examples of synergistic chemicals are: alcohol and certain drugs; Tagamet (ulcer medication) and diazinon (insecticide). Scientists are currently exploring the synergistic effects between DEET (insect repellant), permethrin (insecticide), chlorpyrifos (insecticide), and the anti-nerve gas agent used administered to Gulf War soldiers.

Synergist A chemical that enhances the toxicity of a pesticide to a pest, but that is not by itself toxic to the pest.

Synnema Consisting of a more or less compacted group of erect and sometimes fused conidiophores bearing conidia at the apex only or on both the apex and the sides.

Synnematous With conidiophores borne in tightly packed clusters, as in a synnema.

Synthetic organic pesticides Manufactured pesticides produced from petroleum and containing largely carbon and hydrogen atoms in their basic structure.

Synthetic pesticide A pesticide in which the active ingredient has been manufactured, as opposed to a natural pesticide in which the active ingredient occurs naturally.

Synthetic pyrethroids A diverse class of more than 1,000 broad-spectrum insecticides. Based on the chemical structure and biological activity of pyrethrum (an extract from plants in the genus Chrysanthemum), but more toxic and less rapidly degraded in the environment than natural pyrethrum.

Syntype One of a number of specimens of equal nomenclatural rank that formed all or part of the material the original author had in cases in which the author did not designate or indicate a holotype.

Syphillis An acute and chronic disease caused by the bacteria Treponema pallidum, transmitted by direct contact, usually through sexual intercourse.

Syringe To spray turf with small amounts of water, usually on a hot, dry, windy day, to reduce water loss (transpiration).

System A number of components linked together.

Systematic name A name that fully defines a chemical compound and is derived using a set of rules. The main rules are those produced by IUPAC (International Union of Pure and Applied Chemistry) and CAS (Chemical Abstracts Service).

Systematics Field of biology that deals with the diversity of life. Systematics is usually divided into the two areas of phylogenetics and taxonomy.

Systemic 1. (of a plant pathogen) Occurring throughout the plant.

2. (of a chemical) absorbed into the plant through roots, foliage etc., and translocated elsewhere in the plant.

`Pertaining to a disease in which an infection leads to general spread throughout the plant body. Also, a chemical that spreads internally through a plant.

Systemic acquired resistance (SAR) Reduced disease symptoms on a portion of a plant distant from the area where a hypersensitive response occurred or other stimulus was applied; a rapid and coordinated defense response against a variety of pathogens as a signal travels throughout the plant.

Systemic chemicals Chemicals that enter the plant and move from the point of contact.

Systemic fungicide A fungicide that is absorbed into plant tissue and may offer some curative or after-infection activity; includes fungicides that are locally systemic, xylem-mobile (upward moving), and amphimobile (move in phloem upward as well as downward in the plant).

Systemic infection In virology, an infection resulting from the spread of virus from the site of infection to all or most cells of an organism.

Systemic invasion Micro-organisms that enter plants and penetrate or move throughout the plant.

Systemic lupus erythematosus An inflammatory disease of connective tissue occurring predominantly in women (90%). It is considered to be an autoimmune disease.

Systemic pesticide A pesticide that causes death or injury to pests when it is injested by them during feeding. Systemic materials are applied to plants or animals and are translocated to other parts of the plants or animals than those to which the material is applied.

T

T 2 toxin A mycotoxin of the trichothecene group, produced by several Fusarium spp.; the cause of alimentary toxic aleukia.

T lymphocyte, T cell The most common type of lymphocyte, itself divided into at least three subpopulations on the basis of function – cytotoxic, or killer T lymphocytes, helper T lymphocytes, and suppressor T lymphocytes. T cells play a cardinal role in regulating the immune system.

Tachinid flies A family of flies. In this family the adults are free-living, and larvae are parasitoids of insects. There are four developmental phases - eggs, larva, pupa, and adult. These flies do not harm people or plants.

Tadpole The larva stage of frogs and toads. Tadpoles hatch from eggs laid in water. They breathe through gills and resemble fish. Later they will grow legs and lungs, lose their tails, and become adults.

Tail In nematodes: The portion of the body between the anus and the posterior terminus.

Tail water Irrigation water that has drained from a field.

Takakia sporophyte. renzaglia et al. (1997)vaginula In mosses, a derivative of the archegonium forms a hard, protective epigonium around the developing sporophyte. Later the epigonium breaks up, but the sporophyte retains a conical "dunce cap" or calyptra over its distal end and sometimes a ring around the base. The term calyptra refers to the "dunce cap," and the piece of the epigonium retained around the base (if any) is the vaginula.

Take all A cereal disease caused by Gaeumannomyces graminis (Diaporthales: Ascomycetes).

Talon Bird claw; especially a bird of prey, such as an owl, hawk, or eagle.

Taproot Primary root that grows vertically downward and from which smaller lateral roots branch.

Target organism The organism against which a treatment is directed.

Target pest A pest species that a control action is intended to destroy.

Target spot A lesion in which there is a series of concentric rings, usually necrotic.

Taxon Any named group of organisms, not necessarily a clade; a taxon may be designated by a Latin name or by a letter, number, or any other symbol,

Taxonomic categories Most taxonomic systems are hierarchical and are comprised of a series of taxa of different categories or ranks. Codes of nomenclature do not legislate for taxonomy, they only regulate the naming of the taxa after taxonomic decisions have been made. However, the Codes list the categories or ranks which have been traditionally accepted by taxonomists and so, although the Codes make it clear that a taxonomist may use additional categories provided confusion is not thereby introduced, they help to standardize this aspectof taxonomic practice.

The basic category (or rank) is that of species. Species are grouped into genera, genera into families, families into orders, orders into classes, classes into divisions, divisions into kingdoms, or into intermediate categories, e.g. subgenera, subfamilies etc. Under the Botanical Code species are subdivided into subspecies, varieties and forms. Traditionally these categories are all distinguished on morphological grounds although in some groups, such as yeasts and bacteria, morphology must be supplemented by biochemical or other non-morphological characters.

The Botanical Code makes one provision of particular interest to plant pathologists by allowing ' taxa characterised from a physiological standpoint but scarcely or not at all from a morphological standpoint ' to be distinguished as 'special forms ' (formae speciales) which are, however, not governed by provisions of the Code. Cereal pathologists and others distinguish within formae speciales physiologic races and biotypes which are also outside the Code and their nomenclature is regulated by agreements between specialist workers.

The lowest category recognised by the Bacteriological Code (and also the Zoological Code) is that of subspecies (which under both these Codes is equated with variety). Below or within subspecies (varieties) bacteriologists commonly use such categories as biotype, phagotype, serotype, form (forma), group, phase, variant, mutant, stage, state and in the limit, isolate. Many of these terms (most of which are listed and commented on in the present compilation) are in varying degrees ambiguous although some, such as biotype, serotype, state, mutant and isolate, have precise meanings.

Taxonomist A scientist especially trained in the classification of plants or animals.

Taxonomy The system that scientists developed to classify (put in order) all living things. There are seven groups (Kingdom, Phylum, Class, Order, Family, Genus, Species). Scientists started by putting all living things (organisms) into Kingdoms. There are five Kingdoms (Animal, Plant, Fungi, Protist, and Moneran). Then, each Kingdom is split into smaller groups, called a Phylum. Each Phylum is then split into smaller groups called Classes. Each Class is split into smaller groups called Orders. Each Order is split into smaller groups called Families. Each Family is split into smaller groups called a Genus. Each Genus is split into Species. Species aren't really groups, because the scientists have finally narrowed it down to one particular organism. The way that scientists decided which organism goes in which group is by comparing them to see what they have in common. For example: All the organisms that can move around by running, flying, swimming, etc. went into the Animal Kingdom. Then, all the animals that have backbones went into the Chordate Phylum. Then all the animals with backbones that had hollow bones and feathers went into the Bird Class. And so on. . . This system allows us to organize and keep track of all living things.

Teeth Narrow, round, spore-producing projections hanging from the bottom of the caps of certain mushrooms such as the hedgehog mushroom.

Telamon In nematodes: Rigid, sclerotized portion of the cloacal wall that apparently guides the spicules from the spicular pouch into the cloaca.

Teleomorph The stage characterized by the production of asci/ascospores, basidia/basidiospores, teliospores, or other basidium-bearing organs.

Teleutosorus An old term for telium.

Telial stage (telium) A stage of the rust fungi; a fruiting structure usually appearing as fine, hairlike projections from lower surfaces of infected leaves.

Teliomycetes Class of basidiomycetes including the orders Uredinales and Ustilaginales. Their teleomorphs consist of teliospores from which basidia develop.

Teliospore Thick-walled resting or overwintering spore produced by the rust fungi (Uredinales) and smut fungi (Ustilaginales) in which karyogamy occurs; it germinates to form a promycelium (basidium) in which meiosis occurs.

Telium The final sorus (stage III) produced in the life cycle of the Uredinales and producing teliospores.

Tempeh An Oriental food made by fermenting soybeans with Rhizopus oligosporus.

Temperate phage A phage that reproduces by passing through a prophage state in a lysogenic bacterial cell.

Temperature coefficient of thermal inactivation The ratio (Q/10) between the rates of inactivation at temperatures separated by 10°C. The loss of infectivity of some viruses is closely associated with denaturation; their Q/10 is high and little infectivity is lost at sub-critical temperatures. Viruses with a low Q/10 lose infectivity over a wide range of temperature.

Temperature inversion A condition in which air above an area is warmer than air near the ground. This warm air forms a cap that may cause a pesticide vapor or droplets to collect and concentrate.

Temporary wilt Wilt due to insufficient soil water from which a plant can recover when water is supplied.

Tenacity The property of a deposit or residue to resist removal by weathering or other physical action.

Tenacity index The ratio of the quantity of residue per unit area at the end of a given amount of weathering to that present at the beginning.

Tendril A twisting, root-like structure that a vine uses to grab onto trees and branches for support.

Tensiometer A device for estimating soil moisture levels by measuring the negative hydraulic pressure of water in soil; a porous, permeable ceramic cup connected through a tube to a manometer or vacuum gauge.

Tentacle Long, flexible sense organ on a slug or snail. These animals use their tentacles to explore their environment.

Teonanacatl The Flesh of the Gods,' magic mushrooms used in Central American curing and divining ceremonies.

Teratogen Anything which is capable of producing birth defects in exposed embryos or fetuses. Examples of chemicals for which there is evidence of teratogenicity: thalidomide (drug), mercury (pesticide, pollutant), rotenone (insecticide), aldrin (insecticide), dieldrin (insecticide), thiram (fungicide), xylene (solvent)

Teratogenesis The development of abnormal structures or functional alterations in offspring during pregnancy. Teratogenic effects are not passed on to subsequent generations.

Teratogenic Causing abnormalities of growth of a fetus.

Teratology The study of gross structural abnormalities, e.g. fasciations, galls, witches' brooms etc.

Terminal The growing tip of a stem, especially the main stem.

Terminal spikelet stage Stage in the development of the wheat spike when the primordia of the terminal spikelet are formed.

Terrestrial biotype A strain of an organism adapted to growing on land rather than in water.

Terrestrial Describes an organism that lives and grows on land (not water).

Territory The area which an animal lives. Some animals are "territorial," meaning they protect their territory from other animals of the same species. This is so they don't have to share food or mates.

Tertiary bloom The third production of flowers that occurs at the end of the growing stem of an indeterminate potato cultivar.

Tertiary The geological era covering the period from 65-1 million years ago: include the Palaeocene, Eocene, Oligocene, Miocene, and Pliocene periods; if the last million years, the Pleistocene, is included, this 65 million year span is often called the Cenozoic era.

Tertiary tiller Branch of a grass plant that develops from the axil of a secondary tiller.

Tessellate In nematodes: Checkered; a type of cuticular pattern in which the longitudinal ridges are broken by transverse striations into rows of squares.

Testis A male reproductive organ in which spermatozoa are produced.

Tetrapolar Describes a system of heterothallism in which sexual compatibility is controlled by many alleles, all occurring at 2 loci.

Tetraradiate Describes the staurospores of some stream-inhabiting fungi; the four (sometimes more) arms ensure a stable three-point landing on the substrate.

Tetraspore In a basidium, one of the spores from a four-spored basidium.

Textura Fruit-body tissue, originally applied to discomycetes, now used for all ascomycetes and coelomycetes.

Texture The arrangement of the components of the different tissues, as compact, loose, etc.

Thallic 1. Of the thallus;

2. of conidiogenesis, one of the two basic sorts of conidiogenesis, in which any enlargement of the recognisable conidial initial occurs after the initial has been delimited by one or more septa; the conidium is differentiated from a whole cell.

Thalloconidium A conidium formed by thallic conidiogenesis; types include arthroconidium, chlamydospore.

Thallophyte An organism not differentiated into shoot and root, the cells of which have a cell wall; originally used for bacteria, Fungi and algae.

Thallospore A conidium that has no conidiophore or is not separate from the hypha or conidiophore that produced it.

Thallus Term given to the body of a lower plant which has no recognizable shoot, root, leaf regions. Most of the things this term is used to describe are not even plants, e.g. lichen.

Thatch A tightly intermingled layer of undecomposed or partially decomposed organic residues situated above the soil surface and usually below the green portions of the turf.

Thatch control The process of (a) preventing excessive thatch accumulation by cultural manipulation and/or (b) reducing excess thatch from a turf by either mechanical or biological means.

Therapy Principle of plant disease control marked by the cure of disease, as with heat or systemic chemicals.

Thermal death point The lowest temperature at which heating for a limited period (usually 10 min) is sufficient to kill a microorganism.

Thermal inactivation point The lowest temperature at which heating for a limited period (usually 10 min) is sufficient to cause a virus to lose its infectivity or an enzyme its activity.

Thermolabile Tending to break down when heated.

Thermotherapy Use of heat to reduce or eliminate pathogens in plant tissue; often used on plants prior to meristem culture to produce pathogen-free plants.

Thermotolerant Capable of growing at high temperatures (up to 60C, opt. 40-50C).

Thicket An area filled with close-growing plants, including shrubs, small trees, vines, grasses, ferns and/or others. A thicket is usually difficult to walk through, and provides great shelter for birds and other animals.

Thoracic Belonging or attached to the thorax.

Thorax The middle section of the three sections (head-thorax-abdomen) of an insect's body. The thorax is packed with muscles for the wings and legs which arise from that section. The thorax itself consists of three parts named in order: prothorax, mesothorax, and metathorax.

Threadworm Nematode.

Thrips These are small (0.5 - 5.0 mm), slender insects with rasping-sucking mouthparts. Wings, when present, are very long, narrow and fringed with long hairs. Females are often larger then males and parthenogenesis occurs in some species. Most species feed on plants and many are serious pests of cultivated plants. A few species are predators of small arthropods. Thrips sometimes occur in large numbers and a few species may even bite people. Thrips is both singular and plural.

Thrush A yeast infection of the mouth and throat, esp. in children, caused by Candida albicans.

Thurberia weevil A race of the boll weevil that feeds on wild cotton.

Thyroid A two-lobed structure located in front of and on either side of the trachea, producing the hormone thryoxin; of or relating to the thyroid gland.

Thyroid-stimulating hormone A hormone of the anterior pituitary gland that stimulates and regulates the development and secretory activity of the thyroid gland.

Thyrotoxicosis Poisoning from hyperthyroidism.

Tibia The fourth part (fourth from the body) of an insect's leg.

Tiger stripe A leaf symptom in which clear yellow areas separate marginal and interveinal areas of dark necrotic tissue from persistently green areas along and adjacent to the main veins, giving a striped yellow-and-black appearance. The term was applied to a symptom of hop verticillium wilt and has since been used for similar patterns in other diseases.

Tillage Shattering soil by using various mechanical implements, such as cultivators.

Tiller Shoot of a grass (or cereal) plant, arising from a leaf axil, normally at the base of an older tiller

Tinea A superficial infection caused by dermatophytes.

Tinsel flagellum A flagellum with lateral flimmers or mastigonemes.

Tissue A group of cells of similar structure which performs a special function.

Tissue culture In vitro method of propagating healthy cells from plant tissues.

Titer The concentration of a substance in a solution, or the strength of such a substance detected by titration. In the current context, the term is most likely to refer to antibody titer, which is a measure of the concentration of specific antibodies to selected microbes that are circulating in an individual's bloodstream.

Titillae In nematodes: Small projections on either side of the distal end of the gubernaculum.

Tobacco mosaic virus The plant disease caused by tobacco mosaic virus is found worldwide. The virus is known to infect more than 150 types of herbaceous, dicotyledonous plants including many vegetables, flowers, and

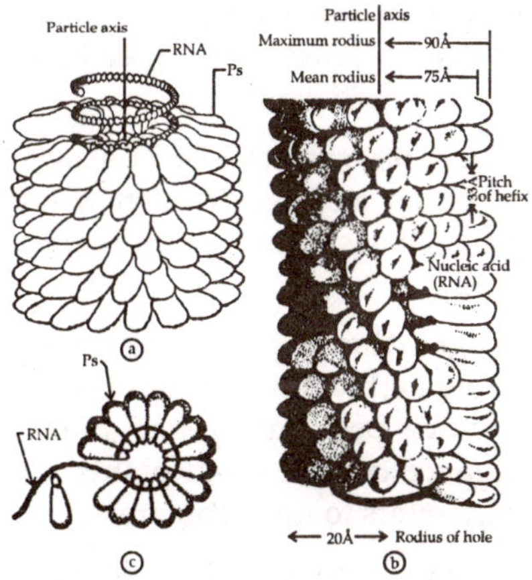

Fig. Structure of TMV: (a) drawing to show the relationship of the RNA and the protein subunits. The RNA shown free of protein could not maintain this configuration in the absence of protein subunits. There are three nucleotides per protein subunit or 49 per turn of this major helix spaced about 5 Å apart (b) photograph of a model of TMV with major dimensions indicated. The structure repeats after 69 Å in the axial direction. The repeat containing 49 subunits distributed over three turns of the helix (c) cross section of (b).

weeds. Infection by tobacco mosaic virus causes serious losses on several crops including tomatoes, peppers, and many ornamentals. Tobacco mosaic virus is one of the most common causes of virus diseases of plants in Minnesota. Many viruses produce mosaic-like symptoms on plants. Mosaic-like symptoms are characterized by intermingled patches of normal and light green or yellowish colors on the leaves of infected plants. Tobacco mosaic damages the leaves, flowers, and fruit and causes stunting of the plant. The virus almost never kills plants but lowers the quality and quantity of the crop, particularly when the plants are infected while young. Virus-infected plants often are confused with plants affected by herbicide or air pollution damage, mineral deficiencies, and other plant diseases. Positive identification of tobacco mosaic virus in infected plants often requires the services of a plant pathologist and the use of an electron microscope. Although it may take a plant pathologist to diagnose tobacco mosaic virus in many ornamental plants, the majority of tomato plants showing mosaic symptoms usually are infected by tobacco mosaic virus.

Tobamovirus Member of a group of plant viruses with rigid, rod-shaped particles containing one molecule of linear RNA, easily sap-transmitted and transmitted in nature by contact and (in the case of some viruses in this group) in seed.

Tolerance level or threshold pest level The population level at which a "pest" species becomes intolerable economically or aesthetically. Under an IPM program, this level determines when control measures are taken - the mere presence of an insect, pest or otherwise, does not warrant pesticide use.

Tolerant Able to endure infection by a particular pathogen, without showing severe disease. Similarly, tolerance: the ability of the affected organism to endure the operation of a pathogenic factor or invasion by a pathogenic organism or virus with little or no reaction, as shown by the more or less complete absence of symptom expression and damage.

Neither of these definitions draws attention to the importance of relating tolerance to the degree of infection to which the host tissues are subjected. A host in which there is apparently unrestricted and extensive colonization by a parasitic organism or virus without symptom development is properly termed tolerant. Absence of symptoms because colonization has been restricted in extent or intensity by host defensive responsive implies resistance, not tolerance.

The terms tolerant and tolerance, like the antonyms sensitive and sensitivity, are applicable also to the response of host or pathogen to toxic chemicals and other damaging agents, and must similarly be used with caution when given quantitative qualification: grades of tolerance must be related to constant and defined degrees of infection by a pathogen, or to constant dosages of chemicals, heat, ultraviolet radiation etc.

Tolerate Able to put up with. A plant may be able to tolerate shade, meaning it can live under other plants. A tree that tolerates drought can go a long time without water.

Tombusvirus Member of a group of plant viruses with isometric (icosahedral) particles containing one molecule of linear RNA, readily sap-transmitted and also transmitted through soil.

Tomentose Covered with soft, matted hairs.

Tooth fungi Members of the family Hydnaceae (Aphyllophorales: Basidiomycetes) in which the hymenium covers downwardly directed teeth.

Toothed margin (leaves) Margin having projections or serrations with pointed apices, indented less than one quarter of the distance to the midvein or long axis of the leaf; compare entire margin and lobed margin leaves.

Top crop Fruit produced in the second fruiting cycle of cotton, mainly on upper branches.

Topdressing A prepared soil mix added to the turf surface; usually incorporated into the soil by raking or irrigating.

Topotype A specimen collected at the type locality.

Tort A legal wrong which causes injury.

Torula sp Torula sp. Reported to be allergenic. Found as a saprophyte. Morphological characteristics - Conidiophores short, dark, simple, branched or absent; conidia 1-to several-celled, round, dark, and in chains. Cultivation - Potato dextrose agar, 24°C.

Toruloid With swollen cells like a chain of pearls.

Tospovirus Member of a small group of plant viruses with spherical particles containing three RNA species, transmitted in nature by thrips in a persistent manner.

Total fertility rate (TFR) The number of children an average women would have assuming that she lives her full reproductive lifetime.

Total vegetation Application of single or multiple herbicides at once or in sequence to provide pre- emergence and/or post-emergence control of all plants. The term usually refers to non-crop areas.

Totipotency The concept that even specialized cells contain all of the genetic information for an organism and, therefore, any cell should be able to regenerate into any tissue or into an entire plant.

Toxic Of, caused by, or acting as, a poison.

Toxicant A poisonous substance; not necessarily synonymous with toxin, which refers usually to poisonous substances produced by living organisms. Used particularly of fungicides, bactericides, insecticides etc.

Toxicity The ability to act as a toxin or toxicant. This is clearly a quantitative expression and distinction should be made between the response of an organism to a toxin or toxicant and to a growth inhibitor or retardant all of which may act at the same cellular site. Toxicity sensu stricto implies an irreversible reaction and is measured in terms of the degree of injury to an organism or to the death rate of a population (e.g. median lethal dose).

Toxicology The study of poisons and their effects.

Toxigenic Producing toxins.

Toxin An organic substance usually acting at a low concentration, which deleteriously and irreversibly affects the normal processes of a living organism. Usually applied to such substances produced by living organisms.

Trabecula A beam-like projection or process; in Basidiomycota, a lamellar primordium or plates of tissue in a young gasteromycete gleba giving rise eventually to a branched columella; in Ascomycota.

Trace elements Elements essential for growth, but required only in minute amounts.

Tracheid Elongated conducting cell of the xylem, with tapering or oblique end walls and pitted walls.

Tracheid types Early plants developed several distinct types of tracheids.

Tracheomycosis A fungal disease in which the pathogen is more or less confined to the vessels in the xylem of the host plant. The term vascular wilt disease is applicable in most cases and is preferred.

Tracheophyte Plants with true vascular tissue, i.e. xylem and phloem;. Includes all land plants higher than bryophytes (mosses) and liverworts.

Trama In hymenomycetes, any sterile tissue within a basidioma, e.g. in lamella, spine, but also in stipe or pileus; context.

Tramal plate In gasteromycetes, layer of tissue partially or entirely devoted to the production of basidia.

Transcript One nucleic acid is the transcript of another if it was synthesized using the original nucleic acid as a template, so that the base sequence of the transcrip is comlementary to the base sequence of the template. By far the most common transcription is the synthesis of RNA in the nucleus by an RNA polymerase, using the organism's DNA as a template. Thus the DNA sequence aaaggttcagt is transcribed as an RNA with sequence uuuccaaguca (RNA contains Uracil bases where DNA has Thymidine). In cell biology and biochemistry, transcription is contrasted with translation, the process in which the base sequence of messenger RNA (mRNA) is used as a template to create proteins.

Transduction The virus-mediated transfer of host DNA (chromosomal or plasmid) from one host cell (the donor) to another (the recipient). Transduction was first observed in bacteriophage/bacterium systems, but has since also been found to be mediated by certain viruses infecting eukaryotic cells.

Transfection The successful virus-infection of cells following their inoculation with viral nucleic acid.

Transfer RNA The RNA that moves amino acids to the ribosome to be placed in the order prescribed by the messenger RNA.

Transformation A process in which exogenous DNA is taken up by a (recipient) cell or protoplast, in which it may be incorporated into the chromosome (or, e.g., into a plasmid) by homologous recombination or converted into an autonomous replicon. The DNA (transforming or donor DNA) may be a fragment of chromosomal DNA from a related strain, a plasmid, or a viral genome.

Transgenic Possessing a gene from another species; used to describe the organisms that have been the subject of genetic engineering.

Transhumance Farmers have a fixed residence but send herds and flocks (with attendant) to pastures

Transitional climatic zone The suboptimal zone between temperate and subtropical climates.

Translation The assembly of a protein on a ribosome, amino-acids being brought to the site by tRNA (transfer RNA) and their order of assembly being specified by mRNA (messenger RNA).

Translocated herbicide A herbicide moved within the plant. Translocated herbicides may move either in the phloem or xylem, but the term is often used to refer to herbicides moved in the phloem.

Translocation The process whereby solutes are transported through the phloem. Unlike xylem, phloem is essentially a chain of living cells. The most widely accepted explanation for translocation of substances in the phloem is the Münch pressure flow hypothesis. The proposed mechanism is as follows: a high concentration of solu.e (e.g., sugars) is loaded into cells of the phloem, presumably by active transport, at locations, such as a leaf, where sugars are produced by photosynthesis. The high solute concentration in the phloem cell creates an osmotic gradient which draws water into the phloem. The resulting high turgor pressure causes flow to occur. If the dissolved sugars or other chemicals carried along with the sugars are removed from the phloem at another place in the tree for use (a sink such as a root or fruit), the decline in concentration of sugar causes water to move out of the phloem cells. Because water is moving in at a source and out at a sink, there is a mass flow of water and substances in the phloem.

Translucent So clear that light may pass through.

Translucent-striate Of a pileus, Appearing striate because the lamellae are visible through the thin pileus.

Transmission breakpoint A critical average worm burden below which the mating frequency is too low to maintain a dioecious parasite species.

Transmission The process by which a pathogen passes from a source of infection to a new host. There are two major types: horizontal and vertical transmission. The majority of transmission processes operate horizontally.

Transmission threshold Occurs for a parasite when the basic reproductive rate Ro is equal to 1. Below this threshold level the disease is unable to maintain itself within the host population. Typically, for directly transmitted parasites there is a transmission threshold for the host population size.

Transmit To spread or transfer, as in spreading an infectious pathogen from plant to plant or from one plant generation to another.

Transpiration The loss of water vapor from the surface of leaves and other aboveground parts of plants.

Transplant shock Nurserymen often prune the roots of trees to be used as bare root transplants or containerized trees. These trees must reestablish an adequate root system in the new location before the entire plant can begin normal growth During this period of system shock, stress symptoms are normal. Symptoms include: overall stunting of leaves and annual shoot growth, wilt, scorch, and abnormally high amount of winter-killed twigs and/or buds. As a rule of thumb, the transplant shock period lasts one year for each one inch of stem caliper.

To minimize transplant shock problems, use correct planting techniques, give trees adequate water for three years after planting, and use no fertilizer for the first two years.

Transposon A discrete piece of DNA that can insert itself into other DNA sequences within the cell. The ends of the transposon DNA are usually inverted repeats.

Transverse section cross section A section perpendicular to the main axis of the plant or organ.

Trap crop Crop planted around a field to protect the inner crop from diseases transmitted by aerial vectors; host crop of a parasitic plant, such as witchweed (Striga spp.), that is planted to stimulate seed germination, and later sacrified by plowing under before the parasitic plant produces new seeds.

Treatment threshold The level of pest population at which a pesticide or other control measure is needed to prevent eventual economic injury to the crop.

Tree A large plant with a sturdy main trunk which lives for many years. Most trees are one of two types: deciduous or conifer. Deciduous trees have broad leaves that die in the Fall. Conifers are evergreen trees, usually with needles for leaves. Most animals depend on trees for food and shelter.

Trehalose A crystalline sugar, $C_{12}H_{22}O_{11}$, characteristically found in fungi.

Tremorgen A mycotoxin inducing a neurotoxicosis (tremor) in man and other mammals; a neurotoxin.

Trenching Physical separation of soil in a vertical plane to sever grafted roots between trees.

Tretic Of conidiogenesis, enteroblastic conidiogenesis in which each conidium is limited by an extension of the inner wall of the conidiogenous cell.

Triarch Consisting of three; for example a triarch stele has three lobes

Trichoderma sp Trichoderma sp. Trichoderma is most commonly found in soil. Trichoderma is often found in litter materials (polluted streams, sewage plants and driftwood). It is found on paper, and in kitchens on many common tableware materials. T. viridae is often isolated from indoor air samples and house dust. Materials such as wood construction and mineral fiber panels can be very affected by this fungus. Trichoderma sp. is reported to be allergenic but are relatively rare. Inhalation of the conidia or teh volatile organic compounds, may cause symptoms similar to those of Stachybotrys reactions. Cultivation - Potato dextrose agar, 24°C.

Trichogyne In some algae, lichens, and fungi, a projection from the female sex organ that receives the male gamete or nuclei before fertilization (karyogamy).

Trichome A plant epidermal hair, of which several types exist.

Trichomycetes An inconspicuous Class of Zygomycota, found attached to the gut walls of insects.

Trichothecenes Mycotoxins (scirpenes) of Fusarium tricinctum, F. sporotrichioides, F. poae, and Trichothecium (Hyphomycetes).

Tricothecium sp Tricothecium sp. Occurring in soil and decaying plant matter, often found as an epiparasite on black knot of plum. Spores are 2-celled, colorless to pink, bilaterally symmetrical and are produced in long chains from the unbranched conidiophores. Cultivation - Potato dextrose agar, 24°C. Produces mycotoxins, trichothecin and other trichothecenes.

Trifurcate Bearing three branches or forks.

Trimastix Flagellum cross section, showing 9+2 doubletsFlagellum. A eukaryotic flagellum is a bundle of nine fused pairs of microtubules called "doublets" surrounding two central single microtubules (the so-called 9+1 structure of paired microtubules; also called the "axoneme"). At the base of a eukaryotic flagellum is a microtubule organizing center about 500 nm long, called the basal body or kinetosome. The flagellum is encased within the cell's plasma membrane, so that the interior of the flagellum is accessible to the cell's cytoplasm. This is necessary because the flagellum's flexing is driven by the protein dynein bridging the microtubules all along its length and forcing them to slide relative to each other, and ATP must be transported to them for them to function. This extension of the cytoplasm is called the axosome. The only thing that the bacterial, archaean, and eukaryotic flagella have in common is that they stick outside of the cell and wiggle to produce propulsion.

Trimitic Describes basidiomata constructed of three different types of hyphae, generative, skeletal and binding.

Triplet codon A set of three nucleotide bases in DNA or RNA that code for an amino acid.

Triradiate Having three radiating arms or branches.

TRNA The RNA that moves amino acids to the ribosome to be placed in the order prescribed by the messenger RNA.

Troop A group of hundreds or even thousands of fruit-bodies, especially basidiomata, in a few square metres.

True leaf Any leaf produced after the cotyledons.

Truffle An edible hypogeous ascoma of the genus Tuber (Pezizales: Ascomycetes). Can also be applied to the basidiomata of many hypogeous basidiomycetes, though these are not generally edible.

Truncate Ending abruptly as though the end had been cut off.

Trunk The main, central, mostly straight part of a tree or shrub, which the branches grow from. Some trees have more than one trunk.

Tuber An enlarged, fleshy, underground stem with buds capable of producing new plants.

Tuberculate Bearing small, wart-like processes or nodules; of a fruit-body, raised or wart-like in shape.

Tuberiform Tuber-like.

Tuft Bunching of twigs or needles.

Tumefaction A plant tumor or gall.

Tumor Abnormal swelling or localized outgrowth, often roughly spherical, produced by a plant as a result of attack by a fungus, bacterium, nematode, insect or other organism.

Tunica A coat; in conidia, a rigid layer, exterior to the cell wall, as opposed to a mucilaginous sheath.

Tuning fork basidia The typical basidia of the order Dacrymycetales (Holobasidiomycetes), which have only two spore-bearing extensions.

Turbinate Top-shaped; flattened at the apex and abruptly tapered at the base.

Turf-grass community An combination of individual turf-grass plants that have mutual relationships with the environment and individual plants.

Turgidity State of being rigid or swollen as a result of internal water pressure.

Turgor pressure Pressure on a cell that maintains it in a inflated or swollen state.

Twig A small branch on a tree or shrub that is connected to a larger branch. Leaves, flowers, and fruits grow from buds on a twig. Most trees and shrubs can be identified by thier twigs, especially in Winter.

Twig borer A small caterpillar that enters the tender shoot tip as it starts growing, causing the leaves to die and the tip to die back.

Tylose A balloon-like intrusion into the lumen of a vessel, formed by enlargement of an adjacent xylem parenchyma or medullary ray cell, and its extrusion through a pit in the vessel wall. In the older literature a single such structure was termed a 'tylosis'. However, it is a current convention that the suffix -osis denotes a process or condition, and it is therefore recommended that the structure should be termed a tylose.

Tymovirus Member of a group of plant viruses with isometric (icosahedral) particles containing a single molecule of linear RNA, easily transmitted mechanically and transmitted in nature by beetles.

Type Type has several different senses in biological writing.

As a noun 1. in the ordinary dictionary sense of' kind ' or' sort ', frequently with the implication that the taxon specified is a good representative (is typical) of the taxon or higher taxon to which it belongs, e.g. as in didactic writing when a group of organisms or a topic is elucidated by the study of a selection of types;

2. in taxonomy, in the technical sense of nomenclatural type, i.e. the element to which the scientific name of a taxon is permanently attached, e.g. a family based on a type which

is a genus, a genus on a type species, a species on a description accompanied by a type specimen. While it is desirable that a nomenclatural type should be typical of the taxon to which it belongs this is not necessarily so. As an adjective, type is used in both the above senses, e.g. type specimen, which indicates that the specimen is a nomenclatural type, and type culture collection, in which most cultures are not nomenclatural types but merely (or hopefully) cultures typical of the taxa to which they belong.

As a verb, bacteriologists use ' to type ' in the sense of to ascertain the serotype or phage type of an isolate; it should not be used as a synonym of ' to identify ' or ' to diagnose '.

As a suffix, ' -type ' is used in a series of diverse compound terms applied to intraspecific categories.

U

Ulocladium sp Ulocladium sp. - (Aw 0.89). This fungus is reported to be allergenic and considered cosmopolitan. It is commonly found as a saprophyte on plant materials and soils. Some species can be also found on dead herbaceous plants, rotten woods, paper, textiles, and other organic substrates (cellulose,)such as water-damaged building materials. Ulocladium is also found in dust and air samples. Ulocladium is known to be a common airway allergen. Cultivation - Potato dextrose agar, 26°C.

Ultrastructure Submicroscopic structure of a macromolecule, cell, or tissue.

Umbilicate Of the pileus: Having a central navel-like depression, sunken in the center, somewhat funnel-shaped.

Umbo Of the pileus: A raised, conical to convex knob or mound on the center.

Umbonate A mushroom cap having a rounded mound in the center.

Uncinate Hooked.

Uncinula necator Uncinula necator is a fungus that causes powdery mildew on grape. It is a common parasite of grape wine (Vitis vinifera). The fungus originated in North America. European wine varieties of Vitis vinifera are more or less susceptible to this fungus. Uncinula necator infects leaves and young berries, and can cause large damage if untreated. This parasite does not need drops of water to infect the plant leaves; high atmospheric humidity is enough. It is also called Oidium. This mildew can be treated with sulfur or fungicides, especially Strobilurines. Uncinula can develop resistance to organic fungicides, so it is important to use different (rotate) fungicides during one season (for example, using two or three different types of organic fungicide and sulfur). The parasite can not develop resistance to sulfur, but efficacy of this fungicide is not as high as the organic ones. But sulfur can be used as a prevention for this plant disease.

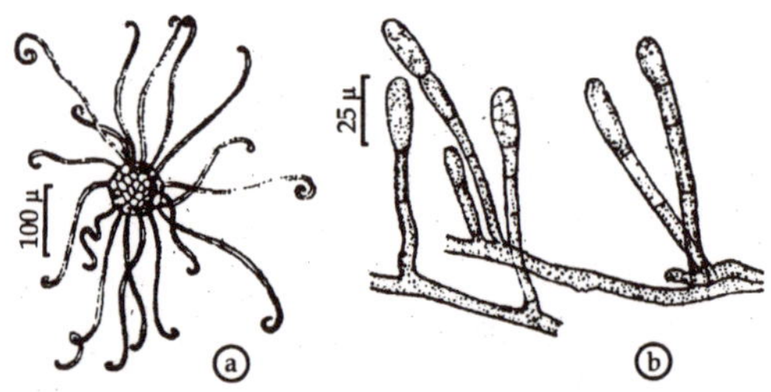

Fig. Uncinula necator: (a) cleistothecium (b) conidiophores and conidia.

Understory An understory tree is a tree that can grow underneath taller trees. An understory tree must be able to tolerate shade. American Beech and Flowering Dogwood are good examples of understory trees. They grow very well underneath taller oaks or hickories.

Undulate Wavy, not flat.

Undulipodium Cilium; sperm tail; motility organelle covered by cell membrane; the term preferred by Margulis et al. for the flagellum in eukaryotes.

Unfurled Unopened.

Uninucleate Having one nucleus.

Unipolar At one end only.

Uniseptate Having one septum.

Uniseriate Of spores in an ascus, in one row.

Unitunicate An ascus without differentiated inner and outer walls is said to be unitunicate. "The unitunicate ascus sometimes has an operculum (a small lid), which opens to liberate the ascospores when the ascus is mature. These asci are called unitunicate-operculate. ... Other asci have no operculum but have an apical pore and/or a ring-like structure at the tip, acting as a valve, or sphincter, through which the ascospores are violently discharged and are dispersed by the air. Such asci are called unitunicate-inoperculate ... There are still other perithecial fungi with unitunicate asci which have no obvious mechanism for ascospore release. In this case, the ascospores are released into the cavity of the ascomata, when the ascus wall disintegrates."

Universal veil A sheet of tissue surrounding the entire fruiting body until it is ready to expand and disperse spores, seen especially well in the genus Amanita . Portions of the universal veil may remain as wartlike masses on the surface of the cap. The bottom portion may persist as a cup, or volva.

Urceolate Pitcher-like, hollow and contracted at the mouth like an urn.

Urea The chief product of nitrogen metabolism in mammals, excreted in the urine. Carbonyl diamide.

Urea formaldehyde A synthetic, slowly soluble nitrogen fertilizer consisting mainly of methylene-urea polymers of different lengths and solubilities; formed by reacting urea and formaldehyde.

Uredinales The rust fungi; obligate plant parasites in the Basidiomycotina.

Urediniospore (also urediospore, uredospore) The asexual, dikaryotic, often rusty-colored spore of a rust fungus, produced in a structure called a uredinium; the "repeating stage" of a heteroecious rust fungus, i.e. capable of infecting the host plant on which it is produced.

Uredinium The structure in rust fungi that gives rust to urediniospores. Sometimes called uredium.

Urediospore Spore produced by a rust fungus that spreads the infection to other hosts.

Uredospore Dikaryotic spore of rust fungi produced in the second host and capable of reinfecting it.

Urocystis tritici fungus affecting leaves and stems of wheat. A smut fungus causing a smut in cereals and other grasses that chiefly affects leaves and stems and is characterized chains of sori within the plant tissue that later rupture releasing black masses of spores.

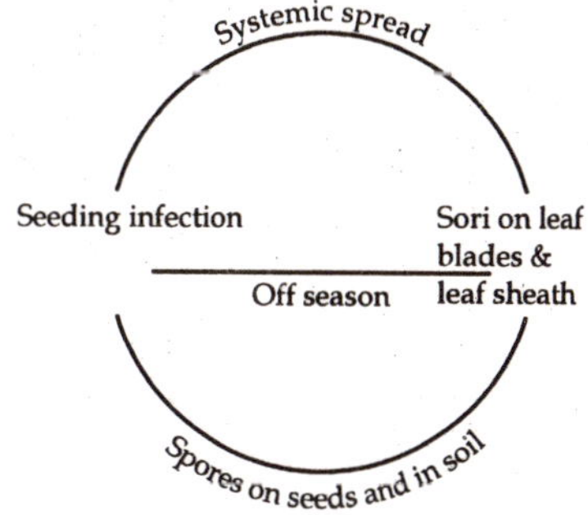

Fig. Disease cycle of Urocystis tritici causing flag smut of wheat.

Urn A term used for anything even vaguely vase- or gourd-shaped, e.g. the sporangium in mosses.

USDA United states department of agriculture.

Ustidium Term used for the basidium of Ustilaginales and Tilletiales (Moore, 1985).

Ustilaginales The smut fungi; plant parasites in the Basidiomycotina.

Ustilaginomycetes The class Ustilaginomycetes comprises more than 1400 species of basidiomycetous plant parasites, which are distributed in approximately 70 genera.

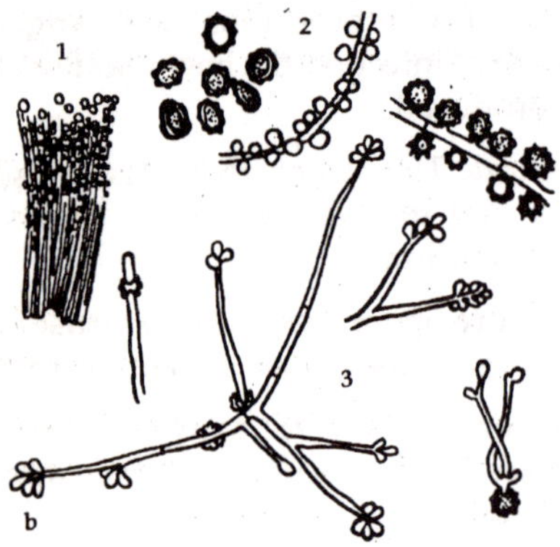

Fig. False smut of rice (Ustilaginoidea virens): (b) 1: section of outer part of sclerotium, 2: formation of spores, 3: spores germinating.

They occur throughout the world, although many species are restricted to tropical, temperate or arctic regions. Some species of Ustilago and Tilletia, e.g. the barley, wheat or maize smut fungi, are well known because they are of economic importance. For example, from 1983 to 1988 the barley smut fungi reduced annual yields by 0.7% to 1.6% in the prairie provinces in central Canada, causing average annual losses of about U.S. $8,000,000. Tilletia contraversa is important in the international wheat trade, and 2-5% of the plants in a corn field are generally infected by Ustilago maydis, while up to 80% of a field can be infected if conditions are good for the smut fungus. On the other hand, the galls of U. maydis are considered a delicacy in the Mesoamerican tradition. They are known in Mexico as "Huitlacoche" and in the U.S.A. as "maize mushroom", "Mexican truffles" or "caviar azteca".

Ustilospore The spore of a smut (Ustilaginales) or bunt (Tilletiales) which, on germination, gives rise to a basidium or, more rarely, a hypha; formerly called chlamydospore or teliospore.

Utriform Bag-like; sac-like.

V

Vaccine A drug intended to induce active artificial immunity against a pathogen.

Vaccines may be live or dead Live vaccines are usually attenuated versions of the wildtype pathogen, such as the MMR vaccines, which are strains of measles mumps and rubella viruses repeatedly passaged through cell lines until non-pathogenic. Typically, live vaccines need only be given as a single dose to induce a full immunological response, inducing specific memory.

Dead vaccines are either killed whole parasite, as with the Salk polio strain and pertussis vaccine, or some highly immunogenic fraction of the parasite, as in toxoid vaccines. Killed vaccines and toxoids which do not multiply in the host must usually be administered in multiple doses to induce a full immunological response.

Vaccination should be distinguished from passive immunization in which concentrated specific antibodies which can be used therapeutically to abrogate an ongoing infection or to provide short term protection (of the order of months), for example against hepatitis A. Passive immunization does not induce immunological memory.

Vacuole Generally spherical organelle within a plant cell bound by a membrane and containing dissolved materials such as metabolic precursors, storage materials, or waste products.

Vagina In nematodes: A canal, lined with cuticle, that connects the uterus or uteri with the female gonospore.

Valid Of taxonomic names and epithets: Published in accordance with several articles of the Code of Nomenclature; such names may be legitimate or illegitimate.

Valve In nematodes: A structure that regulates the rate and/or direction of intake of materials (e.g., the esophago-intestinal valve or cardia).

VAM More correctly AM or arbuscular mycorrhizal, since not all fungi producing this kind of mycorrhiza have vesicles.

Vapor drift The movement of chemical vapors from the area of application. Some herbicides, when applied at normal rates and normal temperatures, have a vapor pressure high enough to cause them to change into vapor form; this may cause serious injury to susceptible plants away from the site of application.

Var Variety.

Variability The property or ability of an organism to change its characteristics from one generation to the other.

Variable A quantity able to assume different numerical values

Variable rate technology The ability to tailor application of seeds, pesticides, fertilizers or other amendments to the needs of specific small areas within a larger field or other management unit.

Variant A general term for any variant from the typical. A culture that is not identical with the parent in all its characters. A loose term now largely replaced by adapted strain or mutant. A useful name for any unnamed variation within a species which avoids using the term variety which has a nomenclatural connotation.

Variegation Pattern of two or more colors in a plant part, as in a green and white leaf.

Variety 1. A subdivision of a species below the rank of subspecies and above the rank of form; subspecies.

2. cultivar 1. One or more races of a pathogen that are characterized by the limitation of their host range to a certain genus or genera. Also, a group of closely related plants of common origin and similar characteristics within a species.

Vas deferens In nematodes: A slender, tube-like gonoduct in the male; unites posteriorly with the rectum to form the cloaca.

Vascular bundle Strand of conductive tissue, usually composed of xylem and phloem (in leaves, small bundles are called veins).

Vascular cylinder The cylinder of vascular tissue in stems or roots.

Vascular discoloration Plants infected with a disease of the vascular tissues (the part of the plant that carries water from the roots to the leaves) will often have brownish streaks in the vascular tissues.

Picture: Brown vascular tissue in a woody ornamental plant.

Vascular mycosis A fungal disease in which infection is more or less limited to the vascular system of the host plant. The term vascular wilt disease is applicable in most cases, and is preferred. However, in cases in which wilting is not a characteristic symptom vascular mycosis may be acceptable.

Vascular pathogen A pathogen that resides in and moves through the xylem and/or phloem.

Vascular Relating collectively to the conductive plant tissues, xylem and phloem. Analogous vascular tissues in bryophytes are referred to as hydroids and leptoids, respectively.

Vascular ring A thin area of potato tuber tissue between the cortex and the medulla in which vascular tissue is concentrated.

Vascular system The system of plant tissues that conducts water, mineral nutrients, and products of photosynthesis through the plant, consisting of the xylem and phloem.

Vascular tissue Plant tissue that conducts water and nutrients throughout the plant.

Vascular wilt disease A disease in which the pathogen is almost entirely confined to the vascular system of the host during pathogenesis, and in which wilting is a characteristic symptom. If wilting is not a characteristic symptom the term is not strictly applicable.

Vasoconstriction Constriction of blood vessels.

Vector A living organism (e.g., insect, mite, bird, higher animal, nematode, parasitic plant, human) able to carry and transmit a pathogen and disseminate disease; (in genetic engineering) a vector or cloning vehicle is a self-replicating DNA molecule, such as a plasmid or virus, used to introduce a fragment of foreign DNA into a host cell.

Vector efficiency The relative amount of transmission (usually expressed as a percentage) that can be expected from a specific vector-virus-host combination.

Vector specificity The comparative range of vector species able to transmit a given virus.

Vectorial capacity In vector-borne infections such as malaria, the vectorial capacity is a concept analogous to the contact rate in directly-transmitted diseases. It is, thus, a function of (a) the vector's density in relation to its vertebrate host, (b) the frequency with which it takes blood meals on the host species, (c) the duration of the latent period in the vector, and (d) the vector's life expectancy.

Vegetable catepillar A mummified lepidopteran larva from which arises the stromatic teleomorph of a species of Cordyceps (Clavicipitales: Ascomycetes).

Vegetation Plant life. Many small and large animals need vegetation to hide and live in.

Vegetative reproduction Asexual reproduction.

Vein Small vascular bundle in a leaf.

Vein banding A change of colour in a narrow zone of leaf tissue alongside the main veins.

Symptom of a virus disease in which regions along the veins are darker green than the tissue between the veins.

Vein clearing An increased translucency of the veinal system in a leal, making the pattern more pronounced (light against dark) by transmitted light.

Veination The arrangement of veins in a leaf.

Vein-delimited The leaf veins from the boundary of the affected areas of a plant.

Velum Veil.

Velutinous Thickly covered with delicate hairs; velvety.

Venom A liquid injected by one animal into another which causes pain or destruction of body tissue. Some venoms are lethal (cause death), depending on the animal (or person) being injected. Honey Bees, Channel Catfish, Copperheads, some caterpillars, and most spiders have venom, but most of these are mild to humans and only cause a stinging sensation. Venom can be injected by stingers, fangs, or spines.

Venter Swollen basal region of the archegonium where the sporophyte develops in bryophytes.

Ventral Front, or lower surface.

Ventricose Swollen, often in the middle or to one side; of a lamella, broader midway between the stipe and the margin of the pileus.

Véraison Beginning of fruit ripening, recognized by berry softening and beginning of pigmentation in colored varieties.

Vermicular Worm-shaped, thickened and bent in places. (Also vermiculate.)

Vermiform Worm-like.

Verrucose 1. With the surface ornamented by small rounded processes or 'warts';

2. distinctly and regularly rough-walled.

Verruculose Delicately verrucose.

Versicoloured Of different colours.

Vertebrates Animals with backbones

Vertical mower A mechanical device whose vertically rotating blades cut into the face of a turf for the purpose of reducing thatch, grain, and surface compaction.

Vertical resistance Resistance which is effective against some biotypes or races of the pathogen, but not others, usually inherited monogenically and expressed qualitatively.

Vertical transmission Vertical transmission occurs when a parent conveys an infection to its unborn offspring, as occurs in syphilis in man or in many arboviruses of arthropods. Perinatal infection is a special form of vertical transmission.

Verticillate Whorled.

Verticillium sp Verticillium sp. Commonly found in soil and decaying plant matter, known also to cause plant diseases.

Characterized by whorls of phialides produced along the length of undifferentiated filaments of on conidiophores. Conidia are colorless, 1- or 2-celled, collected in small wet masses. Cultivation - Potato dextrose agar, 26°C.

Vertifolia effect A term used to describe the loss of general (horizontal, minor gene, multigenic, polygenic) resistance in a cultivar after several generations of selection during which a major gene confers resistance to the dominant race or biotype of the pathogen; first observed in the potato cultivar Vertifolia with late blight resistance.

Vesicle 1. Small, intracellular, membrane-bounded sac in which substances are transported or stored;

2. swollen, lipid-filled cells produced inside plant roots by most endomycorrhizal fungi (sometimes called intramatrical spores).

Vesicular arbuscular mycorrhizas Endomycorrhizas; plant roots colonized by mutualistic fungi of the Glomales (some of which do not produce vesicles, suggesting that these mycorrhizas should simply be described as arbuscular mycorrhizas).

Vessel A xylem element or series of such elements whose function is to conduct water and mineral nutrients.

Vestigial Pertaining to structures or organs that were well developed in an organism's ancestors but have become rudimentary during the course of evolution.

Viability The ability to live; commonly a measure of survival in a population, e.g. of conidia, germination in seeds, etc.

Vicariance Speciation which occurs as a result of the separation and subsequent isolation of portions of an orginal population.

Vigor The capacity of a strawberry plant for active vegetative growth.

Villous Covered with long, soft hairs.

Vine A plant with a weak stem that relies on other plants or structures for support. These plants use different ways of wrapping themselves around other plants to climb upward. Some vines in our area are Poison Ivy, Virginia Creeper, and Wild Grape.

Viraemia The presence of virus in the blood.

Virescence Greening of tissue that is normally devoid of chlorophyll; the abnormal development of flowers in which all organs are green and partly or wholly transformed into structures like small leaves.

Virescent A normally white or colored tissue that develops chloroplasts and becomes green.

Virgate Streaked.

Viricidal Able to inactivate a virus completely and permanently.

Virion The infectious unit of a virus.

Viristatic Able to prevent multiplication of a virus.

Viroid A pathogenic agent, formerly assumed to be a virus, now known to be a ribonucleic acid of low molecular weight and not a ribonucleoprotein. The term was introduced by T. O. Diener following work with potato spindle tuber 'virus'. Citrus exocortis and chrysanthemum stunt diseases may be caused by similar agents.

An infectious nucleic acid without a protein coat that causes potato spindle tuber or chrysanthemum stunt.

Viroplasm Cellular inclusions that are sites of synthesis of viral components and the assembly of virus particles.

Virosis Infection by a virus, or a disease so caused.

Virucide A pesticide that is used to kill viruses in plants.

(Chemicals that kill viruses for medical or veterinary applications are termed "antiviral agents"; they are classed as pharmaceuticals and so are not included in the Compendium.)

Virulence 1. Relative capacity to cause disease; degree or measure of pathogenicity of a parasitic organism or virus.

2. the observed infective capacity of individual strains of the pathogen on application to the tissues of the host. It is recommended that the term should not be used instead of pathogenicity, but should be used with quantitative adjectives to describe degrees of pathogenicity. J. E. van der Plank has proposed that races of a pathogen which interact differentially with varieties should be said to vary in virulence; this restricted use of the term is not recommended: virulence refers to the observable effects of the pathogen on the host, and carries no implication as to the mode of action, mode of inheritance, etc. of this attribute of the pathogen. The degree of pathogenicity of a given pathogen.

1. The case mortality rate of an infection.

2. The extent to which a pathogen harms its host.

These are different usages: what they have in common is that they refer to the effect on an already infected host, not to the degree of transmissibility to a subsequent susceptible.

Virulent Capable of causing a severe disease; strongly pathogenic.

Virulent phage A phage that multiplies in a susceptible cell, which undergoes lysis and liberates new phage particles.

Virulent races Races of a pathogen that are capable of causing severe disease; strongly pathogenic.

Viruliferous Of a vector which carries or contains virus. The vector may or may not be infective and the terms are not synonymous.

Capable of transmitting a virus.

Virus A virus is an extremely small organism composed of genetic material (either RNA or DNA) surrounded by a protein coat. Viruses replicate only inside living cells. Viruses are able to take command of plant cell metabolism and direct the cell to manufacture more virus particles.

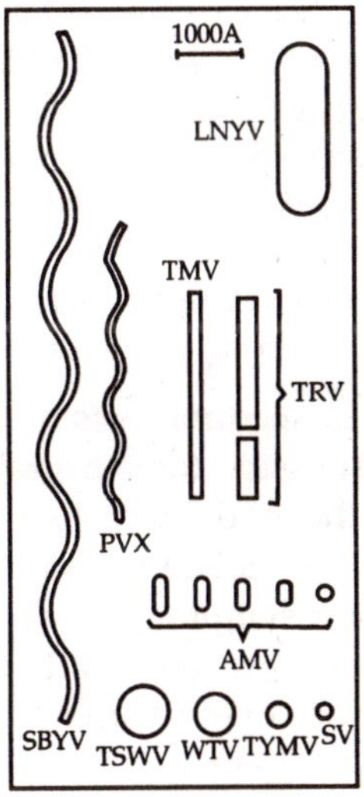

Fig. Figure. Outline drawings to illustrate the range of size and shape found among plant viruses. SV= satellite virus; TYMV= turnip yellow mosaic virus, WTV= wound tumour virus, TSWV= tomato spotted wilt virus, SBYV= sugarbeet yeelow virus, AMV= the five particles of Alfalfa mosaic virus, TRV= tobacco mosaic virus, PVX= potato virus X,LNYV=lettuce nectrotic yellows virus.

An average virus is 100,000 to 1,000,000 times smaller than a plant cell. Due to their size and biology, plant viruses traditionally have been difficult to identify. Symptoms give important clues, but yield no real proof of the exact pathogen involved. Imagine the difficulty of fishing for minute virus particles in a vast sea of plant cell contents. Then, imagine trying to tell these particles apart from other similarly-shaped particles that are entirely different viruses causing other diseases!

In recent years, many advances in molecular biology techniques have allowed plant pathologists to "label" viruses in order to identify them. Virologists have developed many different labeling techniques for virus identification. The following is a simplified example of one such techniques. To detect tobacco mosaic virus (TMV) in tomato, virologists have developed a substance that turns blue only when a solution of crushed plant material containing TMV is present. If the tomato spotted wilt virus (TSWV) is present instead, the solution remains clear. To confirm the presence of TSWV, one must conduct a second test containing a substance that turns color in the presence of TSWV.

It would be too time consuming to run tests for the 30 possible tomato viral diseases. This is where the art of plant disease diagnosis is important. A skilled diagnostician can look at the plant symptoms, make guesses as to the viruses present, and then run tests specific for these viruses to scientifically identify the disease.

Viruses enter plants in specific ways that vary tremendously depending on the type of virus involved. For example, wheat streak mosaic virus (WSMV) always is carried in the saliva of wheat curl mites feeding on wheat plants. Other species of mites do not carry WSMV. Barley stripe mosaic virus (BSMV) always is transmitted inside barley seed.

Means of viral transmission include the following:

feeding by mites, aphids, leaf hoppers, trips, white flies, and nematodes,

on the surface or inside seeds

mechanically in plant sap- on hands, tools, and other equipment,

vegetative propagation, such as cuttings,

grafting,
in certain root-infecting fungi.

Viruses commonly cause dwarfing, odd color patterns on leaves, or stunted plants. Other symptoms include enation (tissue overgrowth), flattening and distortion of the stem, leaf roll, mosaics, and ring spots.

Virus structure A virus has been defined as 'a nucleo-protein entity able to pass through bacterium-retaining filters and to replicate within the living cells of the host'. The nucleic acid component of a virus particle, together with its surrounding protein shell, the capsid, comprise the nucleocapsid. The capsid is formed from a number of smaller structural protein units, the capsomeres. The building of virus particles from their nucleic acid and protein components is described as assembly; the association of virus particles in dense masses, which may be of crystalline or paracrystalline form, is termed aggregation. A virion is a single mature virus particle which, for plant viruses, is usually equivalent to the nucleocapsid, but sometimes includes also an additional surrounding envelope. The enclosure of the nucleic acid (the genome) of one virus within a complete protein capsid of a second virus is described as transcapsidation. A virus in which the genetic information necessary for multiplication is divided between two or more physically separable particles is said to have a divided genome.

Virus transmission The transfer of a virus from an infected host to a non-infected one, commonly through the agency of another organism, the vector, which is often an insect or a nematode, but also may be a mite, fungus, parasitic flowering plant etc. Transmission may also be effected by direct contact, e.g. by rubbing together of leaves or stems, through root grafts etc. The terminology of transmission relates both to the properties of the virus and to its relations with the vector. Hence, the virus may be persistent, semi-persistent or non-persistent, according to the length of time that it remains infective in the vector; propagative, if it multiplies within the vector, or nonpropagative; stylet-borne or circulative; dependent on the presence of a second (helper) virus, or not dependent.

These sets of terms relate to different attributes of virus-vector relations, and although one attribute may be constantly or frequently associated with another, e.g. stylet-borne viruses are usually non-persistent, the terms are not synonymous; similarly, propagative and circulative are not synonymous, even though a propagative virus is necessarily also circulative.

Virus transmission tests Experiments to determine the virus-vector relations involved in the transfer of a virus from an infected host to a non-infected one, generally entailing some or all of the following experimental procedures: after pre-acquisition fasting a non-viruliferous vector is placed on an infector plant for a given acquisition access time (or period). Part of this time may be taken up with probing; that part taken up with feeding is the acquisition feeding time. The now viruliferous vector may immediately become infective, or there may be a delay, the latent period (or pre-infective period) during which the vector remains non-infective. The experiment may necessitate post acquisition fasting before the vector is placed on a healthy test-host (test plant) for a given inoculation access time (or period) during which probing and inoculation feeding can take place. In a serial transmission test a single acquisition access period is followed by inoculation feeding on two or more test hosts in succession.

Virus, circulative Originally restricted to viruses that have been proved to be transmitted by way of the salivary glands of leafhoppers. Such viruses can be detected in the haemolymph and persist for a long time in the vector and through the moult.

Some circulative viruses are known to multiply in the vector and have been referred to as circulative-propagative or simply as propagative viruses: note, however, that 'circulative' and 'propagative' refer to different attributes of the virus and are not synonymous.

Virus, dependent One which is transmitted by its vector only in the presence of a second (helper) virus. Synonymous with assisted virus, carried virus, helped virus.

Virus, helper One which must be present for the transmission of a second (dependent) virus by its vector. Synonymous with assistor virus, carrier virus.

Virus, multiparticulate One which produces more than one type of nucleoprotein component in infected plants.

Virus, NETU (from nematode tubular) A term introduced by Cadman for nematode-transmitted viruses with tubular particles. These viruses comprise the tobravirus group of Harrison et al.with tobacco rattle virus as the type member.

Virus, non persistent One which is retained for not more than a few hours by vectors transferred from infected to healthy plants. Such viruses share additional transmission characteristics: they fail to pass to the progeny of the vector or through the moult; they seem to be carried ' externally ' on the stylets and do not multiply within the vector or appear in the haemolymph; they are acquired most readily by starved vectors after very short acquisition feeding periods; they have very short transmission threshold periods. These additional characteristics should not be regarded as criteria for applying the term 'non-persistent.

Virus, persistent One which is retained for long periods and in some instances throughout life, by vectors transferred from infected to healthy plants. Such viruses share additional transmission characteristics: they pass through the moult; some also pass to the progeny of the vector and/or multiply within it; they are carried internally, may be detected in the haemolymph and seem to be transmitted by way of the salivary glands;

there is a positive correlation between the duration of both the acquisition and inoculation feeding periods and the probability of transmission; they tend to have a long transmission threshold period and a latent period, although the latter may not be clearly defined. These additional characteristics should not be regarded as criteria for the use of the term 'persistent.

Virus, propagative A virus which is known to multiply in its vector. Such viruses are usually also circulative, but the terms are not synonymous.

Virus, satellite One which multiplies only in the presence of a specific second virus.

Virus, semi persistent One which is retained for a maximum of several days by vectors transferred from infected to healthy plants. The transmission characteristics of this small group of viruses are less clearly defined than those of the non-persistent and persistent viruses. The efficiency of transmission increases with increases in both acquisition and inoculation feeding times. There is no latent period and the greatest likelihood of transmission is immediately after vectors leave the virus source. Pre-acquisition fasting has no effect on the efficiency of transmission. The semi-persistent viruses do not pass to the progeny of the vector or appear in the haemolymph and they do not seem to pass through the moult. They cannot be ascribed with certainty to the stylet-borne or circulative categories of Kennedy et al.

Virus, stylet borne One which has been proved to be transmitted on the stylet of its vector. Such viruses fail to pass through the moult or to appear in the haemolymph and do not persist long in the vector. In this sense the stylet-borne viruses are equivalent to the externally-transmitted viruses.

Viscin Sticky substance produced on the seed coat of parasitic flowering plants in the Viscaceae (mistletoes) that helps attach the seed to the host plant branch.

Vitamin A group of organic micronutrients, present in minute quantities in natural foodstuffs, that are essential to normal metabolism.

Vitamin D deficiency (ricketts and osteomalacia) In children, the condition prevents normal bone development; in adults, a lack of vitamin D causes demineralization of bone, particularly in the spine, pelvis, and lower extremities.

Viviparous Bearing living young.

Vivotoxin A toxic substance produced internally in the infected host or by the pathogen within it, the substance being responsible for some or all of the harmful changes produced in the course of the disease.

Volatile Quality which makes a compoun devaporate or vaporize (change from a liquid to a gas) at ordinary temperatures on exposure to the air.

Volunteer crop The undesired emergence of a significant stand of a self-seeded, previously planted crop in a field purposely planted with another crop.

Volunteer Self-set plant; plant seeded by chance.

Volva In agarics and gasteromycetes, the cup-like remains of the lower part of the universal veil attached to the base of the stipe or receptacle.

Vomitoxin 4 deoxynivalenol A trichothecene; a mycotoxin produced by Fusarium graminearum (Hyphomycetes); esp. in barley and winter wheat; has an emetic effect livestock.

Vulva In nematodes: Exterior opening of the mature female's reproductive system (female gonopore); generally appears as a transverse slit on the ventral portion of the nematode.

VZV (varicella zoster virus) A herpes virus which causes chickenpox (varicella) and shingles (herpes-zoster).

W

Walling off Separation of diseased from healthy tissues by barrier tissues produced by a diseased plant.

Warm-season turf grass Turf-grass species adapted to an optimal growth during warm portions (80 to 95 degrees F.) of the growing season.

Wasps Many families of 4-winged insects (including hundreds of thousands of species worldwide) which are related to ants and bees. These families comprise the order Hymenoptera. Many wasps are smaller than a pin-head while a few are a couple of inches from wing-tip to wing-tip.

Waste place Area where plants have been removed and the soil disturbed. Common waste places include roadsides, areas with buildings, deserted lots, and areas where fire occurred.

Water activity (aw) Expresses the available water in a substrate as a decimal fraction of the amount present when the substrate is in equilibrium with a saturated atmosphere (an equilibrium relative humidity of 70% around the substrate means that the substrate has a water activity of 0.70).

Water moulds Members of the Order Saprolegniales (Oomycetes).

Water soaked A dull green coloration of diseased tissues due to membrane leaking of cellular contents into intercellular spaces.

Describing plants or lesions that appear wet and dark and are usually sunken and translucent.

Water solubility Expresses the ability of a chemical to move into water or dissolve at a specific temperature and pH.

We might distinguish demographic stochasticity which arises from the discreteness of individuals and individual events such as birth, and environmental stochasticity arising from more-or-less unpredictable interactions with the outside world.

Water sprout Small, rapidly growing shoot or branch on a large stem, developed from adventitious tissues.

Waterfowl Birds that swim and live near water, including ducks, geese, and swans.

Watershed An area of land that drains its surface water into a defined watercourse or body of water.

Waypoints Locations programmed into a GPS unit by the user that can be recalled to guide return to a specific location or obtain a reference to the current location.

Weaning Removal of young mammals from their source of milk

Weed Any plant that is considered undesirable. One person might think a plant is beautiful and grow it in her garden. Another might think the same plant is a weed and pull it out of his garden. Most plants people call "weeds" are plants that cause problems for many people, such as crabgrass, clover, and plantain.

Weevil Small beetles with a snout in the suborder Rhynchophora.

Welfare Consideration for the comfort and well bieng of farm animals. The Farm Animal Welfare Council has issued 'Codes of Practice' covering the treatment and management of farm animals

Western blot A protein blot. (This term originated as lab jargon; not a generally accepted term.)

Wettable powder Another popular residual spray formulation. A wettable powder is an insecticide in which the active ingredient has been added to a fine absorptive clay. The clay is then mixed with a wetting agent in water to form a liquid spray.

Wetting agent Material that decreases the contact angle of a liquid on a surface; a substance added to a spray to reduce the surface tension of the applied droplets.

Whiplash flagellum A flagellum with a smooth shaft.

White blister Characteristic pustular fructifications of Albugo spp. (Peronosporales).

White rot (of wood) Type of wood decay resulting from enzymatic action of fungi; it degrades all components of wood, including lignin, leaving the wood light-colored and spongy.

White rust Disease of crucifers caused by Albuginaceae (Peronosporales: Oomycota).

Whiteflies These are very small insects, seldom more than 2 or 3 mm in length, that resemble tiny moths. The adults of both sexes are winged and are usually covered with a white dust or waxy powder. First instars are active and are called crawlers. Later instars are sessile and look like scales. They have piercing-sucking mouthparts which they use to suck sap from the leaves of plants. They also excrete large mounts of honeydew which sooty-mold grows upon. Some whitefly species are serious pests of cultivated plants.

Whorl A group of three or more leaves branching out from the same spot.

Wildflower Any flowering plant that grows on its own, without the help of people. Some people consider certain wildflowers to be weeds.

Wilt 1. To lose turgidity, to become flaccid.

2. a disease characterised by loss of turgidity and collapse of leaves, typically caused by vascular infection by Verticillium spp., Fusarium spp. and by certain bacteria, e.g. Pseudomonas solanacearum. However, the term is sometimes used to describe any vascular disease caused by these pathogens even though in some hosts there is typically no loss of turgidity, e.g. hop verticillium wilt. In such instances there may be justification for using the terms vascular mycosis or tracheomycosis instead of wilt.

A common symptom of disease due to a loss of turgor and resulting in subsequent drooping and collapse of the foliage or succulent tissues.

Lack of freshness and turgor and drooping of leaves from lack of water; a vascular disease that interrupts the plant's normal uptake and distribution of water.

Wind break A row or other grouping of trees or shrubs used to provide protection against the effects of high velocity winds.

Wind burn Death and browning most commonly occurring on the uppermost leaves of semi-dormant grasses; caused by desiccation.

Windowpane The removal of the epidermal layer of leaf tissue leaving small segments of clear tissue.

Wing Extension of the nut shell at the suture line; varies in size according to cultivar.

Wing membrane The thin, membranous portion of the wings lying between the veins.

Winter annual An annual plant that usually initiates growth in the fall, lives over winter, and produces seed the following spring.

Winter desiccation damage Winter desiccation is a condition that is most common with evergreens. Winter sun and wind cause water loss from the needles, while roots are in frozen soil and unable to replace this water. The usual symptom is purpling or death of needles on the windward side or on the side facing the afternoon sun. Such symptoms may be more severe in newly transplanted plants that have not yet established a good root system. Late fall and winter watering can minimize winter desiccation.

Winter desiccation The death of leaves or plants by drying during winter dormancy.

Winter freeze damage By January 1, most Montana trees have experienced enough cold units to meet dormancy requirements. After this, trees can begin to grow again when conditions are appropriate. Chinooks may cause winter thaw periods wherein daytime temperatures exceed 60 F for several days at a time. Non-adapted trees may begin growing again as though it was spring, only to succumb to winter injury when temperatures drop again. In such cases, only the main trunk and scaffold branches may be viable in the spring.

Planting species adapted to changing Montana conditions can lessen chances of winter freeze damage. Planting marginally-hardy species in protected locations also is important. Examples of protected locations include the north side of a house or the interior of an effective shelterbelt.

Winter overseeding Seeding cool-season turf grasses over warm-season turf grasses at or near their start of winter dormancy, practiced in subtropical climates to provide green, growing turf during the winter period, when the warm-season species are brown and dormant.

Winter planting A system of planting strawberries in mid to late fall that depends on growth during winter months for production of an early spring crop.

Winterburn Foliar necrosis, often marginal, of plants that retain their leaves in winter due to water deficiency because they cannot take up water from frozen soils.

Winterkill Any injury to turf-grass plants that occurs during the winter period.

Witches' broom A massed proliferation of the branches of a woody plant.

An abnormal proliferation of shoots. A single shoot may be replaced by a large number of shoots; if they all grow in one direction they give the appearance of a broom, but frequently in trees there is irregular orientation and loss of apical dominance, giving an interwoven cluster of shoots which resembles a bird's nest rather than a broom, e.g. that caused by Taphrina betulina in birch trees. The same term is less suitably applied to diseased conditions in herbaceous plants, characterised by increased branching, production of large numbers of slender shoots and often by much reduced leaves, e.g. witches' broom of potato.

Abnormal, brushlike development of many weak shoots.

Wood preservative Pesticides intended to kill wood-decaying fungi.30 Northwest Coalition for Alternatives to Pesticides

Wood ray A radially oriented tier of parenchyma cells that conducts food, water, waste products and other materials laterally in stems and roots of woody plants; rays are usually continuous across the vascular cambium between the secondary phloem and xylem.

Wood Secondary xylem.

Worm burden The number of worms an individual host carries. Sometimes directly measurable (Ascaris), sometimes only indirectly (schistosomiasis).

Woronin bodies Woronin bodies are highly refractive, electron dense, membrane bound particles found on either side of septae that divide hyphal compartments in filamentous fungi. One current theory is that Woronin bodies are peroxisomes that function to maintain cellular integrity. The septae separate hyphal compartments are pierced by pores which allow most cytoplasmic constituents (but not nuclei) to travel freely between hyphae. However, if an adjoining hypha is ruptured, the Woronin bodies block the pore to prevent loss of cytoplasm into the ruptured compartment.

Wound Injuries to plant tissues that often breach barriers (cuticle, bark, cell walls) that might otherwise exclude pathogens; some pathogens (e.g. viruses) can enter plants only through a wound; wounds may occur from natural growth processes, physical and chemical agents, animals (especially insects), and many human agricultural activities, such as pruning.

Wound parasite A parasitic organism which can invade a host only if it can first become established in damaged tissues.

X

Xanthosis A collection of symptoms consisting of distorted leaf growth and yellow leaf margins that develops in strawberry plants infected by mottle virus and either crinkle virus or mild yellow edge virus; yellows.

Xenopathic set A population of a pathogen that is characterised by its pathogenicities on a limited series of host species, varieties or cultivars, but may otherwise be genetically heterogeneous, even in its pathogenic potentialities. A term coined by E. R. Wallace as a more precise substitute for physiologic race.It takes into account the incompleteness of the information which provides the basis for the separation of physiologic races; only if it were possible to determine the pathogenicity of the organism or virus to every variety and cultivar of the hostspecies would it be possible to establish the precise limits of a physiologic race. In practice it is possible only to separate ' sets ' of the pathogen in relation to a limited spectrum of potential host varieties or cultivars; hence the proposal to use a term by means of which this limitation is explicitly expressed.

Xerophyte A plant well adapted to withstand prolonged drought. The typical xerophyte is a deciduous shrub with tiny leaves and a shallow root system that extends well beyond the crown of the shrub.

Xerotolerant Able to grow under dry conditions.

XLB (xylem-limited fastidious bacteria) Bacterial pathogens of plants found only in the xylem, causing wilt, scorch, and stunting symptoms; vectored by insects that feed on xylem fluid; not sap transmissible; require complex culture media.

Xylem Unlike phloem, which does not accumulate but instead is shed with the dead bark, the old xylem is retained, may be used to conduct water for a few years, and eventually just provides structural support as the tree grows larger. At least 90% or more of the trunk of a tree is xylem or wood.

Xylose A pentose sugar, $C_5H_{10}O_5$, found in the cell walls of basidiomycetes.

Y

Y leaf The most recently matured leaf.

Yeast Most species of yeast reproduce asexually through a process called budding. The daughter cell of the parent yeast cell is at first much smaller and tends to cling to the parent and often clumps or chains of cells are formed. Ascospores are formed during sexual reproduction of some yeasts.

Yeasts candida sp Candida sp. Part of the normal flora and other mucous membranes in the body. Thrush and other diseases caused by this yeast usually occur after prolonged treatment with antibiotics or steroids. The environment is not a likely source of exposure for this fungus and cells from the organism are usually not airborne. Mucocutaneous candidiasis is one of the most common manifestations of HIV infection. While other yeasts may occasionally cause clinical disease, Candida albicans is the organism isolated from most patients. Most species of yeast reproduce asexually through a process called budding. The daughter cell is at first much smaller and tends to cling to the parent and often clumps or chains cell is at first much smaller and tends to cling to the parent and often clumps or chains of cells are formed. Cultivation - YM agar, 36°C

Yeasts cryptococcus neofornans Cryptococcus neoformans Yeast considered an opportunistic pathogen, although it can cause a disease in an immune competent host. In its environmental state C. neoformans var. neoformans colonizes pigeon and other bird droppings and C. neoformans var. gatti the bark of the red gum tree. Both forms can cause disease in humans (cryptococcosis). Primary infection with C. neoformans follows exposure to an environmental source and inhalation of the fungus leading to an infection of the lung. A transient colonization of the bronchial tree may result, or more extensive pulmonary involvement may occur. Bronchial infection may be self-limiting or chronic and may lead to dissemination to other parts of the body. Meningoencephalitis is the most common manifestation of dissemination and the most severe. Symptoms depend upon the rapidity of onset of disseminated disease. The indolent cases begin with headache. Nausea, dizziness, decreased comprehension, impaired memory and gait ataxia follow and increase in severity as invasion of the cerebral cortex, brain stem, cerebellum and meninges progresses. Although pigeon excreta is considered a significant source of inoculum of this fungus, it is thought that alternative unknown routes of exposure exist. One possible source that is currently being studied is the sexual stage of C. neoformans var. neoformans, Filobasidiella neoformans var. neoformans. The basidiospores of this organism are dry, readily airborne, which may be more readily deposited in the alveoli. Sampling for C. neoformans should only be performed is a route of exposure is suspected. This would include bulk sampling of pigeon droppings in and around fresh air intakes and surface sampling of ductwork and work areas. Most species of yeast reproduce asexually through a process called budding. The daughter cell is at first much smaller and tends to cling to the parent and often clumps or chains of cells are formed. Cultivation - YM agar, 24°C.

Yellows Now used mainly as a name for diseases in which the most conspicuous symptom is a (bold) yellowing of normally green tissues. If only part of an organ, such as a leaf blade, is affected the transition between yellow and green tissue is often sharp, but not related to the structure of the leaf itself. The name tends to be used (a) of virus diseases, (b) of diseases now attributed to mycoplasmas. It is also well-established for certain other diseases, such as that caused in cabbage by Fusarium oxysporum f. sp. conglutinans but its use as a name for any new disease outside the categories (a) and (b) above is to be deprecated.

A disease characterized by yellowing and stunting of affected parts (caused by fungi, virus, bacteria, or deficiency of essential elements).

Yield An aggregate of the products resulting from growth or cultivation.

Z

Zearalenone A mycotoxin produced by Fusarium graminearum (teleomorph in Gibberella zeae); the cause of oestrogenic syndrome (vulvovaginitis and infertility) in pigs.

Zero grazing Where forage is cut and transported to livestock

Zinc chelate Zinc deficiency in plants can be controlled either by applying zinc sulfate to the soil or by spraying the foliage with relatively small amounts of an organic compound to which zinc has been chelated or bound.

Ziram Ziram is an agricultural dithiocarbamate fungicide used on a wide variety of plant

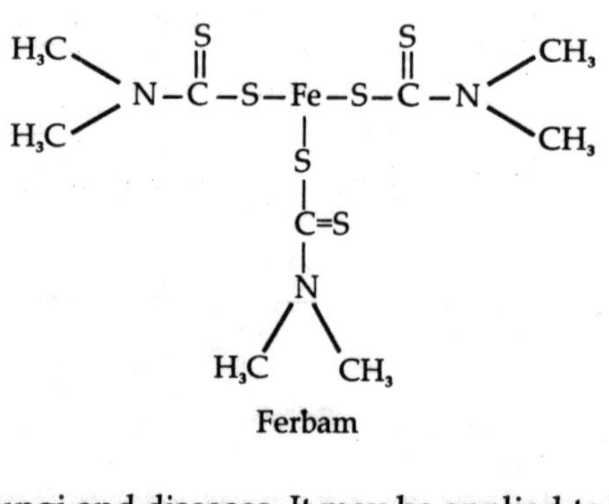

Ferbam

fungi and diseases. It may be applied to the foliage of plants, but it is also used as a soil and/or seed treatment. Ziram is used primarily on almonds and stone fruits. It is also used as an accelerator in rubber manufacturing, packaging materials, adhesives, and textiles. Another use of the compound is as a bird and rodent repellent. Ziram is often marketed as a wettable powder or as granules. Granules or grains are sifted into water and agitated prior to application.

Zonate Targetlike development of tree canker, characterized by successive, perennial rings of callus; any symptom appearing in concentric rings .

Zone line Narrow brown or black lines in decayed or decaying wood.

Narrow dark brown or black lines in decayed wood (especially hardwoods) generally caused by fungi.

Zoned Having concentric bands of color, fibers, or scales, as on the surface of the caps of certain mushrooms such as Lactarius species.

Zoonoses Diseases transferrable between humans and animals.

Zoonosis A parasite naturally transmitted between man and other vertebrate species.

Zoosporangium A sporangium within which zoospores are produced.

Zoospore Flagellated fungal spore capable of locomotion in water.

Zygomycetes A fungus group, also called the Zygomycota, characterized by nonseptate hyphae, sexual zygospores, and sporangiospores produced in a sporangium; common post-harvest pathogens.

Zygomycota Phylum of true fungi; two Classes. 1. Zygomycetes: fast-growing, terrestrial, largely saprobic fungi with no motile cells; produce zygosporangia by fusion of usually similar gametangia; also asexual sporangia containing 1 to many non-motile spores, and borne on simple to complex sporangiophores.

2. Trichomycetes: apparently commensal fungi found attached to the gut wall of arthropods.

Zygomycotina A subdivision of fungi characterized by the formation of a thick-walled resting spore formed after the fusion of two equal gametangia.

Zygosporangium In the Zygomycotina: A usually thick-walled, often ornamented, multinucleate resting sporangium formed following anastomosis of gametangia arising from compatible mycelia (in heterothallic species) or from the same mycelium (in homothallic species).

Zygospore A fungal resting spore produced by the fusion of equal gametes.

Ascospore Spore produced by certain classes of fungi as a result of sexual recombination.

Ascus: Sac-like structure within which the ascospore is produced.

Chlamydospore: Specialized survival structure produced by some species of fungi tolerant of harsh environmental conditions.

Conidiophore: Specialized structure upon which the conidium-producing cell is produced.

Conidium/conidia: Asexual reproductive spore produced by certain classes of fungi.

Forma specialis: A sub species with a very narrow host range.

Hypha/ae: Filaments comprised of fungal cells allowing for growth.

Macroconidium/ia: Large multicellular spore produced by most *Fusarium* species.

Microconidium/ia: Small spore, usually one to a few cells, produced by most *Fusarium* species.

Mycelium/ia: A network of hyphae.

Mycotoxin: A toxic secondary metabolite produced by some species of fungi.

Perithecium/ia: Spherical to flask-shaped structure, often with an opening, within which the asci and ascospores are produced.

Phialide: Specialized cell from which the conidium is produced.

Sporodochium/ia: Specialized structure in which many conidiophores and or phialides are produced.

Zygote A diploid cell or protoplast formed as a result of fusion of two haploid nuclei during sexual reproduction; zygotes often become resting spores, but ultimately germinate to produce either a diploid generation (very few fungi, many algae, all higher plants) or undergo meiosis and, following this genetic recombination, re-establish the haploid phase.

6/9/06 Botany - 15